DATE DUE			
NOV 1 0 '88 S			
JUNE 989			
MAR 1 0 1997 S			

Dynamical Systems
and Cellular Automata

Academic Press Rapid Manuscript Reproduction

Dynamical Systems and Cellular Automata

Edited by

J. Demongeot

Institut IMAG
University of Grenoble
Saint Martin d'Hères
France

E. Golès

Institut IMAG
University of Grenoble
Saint Martin d'Hères
France

M. Tchuente

Institut IMAG
University of Grenoble
Saint Martin d'Hères
France

1985

ACADEMIC PRESS, INC.
(Harcourt Brace Jovanovich, Publishers)
London   Orlando   San Diego   New York
Toronto   Montreal   Sydney   Tokyo

ACADEMIC PRESS INC. (LONDON) LTD.
24–28 Oval Road
LONDON NW1 7DX

United States Edition published by
ACADEMIC PRESS, INC.
Orlando, Florida 32887

BRITISH LIBRARY CATALOGUING IN PUBLICATION DATA
Dynamical systems and cellular automata.
    1. Machine theory
    I. Demongeot, J.   II. Golès, E.
    001.53'5   QA267

    ISBN 0-12-209060-8

LIBRARY OF CONGRESS CATALOGING IN PUBLICATION DATA

Main entry under title:

Dynamical systems and cellular automata.

    Includes index.
    1. Cellular automata—Addresses, essays, lectures.
2. Dynamics—Addresses, essays, lectures.   I. Demongeot,
J. (Jacques)   II. Golès, E.   III. Tchuente, M.
QA267.5.C45D96 •   1985   511   84-46175
ISBN 0-12-209060-8   (alk. paper)

PRINTED IN THE UNITED STATES OF AMERICA

85 86 87 88   8 7 6 5 4 3 2 1

Dedication

This book is dedicated to Professor N. Gastinel from the Université Scientifique et Médicale de Grenoble; he created in 1978 a research group of the Laboratory I.M.A.G. devoted to the study of the theory and applications of cellular automata. M. Cosnard, J. Demongeot, E. Golès, F. Robert, Y. Robert, and M. Tchuente were the first members of this group. In 1979, N. Gastinel organized for the first time at Giens, France a workshop on cellular automata, which brought together the European specialists of this field. We wish to express here our gratitude to our distinguished Maître.

Contents

Applications To Computer Science

Contributors

Numbers in parentheses indicate the pages on which the authors' contributions begin.

J.-P. Allouche (17), *LA 226, Laboratoire d'Informatique et de Mathématiques 351, 33405 Talence, France*

C. André (307), *Laboratoire de Signaux et Systèmes-ERA, 06041 Nice, France*

H. Atlan (171), *Department of Medical Biophysics, Hadassah University Hospital, Ein Karem, Jerusalem, Israel*

Y. Bouligand (75), *EPHE et CNRS, 94200 Ivry/Seine, France*

C. Choffrut (313), *Université de Paris 7, UER de Mathématiques, 75251 Paris 05, France*

C. Cocozza-Thivent (23), *Laboratoire de Probabilités, Université de Paris VI, Tour 56, 75230, Paris, France*

M. Cosnard (17), *Laboratoire I.M.A.G., Université de Grenoble, 38402, Saint Martin d'Hères, France*

K. Culik II (313), *Department of Computer Science, University of Waterloo, Waterloo, Ontario, N2L 3G1, Canada*

D. d'Humières (187), *Groupe de Physique des Solides de l'E.N.S., 75231 Paris 05, France*

J. Davenport[1] (359), *Analyse Numérique, Laboratoire I.M.A.G., 38402, Saint Martin d'Hères, France*

L. Demetrius (87), *Max-Planck-Institut für Biophysikalische Chemie, am Fassberg, 3400 Göttingen, Federal Republic of Germany*

J. Demongeot (1, 99), *Laboratoire I.M.A.G., TIM3, 38402, Saint Martin d'Hères, France*

B. Derrida (147), *Service de Physique Théorique, CEN-Saclay, 91191 Gif-Sur-Yvette, France*

F. Fogelman-Soulié (27), *Laboratoire de Dynamique des Réseaux, 75005 Paris and Université de Paris V., Paris, France*

L. Glass (197), *Department of Physiology, McGill University, Montréal Québec H3G 1Y6, Canada*

E. Golès (1, 47), *Université de Grenoble, Laboratoire I.M.A.G., 38402 Saint Martin d'Hères, France*

J. Hardouin Duparc (57), *Laboratoire d'Analyse Numérique et Informatique, Université de Franche-Comté, 25030 Besançon, France*

B. A. Huberman (187), *Xerox Palo Alto Research Center, Palo Alto, California 94304*

S. A. Kauffman (221), *Department of Biochemistry and Biophysics, University of Pennsylvania, School of Medicine, Philadelphia, Pennsylvania 19104*

M. Kaufman (207), *Service de Chimie-Physique II, Université Libre de Bruxelles, 1050 Brussels, Belgium*

H. T. Kung (321), *Department of Computer Science, Carnegie—Mellon University, Pittsburgh, Pennsylvania 15213*

A. Lissowski (139), *Blackett Laboratory, Imperial College, London SW7 2BZ, England and Department of Psychology, Polish Academy of Science, Warsaw, Poland*

C. Lobry (61), *Université de Nice, Département de Mathématiques, 06034 Nice, France*

H. B. Lück (111), *Laboratoire de Botanique Analytique et Structuralisme Végétal, Faculté des Sciences et Techniques de St.-Jérôme, 13397 Marseille 13, France*

J. Lück (111), *Laboratoire de Botanique Analytique et Structuralisme Végétal, Faculté des Sciences et Techniques de St.-Jérôme, 13397 Marseille 13, France*

Ch. v. d. Malsburg (235), *Max-Planck-Institut für Biophysikalische Chemie, D-3400 Göttingen, Federal Republic of Germany*

L. Melkemi (331), *Université de Grenoble, Laboratoire I.M.A.G., BP 68, 38402 Saint Martin d'Hères, France*

J. P. Nadal (147), *Groupe de Physique des Solides de l'E.N.S., 75231 Paris 05, France*

M. Nakechbandi (339), *Laboratoire d'Analyse Numérique et Informatique, Université de Franche-Comté, 25030 Besançon, France*

R. Occelli (139), *Laboratoire de Dynamique et Thermophysique des Fluides, Université de Provence, 13397 Marseille 13, France*

N. H. Packard (123), *Institute for Advanced Study, Princeton, New Jersey 08540*

J. Pantaloni (139), *Laboratoire de Dynamique et Thermophysique des Fluides, Université de Provence, 13397 Marseille 13, France*

P. Quinton (347), *IRISA-CNRS, Campus de Beaulieu, 35042 Rennes, France*

C. Reder (61), *Université de Bordeaux I, UER de Math-Informatique, 33405 Talence, France*

C. Reischer (369), *Département de Mathématiques, Université du Québec à Trois Rivières, CP 500 Québec G9A 5H7, Canada*

J. Richelle (247), *Laboratoire de Génétique, Université de Bruxelles, Brussels, Belgium*

N. Rivier (139), *Blackett Laboratory, Imperial College, London SW7 2BZ England and Laboratoire de Dynamique et Thermophysique des Fluides, Université de Provence, 13397 Marseille, France*

Y. Robert (359), *CNRS. IMAG, Laboratoire TIM3, BP 68, 38402 St-Martin d'Hères, France*

I. G. Rosenberg (375), *Department of Mathematics and Statistics CRMA, Université de Montréal Succ. "A", Montréal H3C 3J7 Québec, Canada*

P. Schuster (255), *Institut für Theoretische Chemie und Strahlenchemie, Universität Wien, 1090 Wien, Austria*

D. A. Simovici (369), *Department of Mathematics, University of Massachusetts at Boston, Harbor Campus, Boston, Massachusetts 02125*

G. Targonski (65), *Fachbereich Mathematik, Universität Marburg, D3550 Marburg, Federal Republic of Germany*

M. Tchuente (1, 47, 331), *CNRS. IMAG, Laboratoire TIM3, BP 68, 38402 St-Martin d'Hères, France*

R. Thomas (269), *Laboratoire de Génétique, Université Libre de Bruxelles, Rhode-St-Genèse, Brussels, Belgium*

M. Trehel (339), *Laboratoire d'Analyse Numérique et Informatique, Université de Franche Comté, 25030 Besançon, France*

Ph. Van Ham (283), *Université Libre de Bruxelles, Service des Systèmes Logiques Numériques, 1050 Brussels, Belgium*

J. Vannimenus (147), *Groupe de Physique des Solides de l'E.N.S., 75231 Paris 05, France*

G. Weisbuch (293), *Groupe de Physique des Solides de l'Ecole Normale Supérieure, 75231 Paris 05, France*

S. Wolfram (153), *The Institute for Advanced Study, Princeton, New Jersey 08540*

[1]Present Address: School of Mathematics, University of Bath, Bath, England

Preface

This volume constitutes the proceedings of the ''Journées de la Société Mathématique de France'' on ''Dynamical Behaviour of Cellular Automata: Theory and Applications,'' held September 13–17, 1983, at Luminy, France, in the Centre International de Rencontres Mathématiques (CIRM).

The aim of the meeting was to bring together specialists from different fields who were studying cellular automata from a theoretical or practical point of view; the fields of applications covered by the lectures which were given concerned physics, biology, and computer science.

The lectures are grouped into five chapters: the introductory chapter gives an historical background and a review of the recent results about cellular automata theory; the second chapter concerns theoretical approaches of the dynamics of cellular automata; and the last three chapters deal with applications to growth and spread models in physics, biology, and computer science.

We are indebted to all who provided us with their help: The Centre National de la Récherche Scientifique (CNRS), the Société Mathématique de France (SMF), the Association Française pour la Cybernétique Economique et Technique (AFCET), the Direction de la Coopération et des Relations Internationales du Ministère de l'Education Nationale (DCRI), the CIRM, and the Laboratoire d'Informatique et de Mathématiques Appliquées de Grenoble (IMAG).

INTRODUCTION

DYNAMIC BEHAVIOUR OF AUTOMATA

J. Demongeot, E. Goles, M. Tchuente

1.a. FROM THE LOCAL TO THE GLOBAL.

Here, the analysis of dynamic behaviour consists in studing,
beginning with a given local transition function and an
initial configuration $x°$, the sequence $x^{r+1} = F(x^r)$, $r \geq 0$,
where F is the global transition function. The first attempt
in this direction was the article of Schrandt and Ulam (39)
entitled "On some mathematical problems connected with
patterns of growth of figures". In this article, the authors
showed very rich and complex forms obtained by application of
a simple recursive local rule and starting from a simple
geometrical element. It is a very interesting problem whose
formulation is very simple. However, as we shall see, its
solution requires very varied and often sophisticated
mathematical tools.
In the case where the cellular space is infinite, one of the
rare cases, as for as we know, which has been the object of
a deep mathematical analysis, is that of uniform linear
automata. In this class, the set $Q = \{0,1,...,q-1\}$ is
identified with the integers modulo q and the local
transition function is written

$$\delta(x_1,...,x_p) = \sum_{i=1}^{p} a_i x_i; \quad a_i \epsilon Q, \quad i=1,...,p.$$

It has been shown by Waksman (52) and Winograd (53), that if
q is prime, then every finite form reproduces itself
indefinitely during the iteration (see example below).

neighborhood

$$Q = \{0,1\}; \quad \delta(x_1,x_2,x_3,x_4,x_5) = \sum_{i=1}^{5} x_i$$

DYNAMICAL SYSTEMS
AND CELLULAR AUTOMATA

1

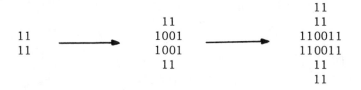

Fig. 1

As this iteration can be written in the matrix form

$$x^{r+1} = Ax^r, \quad r \geq 0$$

these results of self reproduction of figures are obtained by using matrix algebra techniques with finite fields, Gill (14). When the cellular space is finite, the sequence:

$$\{x^{r+1} = F(x^r); \ r \geq 0\}$$

finally becomes cyclic. More precisely, there exists $\bar{p} > 0$ and $\bar{T} \geq 1$ such that

$$F^{r+\bar{T}}(x) = F^r(x) \text{ for every } r \geq \bar{p} \quad \text{and}$$

$$F^p(x) \neq F^q(x) \quad \text{for} \quad 0 \leq p < \bar{p}, \ p \neq q$$

$\bar{p}$ is called the transitory length and $\bar{T}$ the length of the limit cycle. In this framework Martin et al. (31) has recently taken up the study of uniform linear cellular automata, for the case where the cells are connected in a ring (see fig. 2 below).

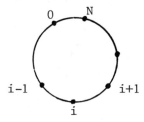

Fig. 2 Each cell i is connected to i-1 and i+1 modulo N.

By identifying every configuration $x = (x_0, x_1, \ldots, x_{N-1}) \epsilon Q^N$ with the polynomial $P(x) = \sum_{i=0}^{N-1} x_i X^i$, he shows that the analysis of the iteration amounts to the study of a transformation on polynomials. This approach allows him to obtain many results concerning transient states and limit

cycles. In addition, he suggests a classification for
cellular automata of dimension 1. He divides them into four
universality classes: the first three are related to fixed
points, cycles and chaotic attractors, whereas the fourth
contains those structures capable of simulating the universal
Turing machine. Also, in their articles appearing in these
proceedings, Wolfram and Packard suggest the use of entropy
as a quantitative statistical measure for studing order and
chaos in patterns generated in the evolution of cellular
automata.

The second class of automata, for which it has been possible
to employ useful and interesting mathematical tools, is that
where each cell is binary and evolves according to a
threshold function

$$\delta(x_1, x_2, \ldots, x_n) = \begin{cases} 1 & \text{if } \sum_{j=1}^{n} a_{ij} x_j \geq b_i \\ 0 & \text{otherwise} \end{cases}$$

Where $x_i \in \{0,1\}$, $a_{ij}, b_j \in \mathbb{R}$ for $1 \leq i, j \leq n$.

a_{ij} is the interaction coefficient between cell i and cell j
and b_i is the threshold associated with cell i.
In the particular case where $a_{ij}=1$ for each i,j and $b_i = \lceil \frac{n}{2} \rceil$,
each cell takes at the instant t+1, the dominant state
in its neighborhood at instant t. It was probably Kitagawa
(27) who was first interested in this type of automata. In
the case of sequential iteration, that is, when at every
instant the local transition function is applied to a single
cell in the network, he showed among other things the
convergence for the following case:

$$Q=\{0,1\} \quad \delta(x_0, x_1, x_2, x_3) = \begin{cases} 1 & \text{if } 2x_0 + x_1 + x_2 + x_3 \geq 3 \\ 0 & \text{otherwise} \end{cases}$$

neighborhood

Fig. 3

In his article, he proposed an analogy with Markov chains,
but this does not allow one to obtain general results. In
fact, it seems today that the most fertile approach, for the
study of finite iterations, is based on the concept of
monotony. The principle consists in associating with every
configuration x, a quantity E(x) which diminishes during the
iteration, i.e.

$$E(x^0) > E(x^1) > \ldots > E(x^P) = E(x^{P+1}) = E(x^{P+2}) = \ldots$$

when E(x) is appropriately chosen, one can deduce the length
of the transient phase as well as the length of the limit

cycle. It should be noted that this idea has been used for a
long time, see for instance Dijkstra (10), in order to
establish formal proofs for the correctness of programs in
computer science.

This approach was used by Tchuente (49) to study parallel
iteration with the majority function in a uniform structure.
By taking for E(x), the number of blocks of cells in state 1
in a finite configuration x, he established results
concerning convergence in dimension 1 and cycles of length
two in dimension 2.

In the case of sequential iteration in dimension 2 with the
majority function on the Von Neumann neighborhood, Robert
(38) showed that the energy function E(x) used in physics in
the context of spin glasses, allows one to prove convergence.
This result has been generalized recently by Fogelman,Goles,
Weisbuch (11) to general threshold functions where.

$$a_{ij} = a_{ji} \quad \text{and} \quad a_{ii} \geq 0 \quad \text{for} \quad 1 \leq 1, j \leq n.$$

This was done with the aid of a function E(x) having an
expression analogous to the hamiltonian used in physics.This
type of functions has previously been used by Hopfield (24)
to study threshold networks introduced as models for
associative memories, and by Hinton (23) to study problems
of visual learning.

The results with we have just recalled show that the approach
by monotony has first been applied especially to sequential
iterations. For the case where one wants to analyse cycles
generated by threshold functions under parallel iterations,
it is also possible to associate a monotone operator in the
case of symmetric threshold networks: Fogelman, Goles,
Pellegrin (12) have introduced the following bilinear
expresion

$$E(x,y) = -x^T A y + (x+y)^T b$$

where A = (a_{ij}) is the matrix of interaction coefficients
and b = $(b_1,...,b_n)$ is the real threshold vector. This
operator permits to give bounds on both the transient
phenomena and the cycles.

Another approach for the study of the periodic behaviour on
threshold automata have been given by Goles, Olivos (16).
Those authors have introduced a purely algebraic technique
which lead to the following result: if $a_{ij} = a_{ji}$ for any i,j,
then the parallel iteration only gives fixed points and
cycles of length two. This approach allows one in addition
to analyze multithreshold functions, Goles and Olivos (17),
threshold functions with memory, Goles (15), and the

generalized functions of majority type, Goles and Tchuente
(18), introduced by Poljak and Sura (30) as formal models for
group dynamics.
To end this paragraph devoted to threshold functions, we draw
attention to the fact that certain authors have obtained
interesting results by making hypotheses on the rank on the
eigenvalues of matrix $A = (a_{ij})$, Accardi (1) or by
interpreting a threshold iteration as the computation of the
sub-differential of a convex functional, Tao (43). All the
approaches which have been developped so far, try to exhibit
monotonous operators adapted to the study of particular class
of threshold automata. But there is a fundamental reverse
question, i.e., the problem of characterizing,the class of
networks which can be studied via a quadratic monotonous
operator. A first important step in this context has been
made recently by Goles (19) who have given a general
mathematical framework for the study of a large class of
networks where the states of the automata are encoded as
vectors and the transition functions verify a positivity
condition. As a particular case this class contain threshold
networks. To end this paragraph, let us note that the analysis
of the dynamical properties of transformations defined on
finite sets can be studied in the context of semi-group
theory (see the paper of Targonski in these proceedings).

1.b. FROM THE GLOBAL TO THE LOCAL

In the preceding paragraph, we were interested in studying
the global transition function, starting from local rules.
Here it is the opposite procedure. We start with a global
behaviour F given a priori and we are interested in the
definition of simple local rules which could lead to F.
The Von Neumann automaton (51) belongs to this category,
because the aim of the author was to conceive simple local
rules which would exhibit on the global level, the property
of self-reproduction. The second example (doubtlessly the
best known) is the "Life" automaton introduced by Conway (8),
see also Bertekamp (5)

$$Q=\{0,1\}; \quad \delta(a_o,a_1,\dots,a_8)=\begin{cases} 1 \text{ if } \begin{cases} a_o=1 \text{ and } 2\leq \sum_{i=1}^{8} a_i\leq 3 \\ \text{or} \\ a_o=0 \text{ and } \sum_{i=1}^{8} a_i=3 \end{cases} \\ 0 \text{ otherwise} \end{cases}$$

neighborhood

Fig. 4 "Life". a_o is the state of the cell under consideration.

The rules are chosen by analogy with what happens in living
environment. A cell is born or survives if an equilibrium
condition, related to the number of living cells in the
neighborhood, is verified; in other cases which represent
insulation or overpopulation, the cell dies or remains in a
"dead" state. This intuition of Conway has turned out to be
brillant because extraordinary configurations such as mobile
figures or gliders guns have been exhibited. More recently,
Hardouin Duparc (see these proceedings) has shown that by
breaking the uniformity of the rules of "Life",it is
possible to obtain glider guns much smaller than those
exhibited by Conway.

The erasing automaton introduced by Toom (49) is another
example where it is desired to go from the global to the
local. However, in the case of erasers the interest is not
in obtaining rich structures as previously, but on the
contrary in characterizing those automata where any finite
configuration evolves towards the empty configuration. It
is clear that these automata, which belong to the first
universality class defined by Wolfram (54), correspond to
the poorest that can be imagined. In these proceedings,
Goles and Tchuente propose for some particular cases of
erasing cellular and threshold automata, an algorithmic
approach which illustrates the complexity of the problem. In
the field of physics and biology, automata networks can be
used as an interesting tool for the modelization of highly
complex global phenomena with very simple local properties;
such as some of the nervous systems properties (see the
papers of Atlan, D'Humieres and Huberman, Fogelman, Von der
Malsburg), Morphogenesis and growth of physical structures
(Bouligand, Lobry and Reder,H.and J. Luck, Rivier and Ocelli,
Vannimenus) and biological evolution (S. Kauffman, Weisbuch,
Schuster).

In several applications, the authors used random automata
networks. The probabilistic aspect is obtained either by
choosing randomly the local transition functions within a
collection of deterministic laws or by assigning a fixed
probabilistic local transition function to all the cells of
the array. For this kind of approach see for instance the
papers of Atlan, Cocozza, Demongeot, Fogelman.

Finally, we bringout the work carried out at the moment by
Fredkin, Toffoli and Vichniac (48,50) at MIT, whose aim is
to bring to the fare, reversible cellular structures which
allow us to calculate without a great dissipation of energy.

1.c. DISCRETE MODELS FOR CONTINUOUS PHENOMENA

It is sufficient to examine very simple examples in dimension

2 to be convinced that the systems regulated by continuous
transformations are very difficult to study analytically,
particularly if one is interested in their iterative
properties.
One way to overcome this difficulty is to make simulations on
a discrete model. Such an approach can clearly not claim to
take completely into account the phenomena of continuity.
However, it can provide very interesting qualitative
informations. In this way, Greenberg et al. (20) have
studied a discrete model based on the observed behaviour of
excitable media. More precisely, the set of states
$Q=\{0,1,\ldots,E,E+1,\ldots,N-1\}$ is partitionned into three classes:
- the excitatory states $q\epsilon\{1,2,\ldots,E\}$
- the refractory states $q\epsilon\{E+1,\ldots,N-1\}$
- the rest state $q=0$
The transition function of any cell i is defined as follows:
- if $1<x_i^t<N-1$ then $x_i^{t+1} = x_i^t+1$ modulo N. In other words, when
a cell is excited, its state evolves according to a fixed
dynamics without any influence from its neighborhood, until
it returns to the rest state
- if $x_i^t=0$ then

$$x_i^{t+1} = \begin{cases} 1 & \text{if i admits a neighbour j such that} \\ & 1<x_j^t<E \\ 0 & \text{otherwise} \end{cases}$$

In other words, excitation can "diffuse" from an excited
region to an adjacent resting region.
The success of this model comes from the fact that
simulations make series of waves appear, a phenomenon
observed in chemical reactions. Combinatorial techniques
have permitted the establishment of many results on the
length of the transient phase and on the structure of the
series of waves, Greenberg et al. (20), Allouche and Reder
(2). For the case where one is restricted to a finite
domain with fixed boundary conditions, Shingai (40) has
established many results on the lengths of cycles.
In these proceedings, R. Thomas and other authors present a
model of genetic control phenomena by means of boolean
representation. This model is proposed to explain the
regulatory properties of the genes. The problem of cross-
regulation in genetic has been already setted by Delbruck
from 1949 (see the discussion of the paper by Sanneborn and
Beale (42)), but the theoretical studies began with Stuart
Kauffmann (25). Thomas and Van Ham describe the theoretical
properties of the asynchronous model; M. Kauffman gives a
very new and exciting application of this model in
immunology; Richelle and Glass are studying the relation
between discrete and continuous approach in these dynamical

system.

2. COMPUTABILITY ON CELLULAR STRUCTURES

Following the pioneering work of Von Neuman (51), the theory
of cumputability on cellular structures was first oriented
towards the study of structures simpler than the 29-state
5 neighborhood cellular automaton of (51), and which are both
computation and construction universal. As explained in
Burks (6), such structures can simulate any Turing machine
and, given any quiescent automaton, they can construct that
automaton in a designated empty region of the cellular space.
For instance, E.F. Codd (7) has shown that there exists a two
dimensional 8-states 5-neighborhood automaton which is
computation-construction universal. In addition, A.R. Smith
(41) has proved that there exist very simple 1-dimensional
computation-universal cellular spaces which can support non-
trivial self-reproduction.

Subsequently, some efforts have been made to find resolutions
to some problems by exploiting the inherent high parallelism
of cellular automata. For instance, it is known from
Barzdin (4) that the time required to recognize a palindrome
of rize n in a one-headed Turing machine is of the order of
n^2, but A.R. Smith has exhibited a one-dimensional array
with nearest-neighbor-interaction, which can do it in time
$\lfloor \frac{n}{2} \rfloor$; another interesting result is due S. Kosaraju who has
shown that any context-free language is $(1+\varepsilon)n$-time
recognizable by a two-dimensional cellular automaton and n^2-
time recognizable by a one-dimensional cellular automata.
However, all the preceding studies were two clearely related
to the sequential model of Turing machines; this fact is well
illustrated by a suggestion of A.R. Smith stating that "for a
first approximation of the complexity of cellular automata,
one can say that they lie somewhere between the completely
serial Turing machine and the highly parallel multi-headed
machines, for certain types of problems". The first
mathematical studies devoted to the characterization of all
the transformations which are realizable on a uniform
cellular structure, were due to G.A. Hedlund (21) and D.
Richardson (37). They showed independently, the following
result: for any finite non empty set of states Q, and for
any, positive integer $n \geq 1$, if the product topology induced
by the discrete topology of Q is assigned to the set of
configurations $\mathcal{C} = \{C:Z^n \to Q\}$, then a mapping F from $\mathcal{C}$ into
itself can be realized by a uniform cellular structure if
and only if it is continuous and commutes with the shift
operation. Subsequently the problem of determining the
capacity for computation of a flexible cellular structure

$A = (Z^n, Q, V)$ where the neighborhood index is fixed but where the transition function may change from time step to time step, has been studied by several authors.

In this context, a mapping $F: \mathcal{C} \to \mathcal{C}$ is said to be computable on A if and only if it can be decomposed into the form $F = F_1 \circ F_2 \circ \cdots \cdots \circ F_n$ where each F_i is a possible global transition function of the system. The first result in this direction was due to Amoroso and Epstein (2) who showed that for $Q = \{0,1\}$ and for any $n \geq 2$, there exist indecomposable maps, i.e. global transition functions of a binary one-dimensional scope-n cellular structure, which are not computable on any binary one-dimensional scope-m cellular structure with $m < n$. This model of computation is very close to the classical completeness problem where one looks for a complete set of functions (see the paper of Rosenberg in there proceedings). This result has been generalized by Nasu (34) to cellular structures of any dimension and finite number of states.

In the case of a finite network $N = (G,Q)$ where Q is the set of states and $G = (V,U)$ is the interconnection graph with $V = 1, \ldots, n$ as set of vertices and $U \subset V \times V$ as set of arcs, the set of transition functions is compared of all mappings $F: x = (x_1, \ldots, x_n) \in Q^n \mapsto (f_1(x), \ldots, f_n(x)) \in Q^n$ such that, for any i, j, $1 \leq i, j \leq n$, if f_i depends on x_j then $(j,i) \in U$. The first studies in this direction were due to N. Gastinel (23), M. Tchuente (45) and K.H. Kim (26) who studied the computation of linear mappings in the particular case where G is a field. Subsequently, M. Tchuente (47) gave a characterization of all Boolean transformations computable on a tree connected network of binary automata.

When computation on a finite network G is defined in terms of factorization of functions, the permutations:
$\sigma: (x_1, \ldots, x_n) \to (x_{\sigma(1)}, \ldots, x_{\sigma(n)})$ play a central role, indeed, such a function can be considered as a formal model for a situation where, for any i, $1 \leq i \leq n$ automaton i wants to send a message to $\sigma(i)$; as a consequence, if $\ell_G(\sigma)$ denotes the minimum length of factorization of σ, then $\ell_G = \max_\sigma \ell_G(\sigma)$ is the maximum delay necessary for the transmission of a collection of messages of the form $(\sigma(1), \ldots, \sigma(n))$. Such a model seems to be appropriate for measuring the communication capacity of a cellular structure. The characterization of networks where any permutation is computable is due to S. Riccard (35), and the comparison of tree-connected structures with respect to the criteria ℓ_G may be found in Tchuente (44). Another aspect of computability on cellular structures, has been developped by Hennie (22) in connection with the

analysis and design of logical circuits. The underlying
idea is that iterative networks composed of identical cells
interconnected to form a regular array, are very suitable
for practical realization. Indeed, such structures are
economical to manufacture and repair and, on the other hand,
they can be enlarged to accomodate more variables by simply
adding more components, i.e. without changing the existing
portion of the network. The fundamental work of Hennie was
concerned with the following general questions: "What kinds
of operations can iterative networks perform, how can they
be analysed, and how can an iterative array be designed to
do a specific job". However this work, as that of many
other authors, did not come to practical realization. This
was due principally to the fact that the efforts were engaged
in establishing very general results and also because this
work was in advance of the technological possibilities.
Today, with the developments in Very Large Scale Integration
(VLSI), it has become clear that the fundamental aspects of
cellular structures, i.e. regularity, modularity and
parallelism, can lead to the concrete realization of high-
performance devices. The major contribution in this
awareness is due to Kung and Leiserson (28). These author,
have brought to the fare the following fact: Initially
cellular networks should be conceived not as universal
machines (see Von Neumann (5)), but rather as procedures on
silicium (Silicon procedures). This means that one should
(and can) from now on design high-performance special-purpose
parallel structures which can be attached to a general-
purpose host computer in order to carry out compute-bound
problems. A second result due to the idea of systolic
arrays introduced by Kung and Leiserson (28) is that pipeline
plus multiprocessing at each stage of a pipeline should lead
to best possible performance. The great success today of
systolic networks is due to the numerous results which have
been obtained in fields as varied as signal processing,
symbolic computation,linear algebra and even the
implementation of some basic devices of computer science
as stacks, counters etc. This large range of applications
is found in the articles in these proceedings.Kung shows in
fact how multilevel pipelining permits one to obtain
efficient implementations. Nakechbandi and Trehel attack
the problem of the solution of a linear program. Davenport
and Robert present an application to symbolic computation,
Melkemi and Tchuente are interested in the computation of
the connected components of a graph and in the
triangularization of band matrices, and Choffrut and
Culik study a problem of folding related to the design
of systolic networks. More recently several authors (Kung

(29), Moldovan (33), Leiserson and Saxe (30), Miranker et al.
(32)) have been interested in using systematic methods for
the construction of systolic networks. This recent tendency
is represented here by Quinton, who proposes us a very
general approach, which, when applied for instance to the
problem of convolution, allows one to obtain a great variety
of networks most of which escape intuition.
Finally, let us note that another important approach to
parallelism has been developped in the context of Petri nets
(see the paper of André in there proceedings).

REFERENCES

(1) Accardi, L. (1971) Rank and reverberation in neural
networks, Kybernetik, 8, 163-164.
(2) Allouche J.P. and Reder C. (1981) Periode des oscillations
d'un automata cellulaire, S.A.N.G., Grenoble.
(3) Amoroso, S. and Epstein I.J. (1974) Maps preserving the
uniformity of neighborhood interconnection patterns in
tessellation structures. Inf. and Control 26, 1-9.
(4) Barzdin V. M. (1965) Complexity of recognition of
symmetry in Turing machines, Problemy Kibernetiki 15.
(5) Berlekamp E.R., Conway J.H. and Guy, R.K. (1982),
Winning Ways, For Your Mathematical Plays, 2, Academic
Press chap. 25.
(6) Burks, A.W. (1970) Essays on Cellular Automata,
University of Illinois Press.
(7) Coddn E.F. (1968) Cellular Automata, Academic Press.
(8) Conway, J. H. (1970) Mathematical Games, M. Gadner Ed.,
Scientific American 120-123.
(9) Cosnard M. and Goles E. (1984) Dynamique d'un automata a
memoire modelisant le fonctionnement d'un neurone,
C.R.A.S. to appear.
(10) Dijkstra, E.W. (1976) A Discipline of Programming,
Prentice-Hall, New Jersey.
(11) Fogelman, F., Goles, E. and Weisbuch, G., (1983),
Transient length in sequential iteration of threshols
functions, Discrete Applied Math, 6, 95-98.
(12) Fogelman, F., Goles, E. and Pellegrin, D. (1984),
Decreasing energy functions as a tool for studing
threshold networks, submited to Disc. Applied Math.
(13) Gastinel, N. (1978) Réalisation du Calcul d'une
Transformation lineaire aux noeuds d'un graphe, Colloque
sur les Méthodes de Calcul pour des systemes de type
cellulaire, Giens, France.
(14) Gill, A. (1966) Linear Sequential Circuits, Mc. Graw-
Hills Series in System Science.

(15) Goles, E. (1983), Dynamical behaviour of neural networks, R.R. N°386, IMAG, Grenoble.

(16) Goles E. and Olivos J., (1982) Comportement periodique des functions a seuil binaires et applications, Discrete Applied Math. 3, 93-105.

(17) Goles, E. and Olivos, J. (1980) Comportement itératif des fonctions a multiseuil, Inf. and Control, Vol.45, N°3, 300-313.

(18) Goles, E. and Tchuente, M. (1983) Iterative behaviour of generalized majority functions, Math. Social Sciences, 4, 197-204.

(19) Goles, E. (1984), Dynamics on positive automata networks, Research Report, Dep. Math., U. de Chile.

(20) Greenberg, J.M. et al. (1980), A Combinatorial problem arising in the study of reaction-diffusion equations, Siam J. Alg. Disc. Meth. 1.

(21) Hedlung, G. A. (1969), Endomorphisms and automorphisms of the shift dynamical systems, Math. Syst. Theory 3, 320-375.

(22) Hennie F.C. III (1961), Iterative Arrays of Logical Circuits M.I.T. Press, Wiley and Sons Inc., New York and London.

(23) Hinton G. E. and Sejnowski, J. T., (1983) Optimal perceptual inference, in Proc. IEEE Conf. in Computer Version and Pattern Recognition.

(24) Hopfield, J.H. (1982) Neural networks and physical systems with emergent collective computational abilities, in Proc. Nat. Acad. Sci. Biophysics, U.S.A., 79,2554-2558.

(25) Kaufman, S. (1969) Metabolic stability and epigenesis in randomly constructed genetic nets, J. Theor. Biol. 22, 437-467.

(26) Kim, K.H. and Roush,F.W. Realizing all linear transformations linear algebra and its applications, 37, 97-101.

(27) Kitagawa, T. (1974) Cell space approaches in biomathematics, Math. Biosciences 19, 27-71.

(28) Kung, H. T. and Leiserson, C.E., Systolic arrays for VLSI in Proc., Symp. on Sparse Matrix Computations and their Applications Knoxville.

(29) Kung H.T. and Lin W. T. (1983), An algebra for VLSI algorithm design, Technical Report, Carnegie Mellon University.

(30) Leiserson, C. E. and Saxe J.B., Optimizing synchronous systems, Journal of VLSI and Computer Systems 1, 1, 41-47.

(31) Martin, O., Odlyzko, A.M. and Wolfram, S. Algebraic

properties of cellular automata, to appear in Comm.
Math. Phys.

(32) Miranker, W. L. and Winkler A. (1982), Space-time
representation of computational structures, I.B.M.
Research Report RC 9775,

(33) Moldovan, D. I. (1982) on the analysis and synthesis of
VLSI algorithms, IEEE transactions an Computers C-31,
11, 1121-1126.

(34) Nasu, M., (1979) Indecomposable local maps of
tessellation automata, Math. Syst. Theory 13, 81-93.

(35) Riccard, S. (1946) Sur les Bases de Groupe Symetrique et
les Couples de Substitutions qui Engendrent un Groupe
Alterné, Vuibert, Paris.

(36) Poljak, S.E. and Sura, M., (1983), On periodical
behaviour of societies with symmetric relations, to
appear in combinatorica.

(37) Richardson, D., (1972), Tessellation with local
transformations, J. Comput. and System Sci. 6,
373-388.

(38) Robert, F., (1978), Une approche booleenne du probleme
de la frustration, S.A.N.G., N⁰302, IMAG, Grenoble.

(39) Schrandt, R. G. and Ulam, S. M., (1970), On some
mathematical problems connected with patterns of growth
of figures, 219-231, in Essays on Cellular Automata, A.
W. Burks ed., University of Illinois Press.

(40) Shingai, R., (1979) Maximum period of two-dimensional
uniform neural networks. Inf. and Control 41, 324-341.

(41) Smith, A. R., (1971), Simple Computation-universal
cellular space, J.A.C.M. 18, 3, 339-353.

(42) Sonneborn, T. M. and Beale, G. H., (1949), Determinisme
des caracteres antigeniques, in Coloque International
du CNRS sur las Unités Biologiques douées de Continuité,
CNRS Ed.

(43) Tao, P. D., (1983), Convergence of a subgradient method
for computing the bound norm of matrices, Rapport de
Recherche IMAG N⁰383.

(44) Tchuente, M., (1982), Parallel realization of
permutations over trees. Discrete Mathematics 39, 211-
214.

(45) Tchuente, M., (1979), Parallel calculation of a linear
mapping on a computer network, Linear Algebra and its
Applications 28, 223-247.

(46) Tchuente, M., (1982), Contribution a l'etude des
methodes de calcul pour des systemes de type cooperatif,
There d'etat, Grenoble.

(47) Tchuente, M., Computation on Tree-Connected networks of
binary automata, submitted for publication.

(48) Toffoli, T., (1980), Reversible computing, M.I.T./LCS/ TM-151

(49) Toom, A.L. (1978), Monotonic evolution in real spaces, in Proc. Locally interacting systems and their applications in biology. Lecture Notes in Mathematics 653, 1-14.

(50) Vichniac, Y. G., (1983), Simulating Physics with cellular automata Technical Report, M.I.T., Computer Science Dept.

(51) Von Neumann, J. (1966), Theory of Self Reproducing Automata, A. W. Burks Ed., University of Illinois Press.

(52) Waksman, A., (1966), A model of replication J.A.C.M. 16, 1, 178-188.

(53) Winograd, T., (1970) A simple algorithm for self-reproduction M.I.T. Project MAC, Artificial Intelligence, Memo 198.

(54) Wolfram, S., (1983) Statiscal Mechanics of Cellular automata, Rev. on Modern Physics 55, 3.

Dynamics of Cellular Automata: Theoretical Approach

SEQUENCES GENERATED BY AUTOMATA AND DYNAMICAL SYSTEMS

Jean-Paul Allouche[*]   and   Michel Cosnard[**]

[*] *LA 226, Laboratoire d'Informatique et de Mathématiques 351, Cours de la Libération, 33405 Talence cédex, France.*

[**] *Laboratoire IMAG, Université de Grenoble, B.P. 68, 38402 Saint Martin d'Hères cédex, France.*

1. INTRODUCTION

Let A be an alphabet, i.e. a finite set, and $A^{\mathbb{N}}$ be the set of sequences with terms in A . Put on $A^{\mathbb{N}}$ the topology of simple convergence and define d to be the shift (for a sequence A in $A^{\mathbb{N}}$, $dA(n) = A(n+1)$) .

If A is a sequence in A , a very simple dynamical system is defined by : $(c\ell \{d^k A ; k \in \mathbb{N}\} ; d)$, where $c\ell$ means closure in the above defined topology.

Conversely, let f be a continuous self-map of $[0,1]$. Can we describe the dynamical system $([0,1],f)$ by means of finite-valued sequences ?

More precisely, let f be unimodal, i.e. :
- f is a continuous self-map of $[0,1]$
- For some c in $]0,1[$ f is strictly increasing on $[0,c[$ and strictly decreasing on $]c,1[$,
- $f(1) = 0$ and $f(c) = 1$.

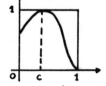

Then, a classical technique to study the iterates of f is the following one :

define γ by $\gamma(y) = \begin{cases} 0 & \text{if} \quad 0 \le y < c \\ 1 & \text{if} \quad c < y \le 1 \end{cases}$, and consider for each x in $[0,1]$ the sequence $(\gamma(f^n(x)))_{n \in \mathbb{N}}$.

Actually, we define $(\gamma(f^n(x)))_n$ to be one or two binary sequences (4) and prove the following theorem :

Put on $\{0,1\}^{\mathbb{N}}$ the lexicographical order, with $0 < 1$, and define for A in $\{0,1\}^{\mathbb{N}}$:

$$\tau A(n) = \sum_{0 \le j \le n} A(j) \quad \text{modulo 2} ,$$

$$\bar{A}(n) = 1 - A(n) .$$

Then (3), (4) :

Theorem 1 :

a) *Let* A *be a binary sequence with* $A(0) = 1$, $A(1) = 0$. *There exists an unimodal function* f *such that*
$$A = (\gamma(f^n(1)))_n , \quad \text{if and only if} \quad \forall k \ge 0 \; \overline{\tau A} \le d^k(\tau A) \le \tau A .$$

b) *Let* A *and* f *be as above, and let* B *be a binary sequence ; then, there exists an* x *in* $[0,1]$ *such that* B *belongs to*
$(\gamma(f^n(x)))_n$ *if and only if* $\forall k \ge 0 \quad \overline{\tau A} \le d^k(\tau B) \le \tau A$ *and* $d(\overline{\tau A}) \le \tau B$.

The interest of this theorem is that the sequences of a) and b) are defined only by combinatorial properties.

2. THE SET Γ

In view of the previous theorem, we define the set Γ by :

$$\Gamma = \{A \in \{0,1\}^{\mathbb{N}} ; \; \forall k \ge 0 \; ; \; \bar{A} \le d^k A \le A\}$$

(a) A first result :

After a straightforward computation we can exhibit the first three elements of Γ :

$$A_0 = 101010... = (10)^{\omega} \quad ; \quad A_1 = (1100)^{\omega} \quad ;$$

$$A_2 = (11010010)^{\omega} \quad\quad ; \quad \text{with } A_0 < A_1 < A_2 .$$

If we carry on, we actually notice that these sequences $A_0, A_1, A_2, ...$ converge to a sequence L , with $L = 1101001100101101...$ and we recognize that

OL = 01101001100101101... is the Thue-Mose sequence (see
(2) for example). Finally we prove (see (1)),

Theorem 2 :

a) L *is in* Γ

b) $\Gamma \cap [0^\omega,L[$ *is discrete and denumerable. More precisely,*
   $\Gamma \cap [0^\omega,L[= \{A_0,A_1,A_2,...\}$ *where the* A_i*'s are* 2^{i+1} *perio-*
   dic and $\lim_i A_i = L$.

c) *For every* R , *with* R > L , *the set* $\Gamma \cap [L,R]$ *has the po-*
   wer of continuum.

(b) q-mirror sequences :

   Let us first recall the definition of a 2-automaton.
A 2-automaton consists of :

- a finite set of states S = $\{i=a_0,a_1,...,a_k\}$, where i is
  called the initial state,

- two functions from S to S labelled 0 and 1,

- a map ϕ from S to $\{0,1\}$.

Indeed we can define a binary sequence in the following way :
the automaton "reads" successively the numbers 0,1,2,3,...
in their binary expansion, giving a sequence with values in
S , and then applies ϕ . We say that this binary sequence is
generated by this 2-automaton ; (for more details see (2)).
For instance :

$$S = \{i,a_1,a_2\}$$

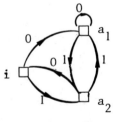

with $\phi(i) = 0$, $\phi(a_1) = 1$, $\phi(a_2) = 0$.

The automaton "reads" a binary word as follows :

$1101i = 110(1i) = 110a_2 = 11(0a_2) = 11i = 1(1i) = 1a_2 = a_1$.

The sequence generated by this automaton is :

$$\phi(0i) = 1 \quad ; \quad \phi(1i) = 0 \quad ; \quad \phi(10i) = 0 \quad ;$$
$$\phi(11i) = 1 \quad ; \quad \phi(100i) = 0 \quad \ldots$$

Let us define now, in order to generalize theorem 2, what we call "q-mirror sequences" :
let q be an integer , $q \geq 3$, and let m be a binary word of length q-1 . We construct successively :

$$B_0 = (m0)^\omega \quad ; \quad B_1 = (m1\overline{m}0)^\omega \quad ;$$

$$B_2 = (m1\overline{m}1 \ \overline{m1\overline{m}1})^\omega = (m1\overline{m}1 \ \overline{m}0m0)^\omega$$

and let Q be the limit of B_i as i goes to infinity. We call Q the q-mirror sequence beginning by m :

$$Q = m1\overline{m}1 \ \overline{m}0m1 \ \overline{m}0m0 \ m1\overline{m} \ 1 \ldots$$

Proposition :

Such a sequence is generated by a 2-automaton. Actually we can prove that the formal series $\sum\limits_{0}^{\infty} Q(n) \ X^n$ is quadratic over $F_2(X)$, the field of rational functions over the two-elements field.

(c) the q-mirror sequences and Γ .

Let q be an integer, $q \geq 3$, m a word of length q-1 and Q the q-mirror sequence beginning by m .

Theorem 3 :

a) *Q is in Γ if and only if the sequence $(m0)^\omega$ is in Γ and is of least period q .*

b) *if Q is in Γ , the set $\Gamma \cap [(m0)^\omega, Q[$ is discrete and denumerable.*
More precisely : $\Gamma \cap [(m0)^\omega, Q[= \{B_0, B_1, B_2, \ldots\}$ where B_i is 2^i q-periodic and $\lim\limits_{i} B_i = Q$.

c) *if Q is in Γ and R such that $R > Q$, then the set $\Gamma \cap [Q, R]$ has the power of continuum.*

(d) More about the structure of Γ :

Theorem 4 :

a) *The periodic sequences of Γ are dense in Γ .*

b) *Γ has a fractal structure ; more precisely : every interval of Γ containing a non-periodic sequence (in particular every non denumerable interval of Γ) has a subinterval isomorphic to Γ as ordered set .*

3. EXAMPLES OF APPLICATIONS :

By means of the results of paragraph 2 and analogous results concerning the sequences described in b) of theorem 1, we can prove :

Theorem 5 :

Let f be a unimodal function and $A = (\gamma(f^n(1)))_n$. Then :

- *If $\tau A < L$, every sequence $(f^n(x))_n$ is asymptotically periodic. More precisely, if τA has period 2^k , then f admits at least one 2^i-cycle for every $i \leq k-1$; f can admit a 2^k-cycle, and has no cycle of different order. For every x , $(f^n(x))_n$ converges to one of these cycles.*

- *If $\tau A = L$, f admits 2^i-cycles for every i , and at least one minimal invariant Cantor-set. For every x in [0,1] the sequence $(f^n(x))_n$ converges either to a 2^i-cycle or to a Cantor-set.*

- *If $\tau A > L$, f admits other cycles and a non denumerable infinity of turbulent sequences.*

Theorem 6 :

Let f_μ be a family of unimodular functions admitting a complete Feigenbaum's period-doubling cascade for μ in $[0,\mu_\infty]$. Then $\tau(\gamma(f_{\mu_\infty}^n(1))_n) = L$; I_n particular the Thue Morse sequence is universal for all such families of unimodal functions, although the Feigenbaum constant is universal only for a sub class of such functions (5).

BIBLIOGRAPHY

1. Allouche, J.P. and Cosnard, M. (1983). Itérations de fonctions unimodales et suites engendrées par automates,
   C.R. Acad. Sc. Paris 296, 159-162.

2. Christol, G.,Kamae, T., Mendes France, M., Rauzy, G.(1980).
   Suites algébriques, automates et substitutions, *Bull. Soc.
   Math. France* 108, 401-419.

3. Collet, P. and Eckmann, J.P. (1980). "Iterated Maps on
   the Interval as Dynamical Systems". Birkhäuser, Boston-
   Basel-Stuttgart.

4. Cosnard, M. (1983). Etude de la classification topologi-
   que des fonctions unimodales. *Preprint.* Grenoble.

5. Cosnard, M. (1983). "Contributions à l'Etude du Comportement Itératif des Transformations Unimodales". Thèse
   d'Etat, Grenoble.

6. Derrida, B., Gervois, A., and Pomeau, Y. (1978). Iteration of endomorphisms on the real axis and representation of numbers. *Ann. Inst. H. Poincaré* 29, 305-356.

7. Metropolis, N., Stein, P.R. and Stein, M.L. (1973).
   On finite limit sets for transformations on the unit
   interval. *J. Comb. Theory A,* 15, 25-44.

8. Milnor, J. and Thurston, P. (1977). On iterated maps of
   the interval. *Preprint.* Princeton.

ATTRACTIVE MARKOV PROCESS ON N^{Z^d}

C. Cocozza - Thivent

Laboratoire de Probabilités, Université Paris VI
Tour 56, 4 place Jussieu, 75230 Paris Cedex 05, France

INTRODUCTION

The purpose of this paper is the rigorous study of the invariant probabilities of a random automaton. Each point of Z^d (called site) is occupied by a finite number of particules so that the state-space of the automaton is

$$E = N^{Z^d} \; ;$$

if η belongs to E and x is a site then $\eta(x)$ denotes the number of particules which lie at site x. The automaton moves according to the following rule: if, at time t, the state of the system is η, the probability that a particule jumps from site x to site y during the interval $(t,t+\delta t)$ is

$$p(y-x)b(\eta(x),\eta(y)).\delta t + o(\delta t) \; ,$$

p is a positive function on Z^d which sum equals one and b is a positive and bounded function on NxN (such that $b(0,.)=0$).

We can construct a Markov process that satisfies this description(see (2), for example) and we denote by $\eta(t)$ the state of the process at time t. If $b(i,j)=0$ for $j\neq 0$ and $b(i,0)=1$ for $i\neq 0$, the process is called the simple exclusion process, it has been studied in details by Liggett (3), Spitzer (4) and recently by Andjel. If $b(i,j)=c(i)$ does not depend on j, we have the zero-range process studied by Andjel (1).

In this paper we intend to determine the space and time invariant probabilities of the process:

DYNAMICAL SYSTEMS
AND CELLULAR AUTOMATA

23

we try to find **all** the space invariant probabilities ν on E such that if ν is the law of distribution of the particules at time zero then ν remains
the law of distribution of particules at any time
t.

INVARIANT PROBABILITIES OF ATTRACTIVE PROCESS

A family of invariant product measures

The first step is to find a family of invariant
probabilities; the simplest ones are the product
measures i.e. probabilities ν on E such that:

(A) $\nu\left(\eta(x_1)=k_1,\ \eta(x_2)=k_2,\ldots,\ \eta(x_n)=k_n\right) =$

   $\mu(k_1)\mu(k_2)\ldots\mu(k_n)$

where μ is a probability on N.
   In order to avoid technical problems, we suppose that $b(i,j)$ is strictly positive if $i\neq 0$.
   The following conditions (C_1) and (C_2) are necessary and sufficient for the existence of invariant product probabilities:

(C_1) $\dfrac{b(i,j)}{b(j+1,i-1)} = \dfrac{b(i,0)b(1,j-1)}{b(j+1,0)b(1,i-1)}$ for $i\neq 0$

   (C_2') $p(u) = p(-u)$ for any u

(C_2) or

   (C_2'') $b(i,j)-b(j,i)=b(i,0)-b(j,0)$ for $i\neq 0$

(Note that these conditions are satisfied for the
zero-range process).
   When these conditions are satisfied, the invariant product probabilities are the probabilities
described by formula (A) and for which we have:

$\dfrac{\mu(i+1)}{\mu(i)} = \dfrac{\mu(1)}{\mu(0)}\,\dfrac{b(1,i)}{b(i+1,0)}$ for $i\neq 0$.

The set of these product probabilities is denoted
by I.

Attractive process

The **process is said to be attractive if it is an**
increasing function of the initial state that is
to say: **if** η **and** ξ **belong to E and** η **is greater**
than ξ then **we can construct the Markov processes**
$\eta(t)$ **and** $\xi(t)$ **with initial states** η **and** ξ **respec-**
tively such that $\eta(t)$ **is greater than** $\xi(t)$ **for any**
t.

This **property is equivalent to the fact that**
$b(i,j)$ **is an increasing function of i and a decrea-**
sing **function of j. This condition will be suppo-**
sed satisfied in the following.

Coupling technics

The **study of the process uses a coupling method.**
We **construct simultaneously two Markov processes**
$\eta(t)$ and $\xi(t)$ **with different initial states which**
jump **together as far as possible. If at time t the**
states of the **two processes are** η **and** ξ **respecti-**
vely, **then the probability that, during the inter-**
val $(t,t+\delta t)$, **a particule at site x jumps to site**
y
for the two **processes simultaneously, is:**

$$p(y-x)\min(b(\eta(x),\eta(y)),b(\xi(x),\xi(y))).\delta t + o(\delta t)$$

for the process in state η only, is:

$$(b(\eta(x),\eta(y))-\min(b(\eta(x),\eta(y)),b(\xi(x),\xi(y)))).\delta t.$$

$$p(y-x) + o(\delta t)$$

for the process in state ξ only, is:

$$(b(\xi(x),\xi(y))-\min(b(\eta(x),\eta(y)),b(\xi(x),\xi(y)))).\delta t.$$

$$p(y-x) + o(\delta t) .$$

Tedious **calculus on these processes** (details
can be foud in (2)) show that, **if the initial dis-**
tributions are **translation invariant, the probabi-**
lity that the **number of particules of** $\eta(t)$ **at site**
x is strictly **greater than the number of particu-**
les of $\xi(t)$ at x <u>and</u> that the **number of particules**
of $\xi(t)$ at y is stictly **greater than the number**

of particules of $\eta(t)$ at y, tends to zero when t
tends to infinity.

This comparison is the key to proove that all
the space and time invariant probabilities of the
process can be represented with the family I.

Results

Under conditions (C_1), (C_2) and an irreductibili-
ty condition on p, a probability on E is a space
and time invariant probability of the attractive
process if and only if it is a convex mixture of
elements of I.

CONCLUDING REMARKS

We have been able to determine all the space and
time invariant probabilities; in the symetric case
(condition (C_2')) there is certainly no time-inva-
riant probabilities which are not space invariant.
In the non-symetric case, if $b(i,j)$ tends to zero
when j tends to infinity, there is time invariant
probabilities which are not space invariant, but
we think that it is not the case if $b(i,j)$ does
not tend to zero when t tends to infinity.

ACKNOWLEDGEMENTS

I would like to thank M. Roussignol for many helpful discussions and F. Ledrappier for his constant
encouragements.

REFERENCES

1.   Andjel, E.D. (1982). Invariant measures for
      the zero-range process, *Annals of Probability*
      10, 881-895.
2.   Cocozza-Thivent, C. (1983). Processus attrac-
      tifs. *In* "Thesis". Université Pierre et Marie
      Curie, Paris.
3.   Liggett, T.M. (1976). The stochastic evoluti-
      on of infinite systems for interacting parti-
      cules. *In* "Ecole d'Eté de Probabilités de
      Saint-Flour VI". 598. Springer-Verlag. Berlin
4.   Spitzer, F. (1974). Recurrent random walk of
      an infinite particle system, *Transactions of the*
      A.M.S. 198, 191-199.

STABLE CORE IN DISCRETE ITERATIONS

Françoise Fogelman-Soulie

*Laboratoire de Dynamique des Réseaux
GSC, 1 rue Descartes, 75005 Paris
and Université de Paris V.
France*

1- INTRODUCTION

Theoretical results on the dynamics of automata networks are
not very many (9,12) and usually rely on very strong assumptions: small number of inputs per element, symmetry of the
network, restrictions on the mappings used by the elements...
When these results do not apply-which may happen in many
practical applications-one would need a method to provide
some information on the dynamics.
In this paper, we present two examples of automata networks
much studied in the litterature: boolean networks and threshold networks and we give a method based on the notion of
forcing mapping first introduced by Kauffman(11), which
allows the effective computation of an "estimate" of the
"typical" limit cycle of the network.
In section 2, we introduce some definitions, in section 3
present simulations of the dynamics. In section 4, we study
boolean networks and give a characterization of their stable
core, and in section 5 perform the same job for unimodal
threshold networks.

2- DEFINITIONS (*)

Definition 1:
A boolean mapping in n variables is a mapping $f:\{0,1\}^n \to \{0,1\}$
A boolean network of n elements is a mapping $F:\{0,1\}^n \to \{0,1\}^n$
such that $F=(f_1,\ldots,f_n)$ where all f_i (i=1,...,n) are boolean
mappings.

_____

(*) Notations are mostly taken from (12) where the reader
can find more details.

DYNAMICAL SYSTEMS
AND CELLULAR AUTOMATA

27

Definition 2

Let $f:\{0,1\}^n \rightarrow \{0,1\}$ be a boolean mapping, f is called a
threshold mapping of coefficients (a_j) and threshold θ iff:

$$f(x) = \begin{cases} 1 \text{ iff } \quad \sum_{j=1}^{n} a_j x_j - \theta \geqslant 0 \\ 0 \text{ otherwise} \end{cases}$$

A threshold network is a boolean network $F=(f_1 \ldots, f_n)$ such
that all f_i (i=1,...,n) are threshold mappings.

Definition 3

A unimodal threshold mapping is a threshold mapping such
that all coefficients are in $\{-1,0,+1\}$
A unimodal threshold network is a threshold network $F=(f_1,..,f_n)$
such that all f_i (i=1,...n) are unimodal threshold mappings.
    Although unimodal threshold networks are indeed a parti-
cular case of threshold networks, it has been shown(7) that
their behavior may be as complicated as in the general case
and thus their study may help understand the general case.

Definition 4

Let $F=(f_1,\ldots,f_n)$ be a boolean network. If all mappings f_i
really depend on $k \leqslant n$ variables only, network F is said
of interconnectivity   k.
    In the following, we will restrict our study to boolean
networks of interconnectivity k=2. Although networks with
larger interconnectivity (k >2) exhibit different behaviors
(10), it can be shown that networks with large k, but mappings
restricted to the class of "forcing"mappings(11) are very
similar to networks of interconnectivity k=2. Also, networks
with interconnectivity k=2 are not fully understood until
now and so any study on boolean networks should begin by
making their case clear.

Definition 5

Let $F=(f_1,\ldots,f_n)$ be a boolean network. The connection graph
associated with F is the graph $G=(X,U)$ with $X=\{1,\ldots,n\}$ and
$(i,j)\in X\times X$ is in U iff f_j really depends on variable i.
(See fig.1)

Definition 6

Let $F: \{0,1\}^n \rightarrow \{0,1\}^n$ be a boolean network. A parallel
iteration on F is defined by: $\forall\ t \geqslant 0, x(t+1)= F\ (x(t))$ where
$x(t)$ is the state of the network at time t.

Fig. 1: Boolean network of interconnectivity k=2.

We show a boolean network of interconnectivity k=2,with regular connections,each element receives its 2 inputs and sends its 2 outputs to its nearest neighbors on a regular grid drawn on a torus (lines 1 and n are connected,as columns 1 and n,for an nxn elements network).This allows a more convenient geometrical representation and does not change the results.

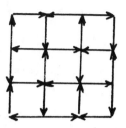

```
 1 11 10 13  9  5 10  7 12  4  9 12  1  9 12  3
 3  7 12 13 10  5 13  8  1 11  9  9 14  7  6 11
14  8 10  4  4  5  6  5  8  4  8  9  8 13 12  2
 2 12 12  2  4  9  8  5 11 12  3  1 13  8 14  6
11 14  7 14 12 13  2 11 14  3  6 14  9  4  6  4
12 12  2 10  8  5 14 14  8  4  7 10  2 14  7  4
10  6 13  3  2 13  8  9 13 10 10  8  1 14  4 14
12 13  2  8  2  1  9  4  2  6  7  3  4  8  4  3
13 13  5 11 13  6 10  8 11  6 12  4  8 12 11  5
10  6 13  7  2  9 10 10  4  2  6  5  2  8 10 14
 8 11  8  2  4  6 10  4  7 12 13  3  4  3  8  1
 4  8 12  9  6  5  6 14  1 12  3  7 13  7 11  1
 1 14  2 11  2  4  5  7  3  3  4  5 12 10 11 13
 6  9  3 12 14  6  9 12 12  5 11  3 11 11 13  8
12 10 12  8  5 12  1 11  1 13 13  2  6 10 11 11
 5 14  7 12  1  9 10 13 12  9 12  3  4  2  7  2
```

This gives the array of boolean mappings in 2 variables used by the elements of a boolean network with regular connections (see above).The numbers refer to the numerotation of Table 1.

```
5 1 0 1 0 6 6 6 6 5 6 5 0 1 1 6      5 0 0 0 0 6 6 6 6 5 1 0 0 0 1 1
6 1 0 0 0 5 6 5 5 5 6 5 5 6 1 0 6    6 0 0 6 0 5 6 5 5 5 6 0 1 1 0 1 6
0 1 1 0 0 1 0 6 5 5 5 6 1 1 1 5      0 0 6 6 0 1 0 6 5 5 5 1 1 1 1 5
1 1 1 5 0 0 5 5 6 6 6 5 1 1 1 0      1 0 6 5 0 0 5 5 6 6 6 5 1 1 1 1
6 6 6 5 5 1 0 6 6 5 5 5 6 5 0 0      6 6 6 5 5 1 0 6 6 5 5 5 6 5 0 1
6 6 5 5 5 1 1 0 1 5 6 5 5 6 6 5      6 6 5 5 5 0 1 1 1 5 6 5 5 6 6 5
5 5 5 6 5 6 0 0 0 1 5 5 5 6 5 6      5 5 5 6 5 6 0 0 0 1 5 5 5 6 5 6
5 6 5 5 5 6 5 0 5 0 0 6 5 5 5 5 6    5 6 5 5 5 6 5 0 5 0 1 6 5 5 5 5 6
1 6 5 6 6 6 5 5 6 0 0 5 5 5 6 0      0 6 5 6 6 6 5 5 6 1 1 5 5 5 6 0
5 6 6 6 5 6 5 5 5 0 0 6 5 5 1 1      5 6 6 6 5 6 5 5 5 0 0 6 5 5 0 1
5 5 5 5 5 5 5 5 6 1 6 5 5 6 1 0      5 5 5 5 5 5 5 5 6 0 6 5 5 6 0 0
5 5 5 0 0 6 5 5 5 5 5 6 5 6 0 1      5 5 5 1 0 6 5 5 5 5 5 6 5 6 1 1
1 0 0 0 1 0 5 6 5 6 5 6 6 6 0 0      0 1 0 1 0 1 5 6 5 6 5 6 6 6 1 1
0 1 1 1 0 0 6 6 5 5 6 5 1 6 1 1      1 0 0 1 0 1 6 6 5 5 6 5 0 0 1 0
0 6 6 0 1 1 5 6 5 6 6 5 1 1 0 0      0 6 6 0 1 0 5 6 5 1 1 0 1 0 0 0
6 6 1 1 1 0 6 6 5 5 6 5 1 0 6 5      6 6 1 0 1 0 6 6 5 1 1 0 0 1 0
```

This figure shows 2 limit cycles for the above network,corresponding to 2 different initial conditions.Points 5 and 6 were fixed during the limit cycle,points 0 and 1 changed states.

A <u>sequential iteration</u> on F associated to the permutation Π of $\{1,\ldots,n\}$ is defined by: $\forall\, t \geqslant 0, x(t+1) = F_\Pi\,(x(t))$ where F_Π is defined as follows:

$$x_{\Pi^{-1}(1)}(t+1) = F(y^1) \qquad \text{where } y_j^1 = x_j(t)\ \forall\, j(1,\ldots,n)$$

$$x_{\Pi^{-1}(2)}(t+1) = F(y^2) \qquad \text{where } y_{\Pi^{-1}(1)}^2 = x_{\Pi^{-1}(1)}(t+1)$$
$$\text{and } y_j^2 = x_j(t) \ \forall\, j \neq \Pi^{-1}(1)$$

$\ldots$

$$x_{\Pi^{-1}(i)}(t+1) = F(y^i) \qquad \text{where } y_{\Pi^{-1}(j)}^i = x_{\Pi^{-1}(j)}(t+1)\ \forall\, j(1,\ldots,i-1)$$

$$\text{and } y_j^i = x_j(t) \ \forall\, j \notin \{\Pi^{-1}(1),\ldots,\Pi^{-1}(i-1)\}$$

$\ldots$

$$x_{\Pi^{-1}(n)}(t+1) = F(y^n) \qquad \text{where } y_j^n = x_j(t+1)\ \forall\, j \neq \Pi^{-1}(n)$$

$$\text{and } y_{\Pi^{-1}(n)}^n = x_{\Pi^{-1}(n)}(t)$$

Remarks: There are n! different sequential iterations on F, associated to the n! different permutations of $\{1,\ldots,n\}$. As the state space is finite, parallel and sequential iterations on F have trajectories which always end up in limit cycles.

Definition 7

Let $F:\{0,1\}^n \to \{0,1\}^n$ be a boolean network and Φ an iteration on F (Φ is defined by F for a parallel iteration and F_Π for a sequential iteration).
The <u>iteration graph</u> of Φ is the graph $G_I = (\{0,1\}^n, I)$
where: $\forall\, x,y \in \{0,1\}^n$, $(x,y) \in I$ iff $y = \Phi(x)$.
    Any circuit in G_I is called a <u>limit cycle</u>. The length of the circuit is called the <u>period</u> of the limit cycle.
Any limit cycle of period 1 is called a <u>fixed point</u>.

Remark: All sequential iterations on F have the same fixed points: those of the parallel iteration on F (12). But they may have different limit cycles.
    It is the purpose of this paper to show how the limit cycles of the parallel and sequential iterations on a boolean network and a unimodal threshold network can be compared and predicted.

3- SIMULATIONS

Simulations of iterations on boolean networks of interconnectivity 2 and on unimodal threshold networks have been made.

They show that the limit cycles of an iteration on a given network are not very different. We will introduce a "geometrical" representation of the limit cycles which will allow to compare them (see also(3) for a discussion of the different methods of comparison).

Definition 8

Let $F:\{0,1\}^n \rightarrow \{0,1\}^n$ be a boolean network and Φ be an iteration on F. Let $C=(x^1,\ldots,x^T)$ be a limit cycle. Then we define the <u>stable part</u> of C as: $S_C = \{i(1,\ldots,n): \forall t(1,\ldots,T), x_i^t = x_i^1\}$

the <u>oscillating subnet</u> of C as: $O_C = \{1,\ldots,n\}/S_C$

the <u>stable core</u> of the iteration Φ as: $S = \bigcap_C S_C$

the <u>oscillating core</u> of the iteration Φ as: $O = \bigcap_C O_C$

(where the intersection is on all possible limit cycles C in the iteration graph G_I).

Stable parts and oscillating subnets allow to compare different limit cycles: if two different limit cycles have similar oscillating subnets and stable parts they usually are "neighbours", meaning that the minimum Hamming distance between the different states on the limit cycles is small (cf fig.1).

The existence of large stable parts in the limit cycles had been noticed in the pionneering work of Kauffman(10,11). Later on, Atlan and al.(1) have analyzed the stability of stable parts and oscillating subnets of random boolean networks with respect to various perturbations (amputation, noise,...).

More recent work (4,6) have shown that,in fact, all limit cycles were found relatively similar because the stable core and oscillating core were large, covering an important part of the network.

Predicting the stable and oscillating cores of a given network is thus a significant task as it allows a good description of the limit cycles of the iteration on the network: any limit cycle is made up from the stable core (which is part of its stable part), the oscillating core (which is part of its oscillating subnet) and from a "wishy-washy" region (3)where points may be either in the stable part or the oscillating subnet.(see fig.2).

Fig. 2:Cores of boolean networks for parallel and sequential
iterations. (stable : 0 , oscillating : 80)

```
18 80 80 80 80  0  0  0  0  0 30 30 80 80 80 30
18 80 80 80 80  0  0  0  0  0 30 30 30 80 80 18
80 80 80 80 80 80 80  0  0  0  0 30 80 80 80  0
80 80 80  0 80 80  0  0  0  0  0  0 80 80 80 80
 0  0  0  0  0 72 72  0  0  0  0  0  0  0 80 80
 0  0  0  0  0 80 80 80 80  0  0  0  0  0  0  0
 0  0  0  0  0  0 80 80 80 80  0  0  0  0  0  0
 0  0  0  0  0  0 80  0 80 80  0  0  0  0  0  0
80  0  0  0  0  0  0  0  0 80 80  0  0  0  0 80
 0  0  0  0  0  0  0  0  0 80 80  0  0  0 80 80
 0  0  0  0  0  0  0  0  0 80 ?0  0  0  0 80 80
 0  0  0 69 69  0  0  0  0  0  0  0  0  0 80 80
80 80 80 80 69 67  0  0  0  0  0  0  0 30 80 80
80 80 80 80 67 67  0  0  0  0  0  0 80 30 76 80
80  0  0 80 67 67  0  0  0 30 30 30 80 80 80 80
18  0 80 80 77 77  0  0  0 30 30 30 80 80 30 30
```

This array shows the stable core (points marked 0) and the
oscillating core for the parallel iteration of the boolean
network shown in fig.1.Points in the "wishy-washy" region
are marked by the number n (0< n<80) of different initial
conditions in which they were oscillating.

```
80 80 80 80 80  0  0  0  0  0  0  0 80 80 80 80
80 80 80 80 80  0  0  0  0  0  0 80 80 80 80 57
80 80 80 80 80 80 80  0  0  0  0 80 80 80 80  0
80 80 80 80 80 80 41  0  0  0  0  0 80 80 80 80
 0  0 80 80 80 41 41  0  0  0  0  0  0  0 80 80
 0  0  0 14 14 80 80  0 80  0  0  0  0  0  0  0
 0  0 14 14 14  0 80 80 80 80  0  0  0  0  0  0
 0  0  0  0 14  0 80 80 80 80  0  0  0  0  0  0
80  0  0  0  0  0  0  0  0 80 80  0  0  0  0 80
 0  0  0  0  0  0  0  0  0 80 80  0  0  0 80 80
 0  0  0  0  0 41  0  0  0 80  0  0  0  0 80 80
 0  0  0 74 74 41 41 41  0  0  0  0  0  0 80 80
80 80 80 80 74 41 41 41 41  0  0  0  0 80 80 80
80 80 80 80 41 41 41 41 41  0  0  0 80 80 80 80
80 57 57 80 41 41  0  0  0  0  0  0 80 80 80 80
80 57 80 80 34 34  0  0  0  0  0  0 80 80 80 80
```

This array shows the stable and oscillating cores (see above)
for a sequential iteration of the preceding network.

These two arrays are not exactly identical,showing that the
parallel and sequential iterations do not have the same iterations graphs.But they nethertheless are very similar.

Work on perturbations of boolean networks with interconnectivity 2 (sending "noise" on some elements) has shown (3) that the "sensitive part" of the network lies mainly in the wishy-washy region. This sensitivity could be used to build a learning machine from boolean networks (see also(2)in this book).

In the following, we will give methods to find the stable core for boolean networks of interconnectivity 2 and for unimodal threshold networks.

4- STABLE CORE OF BOOLEAN NETWORKS

Kauffman, in his studies of boolean networks had shown that stable elements in parallel iterations were to be found in what he called "forcing structures"(11). We have extended these results(6,8) and are now able to compute an estimate of the stable core for parallel and sequential iterations.

Definition 9

The boolean mapping $f:\{0,1\}^n \to \{0,1\}$ is <u>forcing</u> in its i-th variable iff:

$$\exists\, x_i^*,\; v_i^* \in \{0,1\} : \forall\, y \in \{0,1\}^n, y_i = x_i^* \Rightarrow f(y) = v_i^*$$

Variable i is called a <u>forcing variable</u> of mapping f, x_i^* a <u>forcing value</u> of variable i, v_i^* a <u>forced value</u> of mapping f for variable i.

(See table 1 for examples of forcing mappings in 2 variables)

Definition 10

Let $F: \{0,1\}^n \to \{0,1\}^n$ be a boolean network and $G = (X,U)$ be its connection graph. Arc $(i,j) \in U$ is said to be a <u>forcing arc</u> iff: $\exists\, v_i^*,\; x_j^i\; \{0,1\} : v_i^* = x_j^i$ and v_i^* is a forced value

of f_i and x_j^i is a forcing value of variable i for mapping f_j

Graph $H = (X,V)$ is called the forcing graph of network F iff H is the subgraph of G such that $(i,j) \in U$ is in V iff arc (i,j) is forcing.

Any circuit C in H such that for all arc (i,j) in

$C,\; v_j^* = f_j(v_i^*)$ is called a <u>forcing circuit</u> or <u>non-frustrated</u> circuit of graph G.

TABLE 1.
Boolean mappings in 2 variables.

n°	0	1	2	3	4	5	6	7	8	9	10	11	12	13	14	15
		Nor	⇒	$\bar{t}_2$	⇐	$\bar{t}_1$	Xor	Nand	And	⇔	t_1	⇐	t_2	⇒	Or	
	00 00	10 00	00 10	10 00	01 00	11 00	01 10	11 10	00 01	10 01	00 11	10 11	01 01	11 01	01 11	11 11
v^*	0	0	0	0,1	0	0,1	/	0	0	/	0,1	1	0,1	1	1	1
x_1^*	01	1	0	/	1	1,0	/	0	0	/	0,1	1	0,1	0	1	1
x_2^*	01	1	1	1,0	0	/	/	0	0	/	/	0	0,1	1	1	.01

This table shows the different boolean mappings in 2 variables and, for those which are forcing, their forced value v and forcing values of inputs 1 and 2 (x_1 and x_2). Mappings 6 (Xor) and 9 (⇔) are non-forcing. Mappings 3($\bar{t}_2$), 5 ($\bar{t}_1$), 10 (t_1) and 12 (t_2) are forcing in 1 variable. All other mappings are forcing in 2 variables.

Fig. 3: Forcing domain of a boolean network.

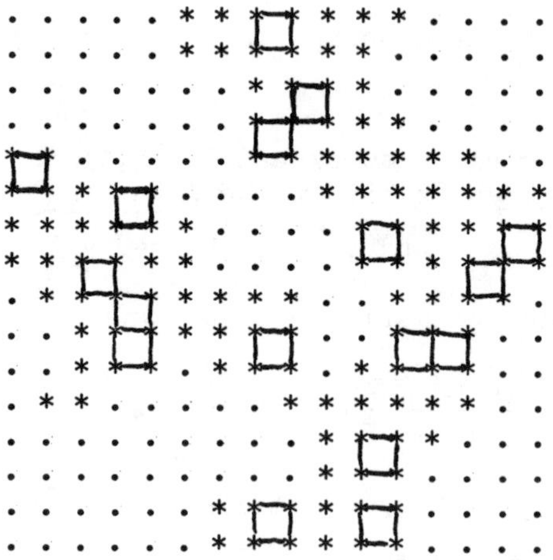

This figure shows the forcing domain of the boolean network of fig. 1, as computed by the algorithm of definition 11. Points marked by a * are in the forcing domain.

Forcing circuits are drawn: □. It is clear on the picture that all forcing circuits are in the forcing domain.

Moreover, when comparing fig. 2 and 3, it appears that the stable cores for the parallel and sequential iterations are well approximated by the forcing domain.

Definition 11:

Let $F:\{0,1\}^n \to \{0,1\}^n$ be a boolean network of interconnectivity k, Let $G =(X,U)$ be its connection graph an $H=(X,V)$ be its forcing graph. $D \subset C$ will be called the <u>forcing domain</u> of network F and $v^* : D \to \{0,1\}$ .the <u>forced value</u> of D iff:

1) $\forall i \in X$, i is on a forcing circuit $\Rightarrow i \in D$ and $v^*(i)$ is the forced value of f_i.

2) $\forall i \in D,(i,j) \in V \Rightarrow j \in D$ and $v^*(j)$ is the forced value of f_j associated to the forcing value $v^*(i)$ of variable i.

3) $\forall t(1,\ldots,k),(i_t \in D,(i_t,j) \in U) \Rightarrow j \in D$

and $v^*(j)=f_j(v^*(i_1),\ldots,v^*(i_k))$

4) $\forall(i,j) \in U, i \in D, v^*(i)=x_i^1 \Rightarrow j \in D$ and $v^*(j)$ is the forced value of f_j associated to the forcing value x_j^1 of i .

The forcing domain of a network F is defined uniquely in terms of mapping F: it is independent of the different iterations that can be done on F. It can be computed through a very fast algorithm (8),by recursively applying the 4 rules in definition 11. (See Fig 3.)
It has been shown through simulations, by Kauffman (11) that forcing structures were generally stable. In (6,8)we have extended these results and proved that the forcing domain was a good approximation of the stable cores of the parallel (6) and sequential (8) iterations on F.

Theorem 1:   Let $F: \{0,1\}^n \to \{0,1\}^n$ be a boolean network of interconnectivity k=2. For any forcing circuit C and any limit cycle of the parallel iteration on F, then:
- either C is stable and in its forced value,
- or C is oscillating, not in its forced value and the following conditions hold:

(cond. 1)$\left\{\begin{array}{l} \forall t(1,\ldots,T),\forall i \in(1,\ldots,n)\cap C: x_i(t)\neq x_i^* \Rightarrow \\ \{\exists k(1,\ldots,n),k\neq i:(k,j) \in U \Rightarrow x_k(t) \neq x_k^*\}\end{array}\right\}$

where:.$x(t) \in\{0,1\}^n$ is the state of the network at time $t(1,\ldots,T)$, in the limit cycle (whose pariod is T).

.x_i^* is the forcing value of variable i for mapping f_j, where j is the successor of i in C($j \in C$ and $(i,j) \in U$).

.x_k^* is the forcing value of variable k for mapping f_j.

(See(6) for a proof)

Corollary 1: The forcing domain D of F is stable under the parallel iteration except if there exists a forcing circuit C which satisfy its (cond.1).

Hence: { $\forall C$ forcing circuit,(cond.1) violated } $\Rightarrow C \subset S$

This theorem proves that forcing circuits—which are part of the forcing domain—must generally be part of the stable core of the parallel iteration: they may also be non stable but the conditions for that -(cond.1)- are very restrictive. Simulations to compute the stable core were made on a small sample of the 2^n possible initial conditions. Hence only an approximation of the stable core was computed:nethertheless this estimate was very similar to the forcing domain,showing -according to corollary 1- that conditions (cond.1) were generally not satisfied (see fig.3).

*Remark:*conditions 2,3,4 in definition 11 explains how to "propagate" the forced value starting from the forcing circuits. Hence, if the forcing circuits are in the stable core (theorem 1), then the whole forcing domain is also in it.

Theorem 2: Let F $\{0,1\}^n \xrightarrow{} \{0,1\}^n$ be a boolean network of interconnectivity k=2. Let Π be a permutation of $\{1,\ldots,n\}$ and $C=(e_1,\ldots,e_k)$ be a forcing circuit of G.Then for the sequential iteration on F associated to Π:
- either C is stable and in its forced value.
- or C is oscillating with a period T such that:

(cond.2) $\{\forall t (1,\ldots,T), \ \sum_{i=1}^{n_\Pi} m_{i+t} = \sum_{i=1}^{n_\Pi} m_{i+t+T}\}$

where:. n_Π is the total number of inversions (* *) presented by couples of successors (e_i,e_j) in C.

　　.e_{m_1} is the first element in C which presents an inversion with its successor.

　　.$e_{m_1+m_2}$ is the second one

　　...

　　.$e_{m_1+\ldots+m_{n_\Pi}}$ is the n_Π -th

(See (8) for more details and a proof).

Corollary 2: The forcing domain D of F is forcing under the sequential iterations on F except if there exists a forcing circuit C which satisfy (cond.2) .

Hence: { $\forall C$ forcing circuit,(cond.2) violated} $\Rightarrow C \subset S$.

(**) $(e_i,e_j) \in U, e_i,e_j \in C$ present an inversion with respect to Π iff $i > j$ and $\Pi(e_i) < \Pi(e_j)$

This theorem thus proves that forcing circuits are generally part of the stable core of the sequential iterations. Simulations show (see fig. 2 and 3) that the different stable cores were both similar among themselves and similar to the forcing domain, thus demonstrating (corollary 2) that conditions (cond. 2) are generally not satisfied.

5- STABLE CORE OF UNIMODAL THRESHOLD NETWORKS;

Various results have been proven on the iteration of threshold networks: they usually rely on the assumption that the matrix A of the coefficients of the networks is symmetric. In (9) Goles-Chacc proved that parallel iterations of thres - hold networks with symmetric matrix A may only have limit cycles of period 1 or 2. In (5) Fogelman and al. proved that sequential iterations of threshold networks with symmetric matrix A with non-negative diagonal may only have fixed points. We have studied (7) unimodal threshold networks in an attempt to better understand the dynamical behaviour of threshold networks with non-symmetric matrix. We also restricted our study to regularly connected networks:

Definition 12:

Let $F: \{0,1\}^{n \times n} \to \{0,1\}^{n \times n}$ be a unimodal threshold network

of n^2 elements. Then F has <u>regular connections</u> iff:
$F = (f_{ij})$ where $\forall i,j (1,\ldots,n) \overline{f_{ij}}$ is a unimodal threshold
mapping depending only on the 4 variables $(i-1,j),(i,j-1)$, $(i,j+1),(i+1,j)$ (where $i-1,i+1,j-1,j+1$ are taken modulo n).
There are only 5 possible such mappings (see table 2). Note that none of these mappings is forcing in the sense of definition 9, which prevents us to use the results of §4. But they are forcing in an extended way:

Definition 13:

Let $f : \{0,1\}^n \to \{0,1\}$ be a boolean mapping. Then f is said to be <u>forcing</u> with respect to the k-tuple $(i_1,\ldots,i_k)$ iff:

$$\exists x_{i_1}^*,\ldots,x_{i_k}^* \in \{0,1\}^k, v_{i_1 \ldots i_k}^* \in \{0,1\} :$$

$$\forall y \in \{0,1\}^n, \{y_i = x_i^*, \forall i (i_1,\ldots,i_k)\} \Rightarrow \{f(y) = v_{i_1 \ldots i_k}^*\}$$

(See table 2 for an example)
We will use this concept to extend our results of §4 to the case of unimodal threshold networks.

TABLE 2.
Unimodal threshold mappings in 4 variables

mapping Variable	4 ++++	3 +++-	2 ++--	1 +---	0 ----
0000	1	1	1	1	1
0001	1	0	0	0	0
0010	1	1	0	0	0
0011	1	1	0	0	0
0100	1	1	1	0	0
0101	1	1	1	0	0
0110	1	1	1	0	0
0111	1	1	0	0	0
1000	1	1	1	1	0
1001	1	1	1	1	0
1010	1	1	1	1	0
1011	1	1	0	0	0
1100	1	1	1	1	0
1101	1	1	1	0	0
1110	1	1	1	0	0
1111	1	1	1	0	0
x^* , v^*	....,1	..1.,1 ...0,1	..00,1 11..,1 .10.,1 001.,0 .011,0	.11.,0 01..,0 .000,1 100.,1	1...,0

*This table gives, for each of the 5 possible unimodal thre-
shold mappings in 4 variables, its values for the 16 possible
combinations of variables, and the forcing patterns x* of the
variables with the associated forced value v*.*

Definition 14

Let $F: \{0,1\}^{n \times n} \to \{0,1\}^{n \times n}$ be a unimodal threshold network with regular connections and $G=(X,U)$ be its connection graph

($X= \{0,1\}^{n \times n}$). Then an elementary chain $(u_1,\ldots,u_k)$,

with $u_\ell = ((i_\ell, j_\ell),(i_{\ell+1}, j_{\ell+1})) \in U, (\ell=1,\ldots,k)$, is said to be

<u>forcing</u> iff there exists a sequence $A_1,\ldots,A_{k+1}$ such that:

(FC) $\left\{\begin{array}{l}
\text{(i) } \forall t(1,\ldots,k+1), A_t \in R(f_{i_t j_t}) \\[2mm]
\text{(ii) } f_{i_1 j_1} = f_{i_{k+1} j_{k+1}} = 3 \quad \text{(see table 2)} \\[2mm]
\text{(iii) } \forall t(2,\ldots,k+1), \text{if } (x,a) \in A_t \text{ then:} \\[1mm]
\qquad *x= \Pi_{i_t j_t}(i_\ell, j_\ell) \text{for } \ell=t-1 \text{ or } t+1 \ (***) \\[3mm]
\qquad *a= v^*(A\ell) \\[1mm]
\qquad \text{(we will denote a as } v^*_{i_\ell j_\ell}, \ell=t-1 \text{ or } t+1)
\end{array}\right.$

where R describes the structure of forcing patterns (see appendix 1).

Definition 15

Let $F: \{0,1\}^{n \times n} \to \{0,1\}^{n \times n}$ be a unimodal threshold network. Then $D \subset X$ will be called the <u>forcing domain</u> of F and

$v^*_D : D \to \{0,1\}$ the <u>forced value</u> of D iff:

(i) if (i,j) is on a forcing chain, then $(i,j) \in D$ and
$$v^*_D (i,j)=v^*_{ij}$$
(ii) for any elementary chain $(u_1,\ldots,u_k)$ (with
$u_\ell = ((i_\ell, j_\ell),(i_{\ell+1}, j_{\ell+1})) \in U, \ell=1,\ldots,k)$ such that there
exists a sequence $A_1,\ldots,A_{k+1}$ satisfying:

(***) For each element $(i,j) \in X$, there exists an ordering Π_{ij} of its antecedents such that, up to this ordering, the coefficients $a_{(i,j),\ell}$ of $(i,j), \ell=1\ldots4$, satisfy the conditions associated with mappings f_{ij}: for example 1,1,1,-1 for mapping 3. This is called the <u>coding</u> of the antecedents of (i,j) (see appendix 1).

$$(SFC) \quad \begin{cases} (i) \\ (iii) \\ (iv) \\ (ii)' \ f_{i_\ell j_\ell} = 3 \ or \ (i_\ell, j_\ell) \in D, \text{for } \ell = 1 \ and \ \ell = k+1 \end{cases} \quad \text{as in (FC)}$$

then all $(i_\ell, j_\ell) \in D$, $\ell = 1, \ldots, k+1$.
Such a chain is then called a <u>semi-forcing chain</u>.

(iii) if $(i,j) \in X$ and there exists $G \subset R_{f_{ij}}$ such that:

$$\forall (x,a) \in G, \Pi_{ij}^{-1}(x) \in D \ and \ a = v^*_D(\Pi_{ij}^{-1}(x))$$
then $(i,j) \in D$ and $v^*_D(i,j) = v^*(G)$.

(iv) if $(i,j) \in X$ and $\forall (k,\ell) \in X, \{((k,\ell),(i,j)) \in U \Rightarrow (k,\ell) \in D \}$

then $(i,j) \in D$ and $v^*_D(i,j) = f_{ij}(x)$ where:

$$x = (x_{k\ell})_{1 < k, \ell < n} \ and \ x_{k\ell} = \begin{cases} v^*_D(k,\ell) \ if \ ((k,\ell),(i,j)) \in U, \\ 0 \ otherwise. \end{cases}$$

Forcing domains have usually been found stable (see Fig.4):
forcing chains are stable unless very drastic conditions
(which are more intricated to write than in theorems 1 and 2:
see (7)). As in the case of boolean networks, the forcing
domain gives a good approximation of the stable core of the
network, thus allowing a prediction of the "typical" limit
cycle.

Nethertheless, it is important to stress a main difference with boolean networks. In this latter case, the algorithm for computing the forcing domain is fast (polynomial).
In the case of unimodal threshold networks, the algorithm is
much slower: at each point of the network, a search has to
be made so as to find a forcing chain. There is no direct
equivalent to the "forcing graph" in this case: this is due
to the fact that forcing mappings for interconnectivity 4
are less than for 2 (11)-which had lead us to the generalization of definition 13- and thus that the forcing structures
need more than one arc per element.

This problem is to be related to the spin-glass problem
where it is well known that the algorithms for finding
ground states are NP-complete.

Fig. 4: Unimodal threshold networks.

```
 -  +  +  -  -  -  -  +  +  +
+.- -.- -.- +.+ +.- +.+ -.- +.- -.- -.-
 +  -  -  -  -  -  +  -  -  -
 +  -  +  -  +  -  -  -  +  +
-.+ -.- +.- +.+ +.- +.- -.+ +.+ -.+ -.-
 -  +  +  -  +  -  -  -  +  +
 -  +  -  -  +  +  +  +  +  -
+.+ -.- +.- +.+ +.- -.- +.- -.- -.- -.+
 -  -  +  -  +  +  +  -  -  -
 +  +  -  +  +  +  +  +  -  -
-.- -.+ +.+ +.- +.- -.- -.- -.- +.+ +.+
 +  -  +  +  +  +  +  +  -  -
 -  +  -  +  -  +  -  -  +  +
+.+ -.- +.+ +.- +.+ +.- -.- +.- +.- +.-
 -  -  +  +  +  -  +  +  +  -
 +  +  -  +  -  +  -  -  -  -
+.- -.- +.- -.+ +.+ -.- +.+ -.+ +.+ -.+
 +  -  +  -  +  +  +  +  +  +
 +  +  -  -  +  -  +  +  -  -
-.- +.- +.+ +.+ +.- +.+ +.- -.- +.- +.+
 +  +  +  +  +  +  -  +  -  -
 -  -  +  +  +  -  +  -  -  +
+.- +.+ -.+ +.- +.- -.+ +.- -.+ +.+ -.-
 -  -  -  -  +  +  -  +  -  +
 -  +  -  +  +  +  -  -  +  +
-.+ -.- -.+ +.- -.+ -.- +.+ -.- -.+ +.-
 +  +  +  -  -  -  -  +  -  -
 -  -  +  +  -  -  -  +  -  -
+.- -.+ -.- -.- +.- -.- +.- -.+ +.- +.+
 -  +  -  -  +  +  +  -  +  -
```

This figure shows a unimodal threshold network with regular connections drawn on a torus,with 10x10 elements.Mappings 0 (----) and 4(++++) were excluded since their main role is to make the corresponding element stable.

The two figures below show the stable (points marked 0)and oscillating (marked 16) cores of the preceding network for parallel (left) and sequential (right) iterations.16 different initial condions were run and the corresponding limit cycles computed.

The forcing domain, as defined by definition 15 is circled. It appears clearly that this forcing domain gives an approximation of the stable cores for both the parallel and sequential iterations.

1	4	5	5	5	16	16	16	16	15
1	0	16	0	0	3	3	16	16	15
1	0	16	16	0	4	0	4	4	1
16	0	0	0	0	4	4	16	4	16
16	16	0	0	0	0	4	16	0	12
0	15	15	0	0	0	0	0	0	15
16	16	0	0	0	0	0	0	0	16
16	16	0	0	0	1	16	16	16	16
16	5	0	16	0	1	16	16	15	15
0	5	0	16	16	16	16	16	16	16

0	0	5	5	5	15	15	15	15	10
0	0	16	0	0	2	2	15	15	10
0	0	16	16	0	2	0	2	2	0
16	0	0	0	0	0	2	16	2	16
16	16	0	0	0	0	2	16	0	11
0	7	7	0	0	0	0	0	0	7
16	16	0	0	0	0	0	0	0	16
16	16	0	0	0	0	15	15	16	16
16	5	0	16	0	0	15	15	10	10
0	5	0	16	16	15	15	15	15	15

6- CONCLUSION

In this paper, we have presented a method for studying the dynamics of automata networks. For two different cases, we have shown that it was possible to characterize this dynamics by its <u>stable core</u> which was predictible by means of the forcing domain.

As there is no exact theory to predict precisely the structure of the limit cycles of such dynamics, the notion of forcing domain proves to be quite powerful on the two examples presented here.

Nethertheless,on these examples, the importance of the interconnectivity appears clearly: as the interconnectivity increases from to 2 to 4, the complexity for computing the forcing domains increases drastically.

This feature may impose severe restrictions on the use of this method in some practical applications.

The author gratefully acknoledges many helpful conversations with H.Atlan, G.Weisbuch and E.Goles who took part in this research. Part of this research was done under financial support from CNRS (AI n°03 41 68 00).

REFERENCES

1. ATLAN H.,FOGELMAN-SOULIE F.,SALOMON J.,WEISBUCH G.(1981).
Random boolean networks.Cybernetics and systems,12,pp103-121.

2. ATLAN H. (1983) this meeting.

3. BEN-EZRA E.(1983).Random boolean networks.Masters Thesis,
Weizmann Institute for Science.

4. FOGELMAN-SOULIE F.,MILGRAM M.,WEISBUCH G.(1983),Automata
networks as models for biological systems:a survey,in"Rythms
in biology and other fields of application". Lecture Notes in
Biomathematics. Springer Verlag,49,pp 144-172.

5. FOGELMAN F.,GOLES E.,WEISBUCH G.(1983).Transient length
in sequential iterations of threshold functions.Discrete
Applied Mathematics,6. pp 95-98.

6. FOGELMAN-SOULIE F.(1983).Frustration and stability in
random boolean networks,submitted to Discrete Applied
Mathematics.

7. FOGELMAN-SOULIE F.,GOLES-CHACC E.,WEISBUCH G.(unpublished)
Unimodal threshold networks.

8. FOGELMAN-SOULIE F.(unpublished).Stable core of discrete
iterations on boolean mappings.

9. GOLES-CHACC E.(1980).Comportement oscillatoire d'une
famille d'automates cellulaires non uniformes.Thèse IMAG
Grenoble.

10. KAUFFMAN S.(1970). Behaviour of randomly constructed
genetic nets,in "Towards a theoretical biology",vol3 Eds C.H.
Waddington,Edinburgh University Press,pp 18-37.

11.KAUFFMAN S.(1972). The organization of cellular genetic
control systems,in "Lectures on mathematics in the Life
Sciences".,vol,3.Ed. J.D. Cowan. The American Mathematical
Society,pp 63-116.

12.ROBERT F.(1981). Itérations discrètes.Cours polycopié,
IMAG. Grenoble.

Appendix 1:

We define the forcing patterns of inputs for the different unimodal threshold mappings(see table 2):laws 0 and 4 were excluded since they are constant.

Let $L=\{1,2,3\}$be the 3 different laws,
$X=\{1,2,3,4\}$be an ordering of the inputs and $R: L \rightarrow pp \left(\bigcup_{k=1}^{3} (Xx\{0,1\})^k\right)$
$v^*: p\left(\bigcup_{k=1}^{3}(Xx\{0,1\})^k\right) \rightarrow \{0,1\}$ be defined as follows:

$\underline{\ell= 3}$ suppose the ordering is such that the coefficients
satisfy $1,1,1,-1$.   Then $R(3)=\{G_3^i\}_{1\leq i\leq 4}$ with

..1. $G_3^i = \{i,1\}$ for $i=1,2,3$

..0. $G_3^4 = \{4,0\}$ and $v^*(G_3^i)=1, \forall i$

$\underline{\ell= 2}$ the ordering is such that the coefficients are $1,1,-1,-1$
Then $R(2)=\{G_2^i\}_{1\leq i\leq 10}$ with ..00. $G_2^1 = \{(3,0),(4,0)\}$

11.. $G_2^2 = \{(1,1),(2,1)\}$

.10. $G_2^i = \{(j,1),(k,0)\}$

$i=3,\ldots$ $6,j=1$ or $2,k=3$ or 4

001. $G_2^i = \{(1,0),(2,0),(j=1)\}i=7,8,j=3$ or 4

.011 $G_2^i = \{(j,0),(3,1),(4,1)\}i=9,10,j=1$ or 2

and $v^*(G_2^i= \begin{cases}1 \text{ for } i=1,\ldots,6\\0 \text{ for } i=7,\ldots,10\end{cases}$

$\underline{\ell= 1}$ suppose the coefficients are $1,-1,-1,-1$.
Then: $R(1)= \{G_1^i\}_{1\leq i\leq 10}$ with

.11. $G_1^1 = \{(j,1),(k,1)\}$ $i=1,\ldots3,j,k \in\{2,3,4\}$ $j\neq k$

01.. $G_1^i = \{(1,0),(j,1)\}$ $i=4,5,6,j=2,3,4$

.000 $G_1^7 = \{(2,0),(3,0),(4,0)\}$

100. $G_1^i = \{(1,1),(j,0),(k,0)\}i=8,9,10,j,k \in\{2,3,4\}$ $j\neq k$

and $v^*(G_1^i) = \begin{cases}0 & i=1,\ldots,6\\1 & i=7,\ldots,10\end{cases}$

Remark :

R(ℓ) does not depend on the ordering of the neighbours
provided it satisfies the conditions on the coefficients.
For a given network F, we define in each element $(i,j) \in X$
an ordering Π_{ij} which respects the conditions on its coefficients for its mapping f_{ij}.
We then have a family $R(f_{ij})_{1 \leqslant i,j \leqslant n}$ of parts $G^k_{f_{ij}}$ which
describes the different possible forcing patterns in (i,j).
The case where those G have only 1 element is similar to
the situation of the boolean networks. The case where there
are only 2 elements allows the construction of forcing chains
The case of 3 elements is used in (iii) of the construction
of the forcing domain.

ERASING MULTITHRESHOLD AUTOMATA

Eric Golès   and   Maurice Tchuente

Université de Grenoble
Laboratoire I.M.A.G.
B.P. 68
38402 SAINT MARTIN D'HERES   cédex
F R A N C E

1. INTRODUCTION

The one-dimensional multithreshold automata studied here are discrete dynamical systems $A = (Z,[K],b,h)$ in which any cell $x \in Z$ can assume an integer state in the range 0 through K, and where the state $c(x,t)$ of any cell x at time t is determined by the states of the cells in a neighborhood of radius h as follows :

$$c(x,t)=f(c(x-h,t-1),\ldots,c(x+h,t-1)) = i$$

$$<=> b_{i-1} \leq \sum_{j=-h}^{h} c(x+j,t-1) < b_i \quad .$$

f is called the local transition function, and

$b = (b_0,b_1,\ldots,b_{K-1}) \in \mathbb{R}^K$ is called the threshold vector ; $c(\cdot,t) : Z \rightarrow [K] = \{0,1,\ldots,K\}$ is called a configuration

and we write $c(\cdot,t) = F(c(\cdot,t-1))$ where F denotes the global transition function.

A configuration $c : Z \rightarrow [K]$ is said to be finite if there are only a finite number of cells such that $c(x) \neq 0$. In this paper, we are interested in the particular case of erasing multithreshold automata, i.e. those where, for any evolution $\{c(\cdot,t) = F^t(c(\cdot,0)) ; t \geq 0\}$ starting from a finite configuration $c(\cdot,0)$, there exists an integer s such that $c(\cdot,s)$ is the null configuration, i.e. $c(x,s) = 0$ for any $x \in Z$. This problem has previously been studied by Toom (10) and Galperin (1,2), for general monotonic local interactions ; however, the results of Toom

concern only the binary case while those of Galperin are difficult to handle algorithmically. The analysis presented in
this paper for K = 2 and K = 3 , is based on very simple
algorithms and it illustrates the complexity of the problem.
The reader interested in more general results about the dynamic behaviour of multithreshold automata, may refer to
(3,4,5,6,7,8,9).

2. NOTATIONS AND DEFINITIONS

A stair e is a configuration such that

$x \leq y$ implies $e(x) \leq e(y)$; and $0 = \lim_{x \to -\infty} e(x) \neq \lim_{x \to +\infty} e(x)$.

We always assume that $e(0) = 0$; as a consequence, we only
need to specify the states of the cells $x \geq 1$; for ins-
tance , $e = \overleftarrow{0}0011 11\overrightarrow{2} = \overleftarrow{0}0^2 1^4 \overrightarrow{2}$ is such that

$e(x) = 0$ for $x \leq 2$, $e(x) = 1$ for $3 \leq x \leq 6$,

$e(x) = 2$ for $x \geq 7$.

$\tau_+(e)$ is the stair obtained from e by a right-shift and
is defined by $\tau_+(e)(x) = e(x-1)$ for any $x \in Z$.

We shall assume that any constant configuration is invariant , i.e. $f(a,a,...,a) = a$ for any $a \in [K]$; it follows
that $(2h+1)(i-1) < b_{i-1} \leq (2h+1)i$ for any $i \geq 1$.

Clearly, A is erasing if and only if, for any stair e ,
there exists an integer t such that $F^t(e) \leq \tau_+(e)$. As
a consequence, it is sufficient, as suggested in Galperin
(1,2) to study the evolution of stairs.

If K = 1 , then it is trivial that A is erasing if and
only if $b_0 \geq h+2$; let us now turn to the case k = 2 .

3. 2-STATE MULTITHRESHOLD AUTOMATA

Proposition 1 :

> Let $A = (Z,[2],b,h)$; $b \in \mathbb{N}^2$, $h \in \mathbb{N}$ be a one-
> dimensional multithreshold automaton in which any
> constant configuration is invariant (i.e.
> $b_0 \leq 2h+1 < b_1 \leq 4h+2$) . A is erasing if and only if
> $h+2 \leq b_0 \leq 2h+1$, $2h+3 \leq b_1 \leq 4h+2$ and $4h+5 \leq b_0 + b_1$.

Proof : Necessary condition : We are going to prove that if one of these conditions does not hold, then there exists a stair e such that $F(e) \geq e$. If $b_0 \leq h+1$ (resp. $b_1 \leq 2h+2$) then the stair $e = \overleftarrow{01}$ (resp. $e = \overleftrightarrow{02}$) verifies $F(e) \geq e$.

If $h+2 \leq b_0 \leq 2h+1$, $2h+3 \leq b_1 \leq 4h+2$ and $b_0+b_1 \leq 4h+4$ then we can write $b_0 = h+1+t$, $b_1 = 3h+3-s$ where $1 \leq t \leq s \leq h$ and it is easily verified that the stair $e = \overleftarrow{01}^{h+1-t}\overrightarrow{2}$ is such that $F(e) \geq e$.

Sufficient condition : Let $h+2 \leq b_0 \leq 2h+1$, $2h+3 \leq b_1 \leq 4h+2$ and $b_0+b_1 \geq 4h+5$ we can write $b_0 = h+1+t$, $1 \leq t \leq h$ and it is easily verified that the stair $e = \overleftarrow{0}1^{h+1-t}\overrightarrow{2}$ is such that $F^2(e) \leq \tau_+(e)$. Thus A is erasing.

Comment : Since the local transition function of a multi-threshold automaton is defined in terms of linear inequalities, one may expect the set E of threshold vectors associated with erasing automata, to be a convex region. Theorem 1 shows that this convexity property holds when the treshold vectors are restricted to be in N^2 ; however, as shown below, this is not true in R^2 .

- 1 - b' = $(h+1.1,3h+2.1) \in E$

          b'' = $(h+2.1,3h+1.1) \in E$

          b = $\frac{1}{2}(b+b')$ = $(h+1.6,3h+1.6) \notin E$

- 2 - b' = $(h+2,3h+2) \notin E$

          b'' = $(h+3,3h+1) \notin E$

          b = $\frac{1}{2}(b+b')$ = $(h+2.5,3h+1.5) \in E$

Let us now study the case when any elementary cell can assume four states (i.e. K=3).

4. 3-STATE MULTITHRESHOLD AUTOMATA

Lemma 2 : If $A = (Z,[3],b,h)$, $b \in N^3$, $h \in IN$ is such that $h+2 \leq b_0 \leq 2h+1$, $2h+3 \leq b_1 \leq 4h+2$, $4h+3 \leq b_2 \leq 5h+3$ and $b_0+b_1 \leq 4h+5$ then there exists a stair $e = \overleftarrow{0}1...12...2\overrightarrow{3}$ such that $F(e) \geq e$ or $F(e) \leq e$.

<u>Proof</u> : For any stair $e = \overleftarrow{0}1^r2^s\overrightarrow{3}$; $1 \le r,s$ we denote

$e_{(i)} = \min \{x : e(x) = i\}$ for $i = 1,2,3$;

$S(e,x) = \Sigma_{j=-h}^{j=h} e(x+j)$ for any $x \in Z$. It is sufficient to

exhibit a stair e such that

(*) $F(e)(e_{(i)}-1) = i-1$ and $F(e)(e_{(i)}) = i$ for two distinct

values of i . Indeed, if for instance property (*) holds
with $i=1,3$, then $F(e)(e_{(2)}) \ge 2$ (resp. $F(e)(e_{(2)}) \le 1$)

implies $F(e) \ge e$ (resp. $F(e) \le e$).

Let us denote $b_0 = h+1+t$; $s = h+1-t = 2(h+1)-b_0$

$a^{(r)} = \overleftarrow{0} 1^s 2^{t+r} \overrightarrow{3}$ for $r \ge 0$;

$c^{(r)} = \overleftarrow{0} 1^{s+r} 2^{t-2r} \overrightarrow{3}$ for $0 \le r \le \dfrac{t}{2}$

Clearly, $a^{(0)} = c^{(0)}$; $F(e)(e_{(1)}-1) = 0$ and

$F(e)(e_{(1)}) = 1$ for $e \in \{a^{(r)},c^{(r)} ; r \ge 0\}$.

<u>Case 1</u> : $S(a^{(0)},a_{(3)}^{(0)}-1) = b_0 + 3h = b_2-1$

$a^{(0)}$ verifies property (*) with $i=1,3$. Let $e = a^{(0)}$;
$S(e,e_{(2)}) = 6(h+1)-2b_0$. If $6(h+1)-2b_0 \ge b_1$ then $F(e) \ge e$,
otherwise $F(e) \le e$.

<u>Case 2</u> : $S(a^{(0)},a_{(3)}^{(0)}-1) = b_0 + 3h \le b_2-2$

Clearly, $S(a^{(r)},a_{(3)}^{(r)}) = S(a^{(r)},a_{(3)}^{(r)}-1)+2$ for $r=1,\ldots,s-1$

$S(a^{(r)},a_{(3)}^{(r)}-1) = S(a^{(r-1)},a_{(3)}^{(r-1)}-1)+1$ for $r=1,\ldots,s-1$

$S(a^{(s-1)},a_{(3)}^{(s-1)}-1) = 5h+1 \ge b_2-2$

Let $e = a^{(p)}$, where $p = b_2-2-(b_0+3h)$; e verifies property (*) with $i=1,3$ and $S(e,e_{(2)}) = b_0+3s-p = 9h+8-b_0-b_2$.
If $9h+8-b_0-b_2 \ge b_1$ then $F(e) \ge e$, otherwise $F(e) \le e$.

<u>Case 3</u> : $S(a^{(0)},a_{(3)}^{(0)}-1) = b_0+3h \ge b_2$

It follows that $4h+3 \le b_2 \le b_0+3h = h+1+t+3h$, hence $t \ge 2$.
On the other hand it is easily verified that :

$$S(c^{(r)}, c^{(r)}_{(3)}-1) = S(c^{(r-1)}, c^{(r-1)}_{(3)}-1)-3 \text{ for } 1 \le r \le q = \frac{t}{2}$$

$$S(c^{(r)}, c^{(r)}_{(3)}) = S(c^{(r)}, c^{(r)}_{(3)}-1)+3 \qquad \text{for } 1 \le r \le q$$

$$S(c^{(q)}, c^{(q)}_{(3)}-1) = b_0 + 3h - 3q$$

$$= h+1+t+3h-3 \frac{t}{2} \le 4h+1 < b_2$$

Let $e = c^{(p)}$ where $p = \dfrac{b_0 + 3h - b_2 + 2}{3}$; clearly,

$$S(e, e_{(3)}-1) = b_2 - \varepsilon , \quad S(e, e_{(3)}) = b_2 + \varepsilon' \text{ where } 1 \le \varepsilon ;$$
$0 \le \varepsilon'$ and $\varepsilon + \varepsilon' = 3$, and

$$S(e, e_{(2)}) = b_0 + 3s + 3p = 6(h+1) - 2b_0 + 3 \frac{b_0 + 3h - b_2 + 2}{3} .$$

If $S(e, e_{(2)}) \ge b_1$ then $F(e) \ge e$, otherwise $F(e) \le e$.

Comment : Clearly, the proof of Lemma 2 can be written as a sequential program containing only elementary operations on numbers with the same order of magnitude as h ; as a consequence, a stair such that $F(e) \le e$ or $F(e) \ge e$ can be constructed in constant time (if we take as unit, the maximum delay necessary to perform a comparison, addition, multiplication or division).

Lemma 3 :

    For any $A = (Z, [3], b, h)$, $b \in \mathbb{N}^3$, $h \in \mathbb{N}$ such that $h+2 \le b_0 \le 2h+1$, $2h+3 \le b_1 \le 4h+2$, $3h+3 \le b_2 \le 5h+3$, $b_0 + b_1 \ge 4h+5$ the question of the existence of invariant stairs, can be answered in constant time.

Proof : From proposition 1, any invariant stair is of one of the following forms :

(i) $e = \overleftarrow{0} \, 1^r \, 2^s \, 2^t \, \overrightarrow{3}$; $1 \le r, s$; $r+s = h+1$; $0 \le t$; $s+t \le h$

(ii) $e = \overleftarrow{0} \, 1^r \, 2^s \, \overrightarrow{3}$; $1 \le r$; $0 \le s$; $r+s \le h$

e is of the form (i) if and only if

$$b_0 + \varepsilon = h+1+s \qquad\qquad = S(e,e_{(1)})$$
$$b_1 + \varepsilon' = b_0 + \varepsilon + 3r - t \qquad = S(e,e_{(2)})$$
$$b_2 + \varepsilon'' = b_1 + \varepsilon' + 3s + 2t - 1 = S(e,e_{(3)})$$

Where $r,s,t \in \mathbb{N}$; $\varepsilon \in \{0,1\}$; $\varepsilon' \in \{0,1,2\}$; $\varepsilon'' \in \{0,1\}$

e is of the form (ii) if and only if

$$b_0 + \varepsilon = r + 2s + 3(h+1-r-s)$$
$$b_1 + \varepsilon' = b_0 + \varepsilon + 3r$$
$$b_2 + \varepsilon'' = b_1 + \varepsilon' + 3s$$

where $\varepsilon,\varepsilon',\varepsilon'' \in \{0,1,2\}$; $r,s \in \mathbb{N}$

Clearly, for any $h \in \mathbb{N}$, these systems of linear equations can be solved in constant time.

We are now ready τ_0 to prove the main Theorem.

Theorem 4 :

   Erasing multithreshold automata $A = (Z,[3],b,h)$ in which any constant configuration is invariant, can be recognized in constant time.

Proof : If $b_2 \geq 5h+4$, then, for any stair $e = \overset{\leftarrow}{0}\ 1^r\ 2^s\ \overset{\rightarrow}{3}$, $F(e)(e_{(3)}) = 2$; as a consequence, A is erasing if and only if (see Proposition 1),

$h+2 \leq b_0 \leq 2h+1$, $2h+3 \leq b_1 \leq 4h+2$ and $b_0+b_1 \geq 4h+5$.
If $b_2 \leq 5h+3$, then A is erasing if and only if (see Lemma 2, 3)

$h+2 \leq b_0 \leq 2h+1$, $2h+3 \leq b_1 \leq 4h+2$; $b_0+b_1 \geq 4h+5$, $4h+3 \leq b_2 \leq 5h+3$

and there is no stair e such that $F(e) \geq e$; moreover, these conditions can be tested in constant time.

The examples below show that the set E of threshold vectors associated with erasing automata is not convex.

1 - $b' = (h+2,3h+2,5h+4) \in \bar{E}$;
       $b'' = (h+3,3h+1,5h+4) \in \bar{E}$
       $b = \frac{1}{2}(b'+b'') = (h+2.5,3h+1.5,5h+4) \in E$

2 - $b' = (h+1.1, 3h+2.1, 5h+4) \in E$;

    $b'' = (h+2.1, 3h+1.1, 5h+4) \in E$

    $b = \frac{1}{2}(b'+b'') = (h+1.6, 3h+1.6, 5h+4) \in \bar{E}$

3 - $h = 4$; $b' = (8, 16, 21) \in E \cap \mathbb{N}^3$;

    $b'' = (8, 18, 19) \in E \cap \mathbb{N}^3$

    $b = \frac{1}{2}(b'+b'') = (8, 17, 20) \in \bar{E} \cap \mathbb{N}^3$

However, we can prove the following

Proposition 5 :

    Let $A = (Z, [3], b, h)$, $b \in \mathbb{N}^3$, $h \in \mathbb{N}$ be such that

    $h+2 \leq b_0 \leq 2h+1$, $2h+3 \leq b_1 \leq 4h+2$,

    $4h+3 \leq b_2 \leq 5h+3$, $b_0+b_1 \geq 4h+5$

(i)   If $b_0+b_1+b_2 \geq 9h+10$ then A is erasing

(ii)   If $b_0+b_1+b_2 \leq 9h+8$ then A is not erasing.

Proof : (i) $b_0+b_1+b_2 \geq 9h+10$. Let   e   be a stair such that
$F(e) \geq e$.

If $e = \overleftarrow{0} \ 1^r \ \overrightarrow{2}$ then $b_0+b_1 \leq S(e, e_{(1)})$; $S(e, e_{(2)}) = 4h+4$ and
this contradicts $b_0+b_1 \geq 4h+5$.

If $e = \overleftarrow{0} \ 1^r \ 3^s$ then $h+2 \leq b_0 \leq S(e, e_{(1)})$, thus $r \leq h$;
on the other hand, $4h+3 \leq b_2 \leq S(e, e_{(3)}) = r+3(h+1)$, thus
$r = h$ and $b_0 \leq h+3$; as a consequence ,

$b_1 \geq 9h+10-b_0-b_2 \geq 4h+4$ and this contradicts $b_1 \leq 4h+2$.

If $e = \overleftarrow{0} \ 2^r \ \overrightarrow{3}$ then $2h+3 \leq b_1 \leq S(e, e_{(2)})$, thus $r \leq h$;
$b_1+b_2 \leq S(e, e_{(2)}) + S(e, e_{(3)}) = 6(h+1)+r \leq 7h+6$ hence

$b_0 \geq 9h+10-b_1-b_2 \geq 2h+3$ and this contradicts $b_0 \leq 2h+1$.

If $e = \overleftarrow{0} \ 1^r \ 2^s \ \overrightarrow{3}$; $1 \leq r, s$ then

$b_0+b_1+b_2 \leq S(e, e_{(1)}) + S(e, e_{(2)}) + S(e, e_{(3)}) \leq 9h+9$ and
this contradicts $b_0+b_1+b_2 \geq 9h+10$.

It follows that, if   e   is the stair exhibited in Lemma 2 ,
then $\{F^t(e) ; t \geq 0\}$ is a strictly decreasing sequence ;
hence A is erasing.

(ii) $b_0 + b_1 + b_2 = 9h+8$; let e be the stair exhibited in Lemma 2 ; if $F(e) \geq e$, then A is not erasing, otherwise, we are in one of the following situations (we use the notations introduced in the proof of Lemma 2).

Case 1 : $b_0 + 3h = b_2 - 1$; $e = \overleftarrow{0} \; 1^s \; 2^t \; \overrightarrow{3}$; s+t = h+1 ;

$S(e, e_{(1)}) = b_0$, $S(e, e_{(2)}) = b_1 - 1$, $S(e, e_{(3)}) = b_2 + 1$; thus $F^3(e) = F^2(e)$.

Case 2 : $b_0 + 3h \leq b_2 - 2$; $e = \overleftarrow{0} \; 1^s \; 2^{t+p} \; \overrightarrow{3}$

$S(e, e_{(1)}) = b_0$, $S(e, e_{(2)}) = b_1$, $S(e, e_{(3)}) = b_2$; thus $F(e) = e$.

Case 3 : $b_0 + 3h \geq b_2$; $e = \overleftarrow{0} \; 1^{s+p} \; 2^{t-2p} \; \overrightarrow{3}$

$S(e, e_{(1)}) = b_0$, $S(e, e_{(2)}) = b_1 - \varepsilon$, $S(e, e_{(3)}) = b_2 + \varepsilon'$; $\varepsilon' = \varepsilon + 1$; thus $F^3(e) = F^2(e)$.

Comment :   This result shows that the region where $E \cap \mathbb{N}^3$ and $\bar{E} \cap \mathbb{N}^3$ "merge" , is contained in the hyperplane

$$H = \{b \in \mathbb{N}^3 : b_0 + b_1 + b_2 = 9h+9\} \ .$$

REFERENCES

1   Galperin, G. (1976). One-dimensional automaton networks with monotonic local interaction. Problemi Peredachi Informatsii . 12,4, 74-87.

2   Galperin, G. (1977). Rates of interaction propagation in one-dimensional automaton networks. Problemy Peredachi Informatsii, 13,1, 73-81.

3   Goles, E. (1980). Comportement oscillatoire d'une famille d'automates cellulaires non uniformes, Thèse USMG, Grenoble.

4   Goles, E. and Olivos, J. (1981). Comportement périodique des fonctions à seuil binaires et applications. Discrete Applied Mathematics 3, 2 pp. 93-105.

5   Goles, E. and Tchuente, M. (1983). Dynamic behaviour of one-dimensional threshold automata, R.R. n° 350, I.M.A.G., Grenoble.

6   Robert, Y. and Tchuente, M. (1982). Relations entre le graphe de connexion et le graphe d'itération d'une fonction booléenne monotone, R.R. n° 307, IMAG,Grenoble

7   Shingai R. (1976). Periodic behaviour of one-dimensional uniform threshold circuits. Trans. IECE, Japan J59,A6.

8   Tchuente, M. (1977). Evolution de certains automates cellulaires uniformes binaires à seuil. S.A.N.G. N°265, Grenoble, I.M.A.G.

9   Tchuente, M. (1982). Contribution à l'étude des méthodes de calcul pour des systèmes de type coopératif. Thèse d'Etat, Grenoble, I.M.A.G.

10 Toom, A.L. (1976). Monotonic binary cellular automata. Problemy Peredachi Informatsii, 12, 1, 48-54.

GENERALIZATION OF "LIFE"

J. HARDOUIN DUPARC

Laboratoire d'Analyse Numerique et Informatique
Route de Gray, La Bouloie - 25030 BESANCON CEDEX
FRANCE

0. INTRODUCTION

The notion of cellular automaton was introduced around 1950
by Von Neumann (3). Cellular automata can be useful in mathe
matical modelling; the most common applications have been in
biology, in nuclear physics and in pattern recognition.
Different generalizations have been proposed, either by modi
fying the cellular space, or the neighborhood, or the local
transition function; the best-known is the Game of Life
introduced by Conway (1).

1. CELLULAR AUTOMATA

A cellular automation is defined as a cellular space or set
of cells regularly distributed in an n-dimensional space; for
example, if n is equal to two, we can take a plane divided
into contiguous squares or equilateral triangles; another
example is that of the torus cut into orthogonal circles.
There exists a discrete time measure called the sequence of
generations. Each cell can possess, in a given generations,a
state chosen from a finite set.
The set of states of all the cells in a given generation is
called a configuration of the cellular automaton.
The neighborhood of a cell is a finite set of cells, this
neighborhood being defined in the same way for all cells in
the space (uniform automata).
The state of a cell in a generation depends exclusively on
the states of the cells in its neighborhood in the proceding
generation.
The law which enables us to determine the state of a cell

from the states of those in its neighborhood in the
previous generation is called the transition function. If
the neighborhood and the transition function   are
symmetric then the cellular automaton is said to by symmetric.

2. THE GAME OF LIFE

The cellular space is ZxZ where Z is the set of integers.
Any cell $x = (x_1,x_2) \in ZxZ$ is directly connected to its
neighbours $\{(x_1+v_1,x_2+v_2): \text{Max}\{|v_1|,|v_2|\} \leq 1\}$ (Moore
neighborhood). Any cell can assume a state $q \in \{0,1\}$ and the
local transition function is defined by

$$\delta(q_0,q_1,\ldots,q_8) = \begin{cases} 1 & \text{if} \begin{cases} q_0 = 0 & \text{and} \quad \sum_{j=1}^{8} q_j = 3 \\ q_0 = 1 & \text{and} \quad 2\leq \sum_{j=1}^{8} q_j \leq 3 \end{cases} \\ 0 & \text{otherwise} \end{cases}$$

Where q_0 is the state of the cell under consideration. This
automaton is uniform, deterministic and symmetric. One of the
most interesting characteristics of this automaton is the
existence of periodic and moving configurations; the best-
known are the gliders, which are of period four and move
according to the vector $(1,-1)$ at every generation. Let us
also draw attention to a complex configuration called the
glider gun, which is identical to itself and in the same
position after sixteen generations, having released two
gliders in the course of this period.
We set out to find the widest generalization possible
preserving the same uniform cellular space and the same
neighborhood as the game of life and which also has the same
states (0 or 1) and a deterministic local transition rule.

3. GENERALIZED AUTOMATON

3.0. THE DEGREE OF FREEDON OF LIFE

If, as in life, the transition function depends on whether
the cell is dead (state equal to 0) or living (state equal
to 1) and on the number of alive neighbours, the cellular
automaton so defined has eighteen degrees of freedom; in
fact, if we consider the number of living cells in the
neighborhood, if we multiply this number by two and subtract
one if the central cell is living, a number between zero and
seventeen is obtained, each value corresponding to a

different configuration of the living cells of the
neighborhood; for the transition function of life the cell is
living in the following generation if the value found is five,
six or seven.

3.1. FIRST STAGE IN THE GENERALIZATION

To begin with, we can consider that the action of the cells
having a common frontier and those having only a common space is not equivalent. Hence, we calculate the sum of the
living cells among the four cells having a common apex and
multiply by ten, then we add two times the number of living
cells in the set of the four cells having a common frontier,
and finally we add one if the central cell is living; the
value of the result is a number between zero and forty-nine;
thus we have a much more considerable genetic capital with
forty-nine degrees of freedom.

3.2. SECOND STAGE IN THE GENERALIZATION

Let us now designate the cells of the neighbourhood by NE,
NW, SW, SE, N, E, S, W and C, the first eigth of these
letters (or pairs of letters) meaning North-West, South-West,
etc, and the last representing the central cell; in order to
respect the symmetry it is necessary for two arrangements,
which can be deduced from one another by symmetry or
rotation, to give the same result, but the contrary is not
necessary. For example, the living sets (NW NE E), (NE SE S),
(SW SE E) should have the same action, but (NW NE S)
possibly another action: the degree of freedom reached is
then 102.

4. SOME EXAMPLES

–Gliders of different periods and speeds can be exhibited: We
have observed periods from 1 to 28, and speeds between 1 and 1/20.
–Tracer vehicle automata, of which a remarkable example is
the locomotive, which leaves behing it majestic puffs of
smoke.
–Gluttonous vehicle automata which rapidly devour objects
placed in their path.
–Automata combining the two preceding processes.
–Automata of different periodicities, frequently with periods
of the form $2^p - 2^q$, where p and q are integers.
–Different kinds of guns.
For each of these automata it is possible to
determine the minimum genetic support, and thus there
remain a certain number of degrees of freedom. It is

possible either to put together several automata, at the
same time making a study of compatibility, or to play with
the remaining degrees of freedom to improve the automaton in
the way the user wants to: for example, on can improve the
phenomena of collision between gliders or between a glider
and a fixed object. We have also determined an automaton
possesseing successive periods of 2, 3, 4, 6 and 7, and another having three vehicles of periods three, four and five
and being itself of period twelve. We have also an automaton
possessing a little glider which turns back on itself in case
of frontal collision and turns ninety degrees if the obstacle
is on the side of its path.
Finally, exemples of differents configurations of
generalizations of life can be seen in (2).

REFERENCES

(1) Gardner, M., (1971), Mathematical Games, Sci. Amer., N°2
    pp. 112-117.
(2) Hardouin Duparc J., (1981), Autocell. dossier de
    programme, R.R. Greco, N°9. 81, Bordeaux.

CONCERNING THE SUPPORT OF THE

SOLUTIONS OF CERTAIN CELLULAR AUTOMATA

C. LOBRY

C. REDER

Université de NICE
Parc Valrose, Dept. de Math.
06000 NICE
FRANCE

Université de BORDEAUX I
U.E.R. Math. et Informatique
351 Crs de la Libération
33405 TALENCE
FRANCE

INTRODUCTION

Let us consider a grid of the plane of mesh h. We write $C_{i,j}$ for the square (the cell) of coordinates i,j. To each $C_{i,j}$ we associate an integer $E_{i,j}$ belonging to the finite set $[0,1,\ldots,N]$. The integer $E_{i,j}$ is the state of the cell $C_{i,j}$. The states change with time according to the formule:

$$E_{i,j}(k+1) = E_{i,j}(k) + F(E_{1,m}(k); 1,m \in V_{i,j})$$

$$V_{i,j} = \{(i,j);(i-1,j);(i+1,j);(i,j-1);(i,j+1)\}$$

The support of the state at instant k, which we write $S(k)$, is the set of cells whose state is different of 0. We are interested in the time evolution of $S(k)$ from the following point of view:
Do there exist applications F such that, starting from the initial condition:

$$E_{i,j}(0) \neq 0 \Longleftrightarrow i^2 + j^2 = 0$$

The support appears, seen from a distance, as being rounded (smooth)?
Let us explain a little more: Seen from close up the edge S(k) is necessarily made up of segments parallel to the axes. But, seen from a distance, when the separating power of the eye does not allow one to distinguish the small irregularities of order h, this edge can very well seem smooth.

DYNAMICAL SYSTEMS
AND CELLULAR AUTOMATA

FORMULATION

Later we shall see the motivations. For the moment we give
a precise mathematical formulation of the question posed in
the introduction. The problem is to give a mathematical
interpretation of "seen from a distance".
We put ourselves in the context (I.S.T.) of the non-standard
mathematics proposed by Nelson (1). We suppose that the mesh
h of the grid is small (infinitesimal). We introduce the
concept of the shadow °S(k) of the support defined in the
introduction. The shadow is obtained by considering the
standardization of the points of R^2 which are near to a point
of S(k). The notion of shadow translates quite well the
idea of seen from a distance. The question posed in the
introduction becomes:
Find, if any exist, applications F such that, starting from
the initial condition:

$$E_{i,j}(0) \neq 0 \iff i^2 + j^2 = 0$$

for some values of k, the shadow °S(k) has a differentiable
edge.
We shall not attempt here to justify this translation.

MOTIVATION

Some natural environments:
- Chemical reaction of Bélousov in the non-homogeneous phase,
- Unicellular populations,
- Cardiac muscle,
- The cerebral cortex,
These are "excitable" environments likely to maintain series
of solitary waves. How to give an appropriate mathematical
representation of these phenomena is a current question.
- The representation by partial differential equations allows
are through computer simulations to reproduce the
observations quite accurately. On the other hand, it does
not give a simple explanation of the observed phenomena, the
theory of this type of equation (systems of parabolic
equations of diffusion reaction type) being difficult.
- The representation by cellular automata allows are to
realize dynamic systems which maintain series of solitary
waves (2). It is known how to explain their fonctioning by
combinatorial methods. Unfortunately the series of waves so
produced have a square form, which is qualitatively very
different from the observed reality (circular or spiral
waves).

Here we continue the project begun in (3) of reconciling the
two points of view. This leads us to the following
considerations, which are certainly very particular with
respect to the question posed, but which seem to lie of the
heart of the problem.
Let us define P_c and P_d as follows:

$$P_c \quad \left\{ \begin{array}{l} \dfrac{\partial u}{\partial t}(x,t) = k^2 \Delta u(x,t); \quad x \epsilon \mathbb{R}^2 \\ u(0,t) = \delta_o \end{array} \right\} \quad \text{heat equation}$$

and the discrete version of P_c:

$$P_d \quad \left\{ \begin{array}{l} U_{i,j}(t+\tau) = U_{i,j}(t) + \dfrac{k^2 \tau}{h^2}\left(U_{i-1,j}(t) + U_{i,j-1}(t) + U_{i+1,j}(t) + U_{i,j+1}(t) - 4U_{ij}(t) \right) \\ \\ U_{i,j}(0) = 0 \quad \text{if} \quad i^2 + j^2 > 0; \quad u_{0,0}(0) = \dfrac{1}{h^2} \end{array} \right.$$

The results of (3) allow us to assert that "seen from a
distance" the solutions of P_d are the solutions of P_c (This
is not very deep, being the "non-standard" version of an
elementary convergence theorem). This raises a small
problem: how is it that the support of the solution to P_d is
a square (this is evident), Whereas the solution to P_c (the
Gaussian one) is invariant for rotation? This small paradox
is easy to elucidate: it is sufficient to distinguish the
support of the points where the solution to P_d is small or
inappreciable from the support of the points where it is
appreciable, which support is certainly invariant for
rotation.
Here it should be indicated that P_d is not truly what is
treated by a computer because a computer only knows a finite
approximation of $\mathbb{R}$.
In fact, what is treated by a computer is more or less the
following:

$$P_a \begin{cases} E_{i,j}(k+1) = E_{i,j}(k) + \left[\dfrac{k^2\tau}{h^2}(E_{i-i,j}(k)+E_{i,j-1}(k)+E_{i+1,j}(k)+E_{i,j+1}(k)-4E_{ij}(k)\right]_\epsilon \\ \\ E_{i,j}(0) = 0 \quad \text{if} \quad i^2 + j^2 > 0; \quad E_{0,0}(0) = \left[\dfrac{1}{h^2}\right]_\epsilon \end{cases}$$

where $[a]_\epsilon$ is the number of the form n_ϵ which is closest to a and ϵ is the smallest real number which the machine can handle. P_a is then a cellular automaton of the type presented in the introduction. It can be asked what the evolution of the form of its support is like.

RESULTS

The results are, for the most part, only experimental.
1) With time the support remains bounded.
2) For a small number of iterations (of the order of $Log(N)$, $N = 1/h^2$), the support is a square ("orthogonal" to the grid)
3) For a large number of iterations (about 10^3 if $\epsilon = 10^{-5}$), the support is a square.
4) For an intermediate number of iterations (between 100 and 250, if $\epsilon = 10^{-5}$), the support closely approximates a disc.
The observation 1) and 2) have evident explanations.
Observation 3) is proved in the case of an automaton similar to P_a, but which is simpler. Finally observation 4) remains a mystery. Numerical experiments lead us to conjecture that, if the number of iterations is of an order of magnitude related to N (but how?), the shadow of $S(k)$ is a disc.

REFERENCES

(1) Nelson E., (1977), "Internal Set Theory". B.A.M.S. n°83, pp. 1165-1198.
(2) Allouche J.P. et Reder C., "Oscillations Spatio Temporelles engendrées par un Automate Cellulaire" Pub. U.E.R. Math. Info. Université de Bordeaux n°8207, to appear in Applied Discrete Math.
(3) Lobry C. Reder C., (1983) Micro Comparmental Systems in Bifurcation theory, Mechanics and Physics, D. Redeil Pub. Comp.

EMBEDDING OF DISCRETE DYNAMICAL SYSTEMS

György Targonski

*Fachbereich Mathematik, Universität Marburg, Lahnberge,
D-3550 Marburg, Federal Republic of Germany*

0. INTRODUCTION

Denoting by X the "state space" of a "system" (without
further explaining these terms) we say that a "next state
mapping" $f : X \to X$ gives rise to an autonomous discrete
semidynamical system $\{X,f\}$. If f is bijective, we have
a dynamical system; every state has a unique predecessor.

Some important questions to be asked in connection with
a discrete semidynamical system are these.

Can the system be embedded in a continuous-time autonomous semidynamical system, that is, a system which, when
restricted to positive integer times, yields the given
discrete system ("$\mathbb{R}^+$- embedding")?

Can the system be embedded in a discrete dynamical
system, that is, in a group, although f is not invertible? ($\mathbb{Z}$-embedding)?

We shall discuss some aspects of $\mathbb{R}^+$-embedding in section 2 and of $\mathbb{Z}$-embedding in section 3. There are, of
course, several other embedding problems; with a self-
explanatory notation, these include $\mathbb{R}, \mathbb{Q}, \mathbb{Q}^+$- embeddings.

The evolution of the discrete system is described by
$\{f^n\}$, $n \in \mathbb{N}$, the semigroup (under composition) of natural
iterates of f. The semigroup property is given by

$$(0.1) \qquad f^m \circ f^n = f^{m+n} \quad (m,n \in \mathbb{N}).$$

Let us now - to discuss embedding in an important
special case - denote the $\mathbb{R}^+$-embedding system by $\{X,F\}$,
where $F(x,t)$ denotes the state into which the state x
is mapped by the evolution during some time span $t > 0$;
that is, F maps $X \times \mathbb{R}^+$ into X.

Obviously, F must have the semigroup property, analogous to (0.1)

DYNAMICAL SYSTEMS
AND CELLULAR AUTOMATA

65

(0.2) $F[F(x,s),t] = F(x, s+t)$ $x \in X$ $s,t > o$

This is(a special case of) the "Translation Equation". The embedding property of F with respect to f is expressed by

(0.3) $F(x,n) = f^n(x)$ $n \in \mathbb{N}$

It is easy to see, that - using (0.2) - condition (0.3) can be reduced to

(0.4) $F(x,1) = f(x)$

The problem of $\mathbb{R}^+$- embedding is therefore the problem to solve the Translation Equation (0.2)(this is easy) with the "time-one condition" (0.4)(this is, in general, impossible).

The key to most iteration problems, thus to the problem of embedding is a problem of "orbits". We discuss these in section 1. We shall then proceed to the question of iterative roots, to be dealt with in section 2, and to the problem of non-trivial units which is the topic of section 3. Both questions will be treated,of course, in the context of the embedding problem.

1. ORBITS

These are the basic structures investigated in iteration theory. The relation

(1.1) $x \sim_f y \iff \exists m \exists n \quad f^m(x) = f^n(y)$
                        $m,n \in \mathbb{N}$

is an equivalence relation, it decomposes X into disjoint subsets, the orbits; they can be considered as (in general infinite) directed graphs, an edge leading from x to y if and only if $y = f(x)$. These directed graphs are "functional": precisely one edge leaves every vertex. For every bijective self-mapping h of X, $h^{-1} \circ f \circ h$ has the same orbit graph as f; h merely furnishes a relabeling of the vertices of X. Describing the orbits of X under f means describing the "combinatorial essence" of the mapping f. Here the adjective "combinatorial" is important; if, apart from the orbit structure imposed by f, X   also has an "intrinsic" structure, for example a topology, then h may play havoc with the continuity of f and so on.

Orbits may be cyclic or acyclic; a k-cycle is k points mapped into each other cyclically by f. A 1-cycle is a fixed point. An orbit contains at most one cycle (cyclic orbit), and exactly one cycle, if the orbit is finite. Thus an acyclic orbit is necessarily infinite. An acyclic orbit is an infinite directed functional tree; a cyclic orbit is a finite or infinite functional rooted "quasi-tree", that is, a rooted tree with the root replaced by a cycle.

If f is bijective, its orbits are very simple; if finite, they are cycles, if infinite, they are chains (ordered like Z by succession under f).

2. ITERATIVE ROOTS

A very simple, very difficult and until recently unsolved problem is this. Given a self-mapping f of X, and a natural number $r \geq 2$, does there exist a self-mapping g of X such that

(2.1) $g^r = f$

If such a g exists, we call it an r-th iterative root of f and denote it by $f^{\frac{1}{r}} := g$.

After initial pioneering work by Isaacs (1950) and substantial progress by Bajraktarević (1965) and Sklar (1969) the general problem of iterative roots of not necessarily bijective mappings was tackled and, in 1976, solved by Zimmermann. These results - some of them not published elsewhere - are discussed at great length in chapter 2 of [1].

The problem of iterative roots of bijections is simpler, and was solved independently of the sequence of results initiated by Isaacs.

Denoting by L_0 the (possible infinite) number of chains and by L_k the number of k-cycles, we have

(2.2)   *Theorem of Łojasiewicz* [2].

A self-bijection of a set has an r-th iterative root if and only if for every $k \in \mathbb{N}_0$, $L_k = \infty$ or L_k is divisible by d_k. Here $d_0 = r$, and for $k > 0$, $d_k = \frac{r}{r_k}$, r_k being the greatest divisor of r relative prime to k. What is the connection of all this with the question of $\mathbb{R}^+$- embedding?

Assume an $\mathbb{R}^+$-embedding of $\{f^n\}$ exists, call it again $F(x,t)$. Now we put $t = \dfrac{1}{r}$ and claim that we have an r-th iterative root of f:

$$(2.3) \qquad F(x,\tfrac{1}{r}) = f^{\frac{1}{r}}$$

Indeed we can verify this immediately by using (0.2) r times, starting with $s = t = \dfrac{1}{r}$ and consecutively putting $t = \dfrac{1}{r}$.

It follows that an $\mathbb{R}^+$-embeddable mapping has iterative roots of all orders, or, expressed in a form better suited to our purpose:

(2.4) *Proposition.* For the $\mathbb{R}^+$-embeddability of a bijection f it is necessary that f should have iterative roots of all orders. We know, of course, from the Łojasiewicz theorem (2.2) what the conditions for the existence of an iterative root of a given order are. We immediately suspect

(2.5) *Proposition.* For the embeddability of a bijection f it is necessary that $L_k = 0$ or $L_k = \infty$ for every $k \neq 1$. For the proof we first note that for $k = 1$, in (2.2) $r_1 = r$ (since every natural number is relative prime to 1) thus $d_1 = 1$ and L_1 is always divisible by d_1; the Łojasiewicz theorem does not impose restrictions on the number of fixed points f may have. In fact, one can see this directly: on the set of fixed points, we may set the $\mathbb{R}^+$-embedding function equal to the identity mapping for all $t > 0$

$$(2.6) \qquad f(x) = x \implies \forall t \atop t > 0 \quad F(x,t) := x$$

Now let $k \neq 1$, given, and assume, contrarily to the proposition, that $0 < L_k < \infty$. We show that for $k \geqslant 2$ f has no root of order $r = k^{L_k}$ and is therefore, by Proposition (2.4), not embeddable. In fact, now $(r,k) = (k^{L_k},k) = k$, r is a multiple of k and $r_k = 1$ thus $d_k = r$. Now $k \geqslant 2$ and $L_k \geqslant 1$, so $d_k = r = k^{L_k} \geqslant 2^{L_k} > L_k$, thus d_k does not divide L_k, and there is no r-th iterative root. For $k = 0$, there is no root of order $L_0 + 1$.
This proves Proposition (2.5). Incidentally, the condition is also sufficient; in order to prove this, we would have to use the Zimmermann results, described in [1],

chapter 2, more extensively than feasible in this exposi-
tion. At any rate, we can derive from Proposition (2.5)

(2.7) *Corollary*. If a bijective self-mapping of a finite
set X is $\mathbb{R}^+$-embeddable, then it is the identity mapping.
   In fact, if $\mathbb{R}^+$-embeddable, it must have $L_k = 0$ or
$L_k = \infty$ for all k ≠ 1. But $L_k = \infty$ is impossible, since
X is finite, thus for all k ≠ 1, $L_k = 0$. All points of
X are fixed points, f is the identity mapping.
   To finish our consideration of iterative roots, we note
that embedding of non-identity mappings, even on finite
sets, becomes possible in some cases if we drop bijecti-
vity.

(2.8) *Proposition*. Any idempotent mapping f is its
own r-th iterative root for any $r \in \mathbb{N}$, and with the
choice

(2.9) $\forall_x \; \forall t \; F(x,t) := f(x)$
            $t > 0$

it becomes $\mathbb{R}^+$-embeddable.
   Obviously $f^r = f$ by definition, this proves the first
part of our assertion. For the second part, we note from
(2.9) that for t = 1, condition (0.4) is satisfied;
applying (2.9) we find $F[F(x,s),t] = f^2(x) = f(x) =$
$= F(x, s+t)$, thus (0.2) is also satisfied.

3. NON-TRIVIAL UNITS

Given the semigroup of natural iterates $\{f^n\}$ of a self-
mapping f of X, $n \in \mathbb{N}$, one can always adjoin to the
semigroup, as trivial neutral element, or "unit", the
identity mapping on X. This, in many cases, is neither
the only choice, nor - for some purposes - the best one.
There exist, in those cases, additional units.
   We shall call these additional units nontrivial. Recently they have been studied in the 1981 doctoral dissertation of Graw [3] and the 1983 diploma thesis of Weitkämper [4].
   We define a strong unit η for the semigroup $\{f^n\}$ of
natural iterates of a self-mapping f of X to have the
following properties

(3.0)   (a) $\eta \circ f = f$ (b) $f \circ \eta = f$
            (c) $\eta^2 = \eta$ (d) $\eta(X) = f(X)$

Mappings with (a),(b),(c), but not necessarily with (d) will be called weak units e for $\{f^n\}$.

We start with the discussion of weak units. To investigate e, we start with (c); e is idempotent: every orbit consists of a fixed point and a (possibly infinite) number of points which reach it in one step under f ("points of height one over the fixed point").

The behaviour of e with respect to f is given by (a) and (b).

From (a) we note that e is the identical mapping on $f(X)$. In fact, if $x \in f(X)$, then there exists a $z \in X$ with $x = f(z)$. Then by (a) $e(x) = (e \circ f)(z) = f(z) = x$ whenever $x \in f(X)$.

From (b) we infer that e is a mapping within the set of predecessors of a given point. More precisely associating, for a mapping $f : X \to X$, with every $z \in X$ the set P_z of its (immediate) predecessors, we find that X is the disjoint union of the P_z

$$(3.1) \qquad X = \bigcup_{z \in X} P_z \qquad P_z \cap P_w = \emptyset \ (z \neq w)$$

For some z, P_z may be empty; this occurs precisely when z is an "initial point"; $z \in X \smallsetminus f(X)$. Obviously (b) means that e is the union of its restrictions to the P_z

$$(3.2) \qquad e = \bigcup_z e|P_z$$

From c, on the other hand, we see that e is not arbitrary within the P_z; it must be idempotent, that is, every P_z is the disjoint union of subsets P_z^ν such that each P_z^ν contains exactly one fixed point ξ_z^ν of e, and

$$(3.3) \qquad e(P_z^\nu) = \{\xi_z^\nu\} .$$

If f is bijective, the matter becomes extremely simple. $X \smallsetminus f(X)$ is then empty: there are no initial points. $P_z = \{f^{-1}(z)\}$ contains exactly one point and e is the identity mapping on all of X.

Let us now add (d) in (3.0) to our conditions; that is, we consider now what we called strong units; we denote these as said by η. Reviewing what has been said so far, we see that (d) really imposes only one condition, and this in connection with (3.3); we must have

$$(3.4) \qquad \underset{z \in X}{\forall z} \qquad \xi_z \in f(X)$$

because then and only then $e(x) \in f(X)$ for all X.

The significance of (3.4) is that every point not in $X \smallsetminus f(X)$ must have at least one immediate predecessor which is in $f(x)$, that is, which itself has a predecessor. Therefore, there can be just two kinds of points

a)   points with no predecessor, therefore empty P_z, that is points in $X \smallsetminus f(X)$

b)   points which have a predecessor with a predecessor, that is, points in $f^2(X)$. Summarizing, we have the disjoint decomposition

$$(3.5) \qquad X = (X \smallsetminus f(X)) \cup f^2(X)$$

Since of course the disjoint decomposition
$X = (X \setminus f(X)) \cup f(X)$ also holds, this implies

$$(3.6) \qquad f(X) = f^2(X)$$

Thus $f|f(X)$ is surjective; f is surjective on its range.

What is motivating our concern with nontrivial units? One reason (by no means the only one) is this. Assume f is non-bijective but that it is still $\mathbb{Z}$-embeddable. Then there must exist an inverse f^{-1} of the group element f; and $f^{-1} \circ f = f^{-1} \circ f^1 = f^0$. But f^0 cannot be the identity mapping, otherwise f would have an inverse, in contradiction to our assumption that f is not bijective. Therefore, $f^0 = e$ is a unit different from the identity mapping.

In fact, there exists a class of non-bijective mappings which is embeddable in a $\mathbb{Z}$-group.

(3.7) *Definition*. (A.Sklar; [5]). A self-mapping f of a set X is called ultrastable, if it is bijective on its range, i.e. if $f|f(x)$ is bijective.

The orbits of an ultrastable mapping are easy to visualize. Take a bijective orbit, that is, a cycle, or a chain, and add points reaching the cycle or chain in one step. There are, in general, infinitely many weak units, and exactly one strong unit, because there is exactly one ξ_z - value (cf. (3.3) and (3.4)) in $f(X)$. We can write down the explicit formula for this unique η:

(3.8) $\eta(x) = [f|f(X)]^{-1} f(x)$

If, for instance, one orbit of $f|f(X)$ is a cycle, and $x \in X \setminus f(X)$, then we obtain $\eta(x)$ as follows, starting from x. We go from x in one step into the cycle: $f(x)$. From there, we go clockwise along the cycle in one step to $\eta(x)$. (Here we assumed that the mapping f proceeds counterclockwise in the cycle.)

Now we can reduce the $\mathbb{Z}$-embedding of an (in general non-bijective) ultrastable mapping f to the $\mathbb{Z}$-embedding of the bijection $f|f(X)$

(3.9) $f^n := [f|f(X)]^n \circ \eta$

The correctness of (3.9) can be directly verified. Note that without condition (d) (that is, replacing η by a weak unit e) $f|f(X) \circ e$ would not be defined on all of X.

Ultrastability is more than just an interesting example. We have

(3.10) *Proposition.* (Weitkämper [4], Theorem 1.36). A self-mapping of a set is $\mathbb{Z}$-embeddable if and only if it is ultrastable.

By (3.9) we have shown that ultrastability is sufficient. For the (longer) proof of necessity we point to [4] and to a forthcoming publication.

Many additional results on nontrivial units are to be found in [3], pp.13 - 25 and in [4], pp.23 - 34.

References

[1]   Targonski, György: "Topics in Iteration Theory",
      Vandenhoeck & Ruprecht, Göttingen and Zürich, 1981
[2]   Łojasiewicz, S.: Solution générale de l'équation
      fonctionelle $f(f(...f(x)...)) = g(x)$ Ann. Soc.
      Polon. Math. 24 (1951) 88 - 91
[3]   Graw, R.: "Über die Orbitstruktur stetiger Abbildun-
      gen". Ph.D. Thesis, University of Marburg, 1981
[4]   Weitkämper, J.: "Einbettung von Abbildungen in ein-
      parametrige Halbgruppen". Diploma (M.Sc.) Thesis,
      University of Marburg, 1983
[5]   Sklar, A.: Canonical decompositions, stable func-
      tions, and fractional iterates, Aequ. Math. 3 (1969)
      118 - 129.

Applications to Growth Models

SYMMETRY PRINCIPLE AND MORPHOGENETIC GAMES

Y. Bouligand

EPHE et CNRS, 67, rue M.-Günsbourg, 94200 Ivry/S.,F.
FRANCE

1.INTRODUCTION

Symmetry propagation and symmetry breaking are two essential processes in physical and in biological morphogenesis. For instance, crystal growth corresponds to the spatial propagation of a given symmetry group. The phase transition observed at a crystal interface and the unavoidable defects within the crystal are examples of symmetry breaking. Similarly, after fertilization, the symmetries created by cleavage divisions in an egg are broken by cell differentiation. All natural morphogeneses of animate or inanimate systems show such propagations of symmetries and their breakings follow a more or less defined program (1).

The behaviour of networks of cellular automata is of great interest in morphogenesis. A small number of instructions can lead to quite elaborate forms and their growth is easy to observe. Von Neumann and Ulam were among the pioneers of these morphogenetic games(2,3). I intend to analyse here the evolution of symmetries in some examples of morphogenetic games and to compare them with the development of natural systems.

There are definite relations between symmetries of causes and symmetries of effects in physical phenomena, which were established by Pierre Curie after a profound reflexion about his own work in physics of crystals (4) and particularly his discovery of piezoelectricity in collaboration with his brother Jacques (see the historical survey by de Gennes, 5). In this article on symmetry by P. Curie, one finds the well known symmetry principle or Curie principle, which states that symmetries of causes are kept in the effects and that dissymmetries of effects are already present in the causes. On the other hand, certain symmetries of effects are not necessarily present in the causes, and new symmetries can arise, e.g: a cubic crystal shows isotropic optical properties and the optical effect is thus more symmetric than the causes (a cubic system). In symmetry breaking, Curie's principle seems to be violated. One could suppose that symmetry breaking amplifies considerably a weak dissymmetry already present in the causes. Moreover a statistical examination of a given type of symmetry breaking allows one to verify the Curie principle. Curie himself studied an example of symmetry breaking, the paramagnetic-ferromagnetic transition, which occurs in iron at 780°C, the 'Curie point'.

The sequence of symmetry groups observed in natural morphogeneses and in their computer simulations merit discussion in the light of 'Curie's principle'.

2. CONWAY'S GAME LIFE (6)

These games and some variants, well known from computer scientists, simulate some aspects of life (birth , survival,death,reproduction,organization, etc...). In such deterministic games, one observes that time is arrowed (fig.1). The situation at time t determines those observed at times t+1, t+2 etc... Inversely, if one looks for configurations at time t-1 which led to the situation observed at time t, one finds in general an infinity of solutions which form symmetrical pairs. In the example of fig.1, isolated points can be added far enough from the initial pattern, without effect on the following sequence, since such points die immediately. One sees also that, in fig.1, pattern nr.4 can be derived from pattern nr.1 , as well as from pattern nr.1 reflected in the horizontal plane vv'. These two possibilities form a symmetrical pair. Pattern nr.4 going back in time will choose one or the other of the symmetrical pairs and not both. Playing Conway's game against the arrow of time forces choice of patterns which is analogous to symmetry breaking.

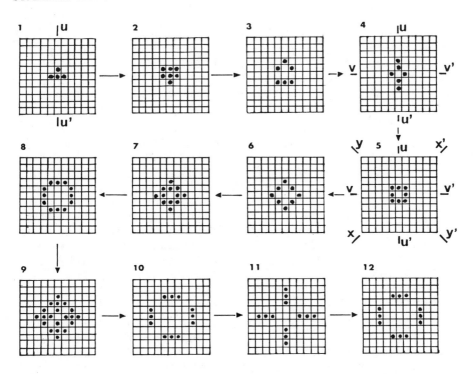

Fig.1. A succession of patterns arising from an initial configuration in Conway's game. One consider an indefinite square lattice. Each cell of the checkerboard has eight neighboring cells (adjacent orthogonally or diagonally).
Birth: each empty cell adjacent to three neighboring counters is a birth cell; a counter is placed on it at the next generation.
Survival: every counter with two or three neighbours survives for the next generation.
Death: in other cases, counters die (are removed) from overpopulation (4 to 8 neighboring counters) or from isolation (0 or 1 neighboring counters). Pattern nr.1 shows a symmetry axis uu', found also in nr.2 and nr.3. A new symmetry axis vv' appears at generation 4 and two other ones xx' and yy' at generation 5. From generation 10, the pattern oscillates and this means that a maximum of symmetry is reached. (Redrawn after ref.6).

In contrast with classical mechanics, where the knowledge of positions and velocities at time t determines how a system will evolve all along the t axis(past and future), the rules of Conway's game lead to a fully determined future, i.e: a unique solution, whereas the past remains largely undetermined.

Curie's principle applies to Conway's game in the direction of increasing time. If present at time t, a symmetry will be kept indefinitely, but new symmetries can appear. Such systems accumulate symmetries, but they never show symmetry breaking. In this sense, they cannot be considered a simulation of natural morphogeneses. On the other hand, Conway's game illustrates remarkably Curie's principle as shown in fig.1. When a pattern undergoes a cycle (oscillating systems and gliders,6), the symmetries are necessarily invariable (see patterns nr.10,11,12 in fig.1).

People who have played Conway's game by hand are well aware of the instability of the observed patterns. One mistake may lead to drastic changes of subsequent configurations. These mistakes, considered as perturbations introduced at given times generally break the symmetry. In some cases, such noises have no effects and this is obvious when they occur at sufficient distance from any particular pattern. As indicated above, an isolated birth immediately followed by death does not change the evolution. If the perturbation touches essential structures, one is often led to a radically different evolution . In natural morphogeneses, on the contrary, regulatory or positive feed-back processes are observed. Regulation is the rule for embryos in development. The shape of growing crystals also can be stable over long periods. These examples indicate how remote the so-called'game life' is from natural morphogeneses and, in particular, from living systems.

3. FOGELMAN'S MODELS

Another example of boolean networks worked out by Fogelman (7) offers some plausible models for the epigenetic stabilization of neuronal circuits. Her networks are derived from those studied by

Kauffman to simulate a set of interacting genes(8).
Fogelman studied a set of deterministic automata
linked into a square lattice drawn onto a torus, as
shown in fig.2a,b. Many other nets with no free
ends behave in a similar way. Each automaton α takes
its imputs i and i' at time t, from two neighbouring automata, and that determines the state of α
at time t+1. Similarly, the state of α at time t
serves as imput for the two other neighbouring
automata, as indicated by the arrows. The state
(zero or one) of an automaton is assigned by the
boolean function associated with α. Clearly, there
are sixteen boolean functions for two variables,
each of which can be in one of the two states,
zero or one. Here, these two variables are the
states of the two automata linked to α through i and
i', at time t-1 (fig.2a). Each automaton is given
a boolean function, which is chosen at random at
time zero and, then, remains constant. The states
of automata are similarly chosen at random, at
time zero. The evolution is then purely deterministic and, after a certain delay, one observes
the emergence of a cyclic behaviour, which is
expected, since the number of states in such a
system is finite. The period of the cycle is
variable and depends on the boolean functions
assigned to the automata.

In systems which have reached one of their
possible cycles, one can distinguish two types of
region: islands of automata which show a cyclic
behaviour and a connected region where the state
of automata is constant (fig 2c). Is this then an
example of symmetry breaking ? Perhaps not since
Curie's principle must apply. In fact, one probably observes an amplification of heterogeneities
already existing at time zero. They are much more
important than the dissymmetric fluctuations which
originate genuine symmetry breaking. The heterogeneities are not visible, since one only represents the states and not the boolean functions
themselves. Fogelman et al.(9) showed that these
functions play an important rôle in determining
the final patterns. For example, some boolean
functions are said to be 'forcing' and stabilize
the corresponding automata and some others in
their vicinity.

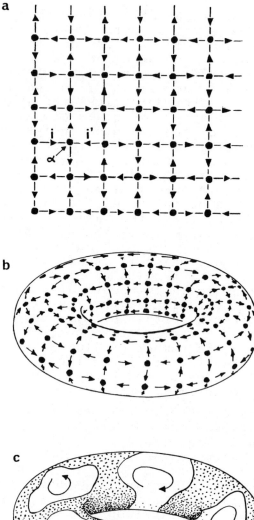

Fig.2. Fogelman's automata network. The square
lattice formed by these automata (a) can be viewed
as drawn onto a torus (b). Each automaton α receives two imputs: the states at time t of the two
automata linked through i and i', from which is
derived the state of α at time t+1. When the

Such complex models, with randomly chosen characteristics are unlikely to exhibit symmetries and their phenomenology is different from what is usual in morphogenesis: an alternation of symmetry propagation and of symmetry breaking. This model is however interesting for one aspect of neural morphogenesis. There is an extremely complex set of synapses in developing nervous system, when cells stop dividing and symmetry is no longer created nor discarded. Fogelman's model suggests that stabilization of certain neuronal circuits is inevitable and that chemical mechanisms supposed by Changeux et al.(10) reinforce patterns which occur in an asymptotic limit.

The organisation into oscillating subnets is stable against different types of perturbations in this model but, in some cases, after introducing noises (one changes arbitrarily the state of one or several automata at a given time), the system reaches a new limit cycle. The length of the delay necessary to come back to the preceeding cycle after perturbation or to reach a new limit cycle depends on the difference between the state of automata at the time of perturbation and the states observed in the cyclic behaviour. For example, if the state of automata at time zero corresponds by chance to one of the states observed in the cyclic behaviour, there is obviously no delay. Since at time zero, the state of automata is chosen arbitrarily, this delay is not negligible in general.

The organization into stable subnets in this model recalls that observed in intracortical neuron networks, when certain preferred circuits are stabilized. These neuron networks are however different , since they present both afferent and efferent axons. The efferent impuses create perturbations. The limit cycle of a subnet can be reestablished after a certain delay or it can be changed. The length of the delay or this qualitative change could provide the bases of a recognition process.

---

network has reached its asymptotic limit (c), one distinguishes islands of automata with a periodic behaviour(white areas with arrows) separated by a sea of automata in a constant state (dotted areas).

4. LUCK'S MODELS

Formal languages are built with an alphabet
and local production rules (11). They are generalizations of the preceeding examples and they allow
one to observe morphogeneses, when one starts from
an initial 'word'. Works by H.B. Lück on the
growth of algae are at the basis of several
theoretical developments, and many of his models
seem to be very close to biological reality (12).
One reason for this closeness is that symmetry
propagation and symmetry breaking are specified in
the production rules themselves. For instance, in

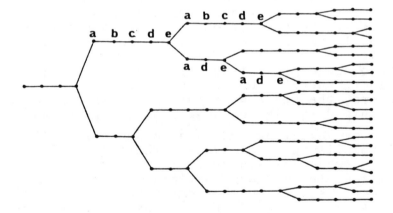

Fig. 3. a. Schema summarizing production rules
in morphogenesis of Chaetomorpha. Each letter
represents a night. Each cell divides always
at a night called e and this occurs either
three nights later, through e→a→d→e or five
nights later, following the path e→a→b→c→d→e.
b. From these rules, one deduces the family-
tree, with the absolute age of cells, each point
representing one night.

algae of the genus <u>Chaetomorpha</u>, cells divide
during the night and a mother cell gives two
daughter cells. One daughter divides three nights
later, whereas the other does so on the fifth night.
This process is represented schematically in fig.3.
In this model, the cell kinetics is based on
geometric and temporal rules, and one adjusts a
small number of parameters to get good agreement
between iteration of the model and the observation
of living material. Building symmetry breaking
into the rules themselves is a way of side-
stepping Curie's principle and an effective way of
tackling the problem. In Lück's studies, the
dissymmetry production is an intracellular process
linked to cell division itself and is present in
all divisions. His example of morphogenesis(fig.3)
is a collective process, involving populations of
cells and is situated at a higher organization
level than a single cell.

Such iterative models often lead to systems
showing structures resembling Mandelbrot fractals
(13), as shown in fig.4. In this example, H.B. and
J. Lück describe the successive partitioning of
cells in a plant tissue (14). Dissymmetries are
clearly visible and are generated by the division
process itself. The question of dimension arises,
though one does not find the perfect self-
similarity of the scaling figures considered by
Mandelbrot, the von Koch snow-flake for instance.
Mandelbrot has published several pictures
resulting from an iterative production of
dissymmetry and one famous example is his 2d-lung
model (15). Branching processes are common in
plants and animals (organs of respiration,
circulation, nerves etc...).

5. CONCLUSIONS

Three types of models are analyzed here. The
main conclusions are: 1. In Conway's game life,
no   symmetry breaking can take place, when it is
played in the same direction as the time arrow.
Symmetries accumulate and stabilize. 2. In automata
networks of Fogelman, one finds again stabilization and the rupture of symmetry is an illusion.
Such stabilizations might occur in certain
developmental processes. 3. In models due to

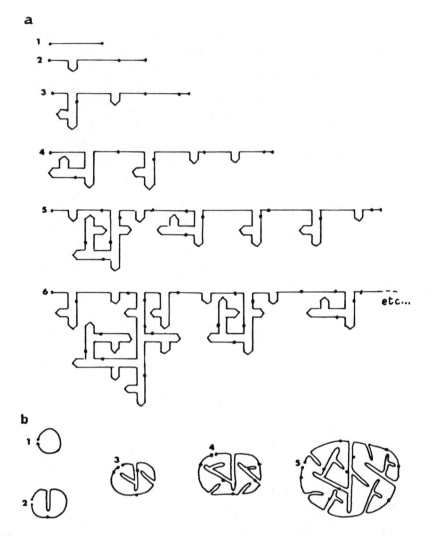

Fig.4. A morphogenesis related to that of a fractal
(a). The fifth pattern is a series of successive
steps in the formation of a self-similar branching
structure. This system is obtained by developing the
partitions which correspond to a plant cell dividing
four successive times. The cell membrane can be
followed along a unique circuit which is developed
in (a), by making straight and horizontal those
parts of the membrane which are in contact with the
environment (after J. and H.B. Lück, ref.14).

H.B. and J. Lück, dissymmetry is built into the
rules of the model, at the intracellular level,
and symmetry breaking is observed at a higher
organization level than the single cell. The
macroscopic dissymmetries originate from
specifie microscopic dissymetries. Other morphogenetic models merit this discussion on symmetry
and ,in particular, asynchronous and stochastic
systems.

Acknowledgements . I thank Dr. P.E. Cladis for
several enlighting discussions.

REFERENCES

(1) BOULIGAND Y. Symétries et brisures de symétrie
Colloque P. Curie, Symmetries and Broken Symmetries
N. Boccara ed., IDSET, Paris,131-140 (1981).

(2) VON NEUMANN J. The general and logical theory
of automata, in 'The World of Mathematics', J.R.
Newman ed., Simon and Schuster, N.-Y., $\underline{4}$,2070-
2090 (1956).

(3) ULAM S. On some mathematical problems connected with patterns of growth figures. Proc. Symp.
Appl. Math., $\underline{14}$,215-224 (1962).

(4) CURIE P. Oeuvres. Gauthier-Villars,Paris(1908)

(5) DE GENNES P.-G. Pierre Curie et le rôle de la
symétrie dans les lois physiques, in Colloque P.
Curie, Symmetries and Broken Symmetries, N.Boccara
ed., IDSET,Paris,1-9 (1981).

(6) GARDNER M.Scientific American, $\underline{223}$(4)120,(5)
118,(6)114; $\underline{224}$(1)105,(2)112-117,(3)108; $\underline{226}$(1)
(1971,1972).

(7)FOGELMAN-SOULIE F.Les réseaux d'automates: une
modelisation de systèmes biologiques? Actes du
premier séminaire de biologie théorique,ENSTA,
(1981). See also Fogelman F. et al., Lecture notes
in Biomath., $\underline{49}$,144-172(1983) and Réseaux d'automates et morphogenèse, in L'autoorganisation, de
la Physique au Politique, P. Dumouchel and J.-P.
Dupuy ed., Seuil, Paris, 101-114 (1983).

(8) KAUFFMAN S. Behaviour of randomly constructed genetic nets, in 'Towards a Theoretical Biology', C.H. Waddington ed., Edimburgh Univ. Pr.,3(1970).

(9) FOGELMAN-SOULIE F., GOLES-CHACC E. and WEISBUCH G.Specific roles of different boolean mappings in random networks. Bull. Math. Biol. and see also this volume.

(10) CHANGEUX J.P., COURREGE P. and DANCHIN A. A theory of the epigenesis of neuronal networks by selective stabilization of synapses, Proc.Nat.Acad. Sci.70,2974-2978(1973). See also Gouzé, Lasry J.M. et Changeux J.P., Biol. Cybern., under the press, and Changeux J.P. L'homme neuronal, Fayard,(1983).

(11) HERMAN G.T. and ROZENGERG G. Developmental Systems and Languages , Amsterdam, Elsevier-North Holland (1975).

(12)LUCK H.B. Sur l'histogenèse et la différencia-tion cellulaire. Ann.Sc.Nat.,12° Sér.,3,1-21(1962) Elementary behaviour rules as a foundation for morphogenesis. J.Theor.Biol.,54,23-34(1974).

(13) MANDELBROT B. Fractals: Form, Chance and Dimension, Freeman (1977).

(14) J. and H.B. LUCK Proposition d'une typologie de l'organisation des tissus végétaux. Premier Séminaire de Biologie Théorique, ENSTA, Paris, 335-371 (1981).

(15)MANDELBROT B. Les facettes fractales de l'anatomie, in 'La Morphogenèse, de la Biologie aux Mathématiques', éd. Y. Bouligand, Maloine, Paris, 83-89 (1980).

EVOLUTIONARY MODELS – CELLULAR AUTOMATA
AND ERGODIC THEORY

L. Demetrius

*Max-Planck-Institut fur biophysikalische Chemie am Fassberg,
3400 Gottingen, West Germany*

1. INTRODUCTION

Evolutionary theory is concerned with understanding the
changes in the diversity of populations in so far as this
diversity is determined by differences in the replication
rates of members of the population. The mathematical models
that have been proposed to describe the dynamics of diversity can be considered to be of two main types.

The first class of models is concerned with the genotypic
and phenotypic diversity. These models, which have their
origin in the work of Fisher (1), Lotka (2), consider a
population as divided into different types. The variables in
the model are the numbers of individuals of each type. The
dynamical equations describe changes in these numbers in
terms of the replication and mortality rates of the distinct
types. The asymptotic properties of the dynamical equation
give information on the relative proportions of the types in
the population. This approach, which was originally introduced to model the evolution of genotypic diversity in
higher organisms, has had considerable influence in understanding the dynamics of the cell cycle and the evolution of
macromolecules.

The second class of models is concerned with understanding what we will call structural diversity. The problem
that arises in this case concerns the understanding of the
evolution of the geometry of morphology of a tissue of cells,
this morphology emerging from the differences in the
replication rates of the cells that constitute the tissue. A
tissue can be considered as made up of an array of cells,
each cell capable of assuming a certain number of states.

State of each cell changes as a result of its interaction
with its neighbours. The morphology of the tissue at any
instant is described by the states of its constituent cells.
The model representing this process, has its origin in the
work of Von Neumann (3). In this model, called cellular
automata, the cells are described by sites and the tissue by
a lattice of sites. Each site can assume a finite number of
values. The state or configuration of the lattice is given
by the values of the variables at each site. These values
evolve synchronously in discrete time, the value of the
variable at one site being affected by the values of the
site in its neighbourhood. The evolution of this automata
leads to correlations between sites and the formation of
structures. A recent account of this class of models
together with extensive references to different areas of
application is given in Wolfram (4).

These two classes of models both lead to the formation of
patterns or organized structures given some initial configuration. The mathematical characterization of this
evolution towards organization is one of the central
problems in the study of these models. This paper shows that
the ideas of ergodic theory and the notion of entropy
provide a unified framework for the understanding of this
evolution. The connection with ergodic theory rests on the
observation that, mathematically, both classes of models can
be represented in terms of an operator acting on a suitable
sequence space. For the models of phenotypic and genotypic
diversity, the sequence space is the space of genealogies
Demetrius (5), and the operator acting on this space is the
shift operator. For the models of cellular automata, the
sequences space is the space of configurations (3), (4). The
operator acting on the space of configurations is the rule
which describes how the configurations evolve in time.

A dynamical system consisting of a sequence space and an
operator acting on it, can be studied from two points of
view : (a) by considering the operator as a measure preserving transformation and studying its asymptotic properties.
(b) by regarding the operator as a continuous map on the
sequence space and studying the properties of the orbits
generated. These two points of view lead to two different
characterizations of the asymptotic properties of the
operator : (a) the measure theoretic entropy which describes
the asymptotic rate of increase of the number of typical
genealogies of a given length, (b) the topological entropy
which represents the asymptotic rate of growth of the number
of approximate orbits of a given length. We will indicate
the significance of these two entropy measures as global

indices of organization for the two kinds of models and we also indicate how these entropy measures can be invoked to develop a statistical mechanics for the models.

2. MODELS OF PHENOTYPIC DIVERSITY

In this class of models, the population is described by a certain number of types. The problem of interest is the evolution in the diversity of types. The basic variable in the model is the number of elements in each type. These numbers evolve as a result of the interaction between the types, the rules of interaction depending on the level of biological organization considered. We treat three levels of biological organization and we consider both the continuous time and the discrete time representations.

2.1 Continuous time representation
a) Macromolecular models
Each type is described by its polynucleotide sequence. Let $x_i(t)$ denote the concentration of the polynucleotides I_i. The evolution of polynucleotides is modelled by a differential equation of the form, Eigen (6), Eigen and Schuster (7),

$$\frac{dx_i}{dt} = \sum_j w_{ij} x_j - x_i (\sum_{rs} w_{rs} x_s) \tag{2.0}$$

or in vector notation

$$\frac{d\bar{x}}{dt} = W\bar{x} - (\bar{1}.W\bar{x})\bar{x} \tag{2.1}$$

on the unit simplex

$$S_m = \{ \bar{x} \epsilon R^m : x_i \geq 0, \sum_j x_j = 1 \}$$

Here $\bar{1} = (1, 1, \ldots 1)$ and
$W = (w_{ij})$.
The coefficients w_{ii} represent the replication rates of I_i, w_{ij} the mutation rates from I_j to I_i. The term
$$\Phi = \sum_{rs} w_{rs} x_s = \bar{1}.(W\bar{x}).$$
is interpreted as an externally controlled dilution flow which keeps the total concentration $\sum_i x_i$ constant.

There exists a connection between (2.0) and the linear

equation

$$\frac{d\bar{y}}{dt} = W\bar{y}$$ (2.2)

Thus if $\bar{y}(0) \in R^m$, then

$$\bar{x}(t) = \frac{1}{\Sigma y_j(t)} \bar{y}(t)$$

is well defined and in S_m for all $t \geq 0$. Moreover $\bar{x}(t)$ is a solution of (2.0). Conversely, as noted in Thomson and Mc Bride (8) one obtains (2.2) from (2.1) by writting

$$g(t) = \int_0^t \Phi(u) \, du$$

and

$$\bar{y}(t) = \bar{x}(t)\exp [g(t)].$$

b) The cell cycle model
The type in this case corresponds to the physiological state or age of the cell. The fundamental variable in this model is the age density function $u(x,t)$ at time t. Here x denotes the age of the cell, that is, the time elapsed since its birth. The total population at time t is expressed in terms of the age-density function by

$$N(t) = \int_0^\infty n(x,t) \, dx.$$ (2.3)

Cell disappearance is assumed to be the sum of two terms one corresponding to cell death and the other to mitosis. We have $\mu = \mu_d + \mu_m$
Assuming that μ is age-dependent but time independent, the changes in the age-density function is given by

$$\frac{\partial u}{\partial x} + \frac{\partial u}{\partial t} = -\mu(x) \, u(x,t)$$ (2.4)

with boundary conditions

$$u(o,t) = 2\int_0^\infty \mu_m \, u(x,t) \, dx.$$ (2.5)

$$u(u,o) = f(a)$$ (2.6)

where $f(a)$ is the initially prescribed age-distribution of the cell populations. The equation (2.4) is usually called the Von Foerster equation.

c) Demographic Model
The state of an individual in the population of organisms is described by its age x. As in the cell cycle model, x is considered a continuous variable and $u(x,t)$ represents the age-density function. Changes in this function arise from $m(x)$, the age-specific fecundity and $\mu(x)$, the age-specific

mortality. The dynamics of $u(x,t)$ is also given by (2.4) with the boundary condition (2.6) corresponding to the initial age-distribution. The boundary condition which reflects the replication rate of the organisms is given by

$$u(o,t) = \int_o^\infty m(x)\ u(x,t)dx. \qquad (2.7)$$

2.2 Discrete time representations

Finite difference approximations can be obtained from the age-structured models and the model of macromolecular evolution given by (2.2). These difference equations give rise to the discrete time representation

$$\bar{x}(t+1) = A\bar{x}(t). \qquad (2.8)$$

For model (a), the element $x_i(t)$ of the vector $\bar{x}(t)$

represent the number of polynucleotides of type I_i in the

population at time t. The diagonal elements of the matrix describe the replication rates of the polynucleotides I_i.

The off-diagonal elements represent the mutation rates between types.

For model (b), the elements $x_i(t)$ describe the number of cells in each phase of the cell cycle. The elements in the first row of the matrix describe the mitotic rates. The elements in the other rows represent the rates at which cells pass from phase (i) to phase (j).

In the dynamical system representing model (c), $x_i(t)$

denotes the number of individuals in age-class (i). The elements in the lower off-diagonal correspond to the mortality rates of each age-class whereas the elements in the first row describe the fecundity elements.

2.3 Genealogies and entropy

In terms of the discrete representation given by (2.8), we can consider, in the case of macromolecular replication, a polynucleotide I_j, as generating at each instant in time a

sequence of types I_{k_1}, I_{k_2}, I_{k_3} ... and so on. Assuming that there are m possible types of polynucleotides, a genealogy is defined as a sequence $(I_{k_t})\ k_t \in \{1,2,...m\}$ such that I_{k_t}

is a direct copy (correct or mutant) of $I_{k_{t-1}}$, for t=1,2...

The set of genealogies can be formally characterized as follows :

Let $S = (1,2, ...m)$

and let $X = \overset{\infty}{\underset{0}{\Pi}} Sn$, where $S_n = S$.

Let $\Omega^+ = \{x \in X, a_{x_i, x_{i+1}} > 0\}$

where $A = (a_{ij})$, correspond to the matrix in (2.8).

A genealogy is an element $x \bar{\in} \Omega^+$.

Let $T : (x_k) \to (x'_k)$, where $x'_k = x_{k+1}$, denote the shift

operator on Ω^+.

The equilibrium measure μ on Ω^+ can be characterized, (5). This is a probability measure μ which is invariant under the shift operator T. This measure extends in a natural way to a measure on the doubly infinite sequence Ω. We will denote the model by the triple (Ω, μ, T).

Now, for a given dynamical system (Ω, μ, T), where $T: \Omega \to \Omega$ and μ is a probability measure preserved by T, we can define a measure theoretic entropy as follows.

Consider a partition $\alpha = (\alpha_1, \alpha_2 \ldots \alpha_k)$. The entropy of the partition α is given by

$h(\alpha) = - \Sigma \mu(\alpha_i) \log \mu (\alpha_i)$.

Write $\alpha^* = \alpha v T^{-1}(\alpha) v \ldots v T^{-n}(\alpha)$ and define

$h(\alpha, T) = \underset{n \to \infty}{\lim} \frac{1}{n} h(\alpha^*)$.

The measure theoretic entropy H (T) is given by

$H_\mu(T) = \underset{\alpha}{\sup} h(\alpha, T)$

The point of view taken here is that a partition corresponds to an experiment and $h(\alpha)$ represents the average amount of information contained in this experiment. The quantity $h(\alpha, T)$ is the average rate of information produced per unit time when the experiment α is performed many times in succession. The number $H_\mu(T)$ is the supremum of this rate over all experiments α. This number which we simply denote by H is a measure of the degree of randomness of the system.

In the model considered in which Ω represents the set of genealogies, H describes the asymptotic rate of increase of the number of genealogies of length t. This property can be characterized as follows.

Let $\Omega^{(t)}$ denote the set of genealogies generated by a single individual during the interval (0,t). Let E denote a subset of $\Omega^{(t)}$.

Consider an element $z \in E$ and write

$[z] = \{x \in \Omega^+, x_i = z_i, 0 \le i \le t-1\}$

Write $\mu_t(E) = \underset{z \in E}{\Sigma} \mu[z]$

Let $N_t^*(\varepsilon) = \min \{\text{card } E : E \subset \Omega^{(t)}, \mu_t(E) > 1-\varepsilon\}$

The number $N_t^*(\varepsilon)$ describes the minimal number of genealogies whose total probability exceeds $1-\varepsilon$. We have shown in (5),

that $\lim_{t \to \infty} \frac{1}{t} \log N_t^*(\varepsilon) = H$

Hence H represents the rate of increase of the typical genealogies. The number H can be computed for each of the models considered. See (9), (10), (11) for models (a), (b), (c) respectively. In this article, we shall only give the interpretations of H for the different models.

Model (a): H represents the extent to which mutation back to the wild type occur. When there are no back mutations to the wild type, H = 0. In the case of single point mutations, the maximum value of H is attained when correct and erroneous mutations occur at the same rate.

Model (b): H describes the degree to which the different cells pass through the cell cycle. When there is complete synchrony of the cycle, that is, all cells age at the same rate and divide at the same instant, we have H = 0

Model (c): H measures the variability in the ages at which individuals reproduce. In the case of organisms which reproduce once in their life time H = 0, otherwise H is positive.

Thus the measure theoretic entropy in the equilibrium state reflects an essential feature of the degree of organization of the biological process.

The entropy H also determines the rate of convergence to the equilibrium state μ. As shown in (12), one can consider for the dynamical system (Ω, μ, T), non-stationary measures ν on Ω, absolutely continuous whith respect to μ. Goldstein and Penrose (12) introduced a non-equilibrium entropy of the state ν and showed that the measure theoretic entropy of the state μ is the rate at which the non-equilibrium entropy increases. This fact, see (13) for an application to the demographic model, shows the significance of H as a measure of organization.

3. MODELS OF STRUCTURAL DIVERSITY

In the models of phenotypic diversity discussed in section (2), the state is described by a vector giving the number of each type in the population ; the relative positions of the different types does not enter into consideration. In the case of models of structural diversity, the spatial structure plays an essential role. For this class of models the population is represented by a regular uniform lattice of dimension $d \geq 1$. Each site in the lattice is capable of a finite number of states specified by integers from the set $S = (1, 2, \ldots m)$. The state of the neighbours at time t determines the state of a site at time (t+1). The rules of transition can be deterministic or have random elements ; the number of sites in the lattice may be finite of infinite. We will assume that the lattice is one-dimensional, the rules of transition deterministic and the number of sites infinite. We denote the sites by the set of integers Z. A configuration of the automaton is now given by assigning an element of S to each site $i \epsilon Z$. Formally, a configuration x is a mapping x : S → Z.

Hence a configuration can be described by a doubly infinite sequence

$$x = (\ldots x_{-1}, x_0, x_1, \ldots)$$

where x_i assumes values from the set S. We let Ω denote the set of all possible configurations. The rule which describes the time evolution of a configuration is given by

$$R: (x_i) \to g(x_{i-\alpha}, x_{i-\alpha+1}, \ldots, x_{i+\alpha}) \qquad (3.0)$$

Here $g: S^{2\alpha+1} \to S$

This rule indicates that the state at time (t+1) depends only on a neighbourhood of at most $(2\alpha+1)$ sites.

A natural metric can be defined on Ω as follows

Write $d(x,y) = \dfrac{1}{2^r}$

where $r = \sup \{m \epsilon Z, x_i = y_i, |i| < m\}$

With respect to this metric, the space Ω is a compact metric space and the operator R is a continuous function on Ω.

We denote the cellular automaton by (Ω, R). Let T denote the shift operator on Ω

$$T: (x_k) \to (x'_k) \quad \text{where } x'_k = x_{k+1}$$

As is well known T is a homeomorphism on Ω and the homogeneity of R means that T commutes with R, that is, RT = TR.

Now, by the Curtis, Hedlund, Lyndon theorem (14), any map that commutes with the shift is of the form (3.0). This fact implies that the transition rules of all one-dimensional cellular automata are characterized by continuous shift-commuting maps. This observation brings the study of cellular automata in close contact whith the study of what are called factor maps in ergodic theory. We exploit this connection in studying the organization in cellular automata.

3.1 Configurations and Entropy

The cellular automaton (Ω, R) can be considered as a topological dynamical system with the special property that Ω is a Cantor set and the transformation R commutes with the shift operator.

Now, for an arbitrary topological dynamical system (Ω, Φ) where Ω is a compact topological space and Φ a continuous map from Ω into itself, a topological invariant, topological entropy can be defined.

Let α be an open cover of Ω and let $N(\alpha)$ be the minimum number of elements of α needed to cover Ω. Write
$$h(\alpha) = \log N(\alpha).$$

Let $\quad h(\alpha, \Phi) = \lim_{n \to \infty} \frac{1}{n} h\left(\bigvee_{i=0}^{n-1} \Phi^{-i}(\alpha) \right)$

The topological entropy $h(\Phi)$ is defined by
$$H(\Phi) = \sup_{\alpha} [h(\alpha, \Phi)] \tag{3.1}$$

where the supremum is taken over all open covers of Ω.

The interpretation of these operations is as follows. The space Ω is considered as the state space and open covers α represent experiments, the elements of a cover describing the outcomes. The quantity $h(\alpha)$ represents the amount of information obtained by performing the experiment.

Experiments are performed many times in succession ; the quantity $h(\alpha, \Phi)$ represent the long time average of the amount of information obtained per performance. The topological entropy $H(\Phi)$ is the largest possible rate at which information can be extracted from the system. The number $H(\Phi)$ is a measure of the degree of randomness of the system.

The topological entropy can also be considered as the asymptotic growth rate of the number of finite length orbits as the length goes to infinity (15). To understand this point of view which we will invoke in characterizing cellular automata, we consider a set $E \subset \Omega$. The set E is called (t, E) separated for Φ if

$x, y \in E$ implies $d(\Phi^k(x), \Phi^k(y)) > \varepsilon$
for some $k \in [0, t]$

Let $N^*(t, \varepsilon) = \max \{\text{card } E : E \subset \Omega \text{ is } (t, \varepsilon) \text{ separated for } \Phi\}$

The number $N^*(t, \varepsilon)$ is the number of finite length orbits, where an orbit is defined as a sequence
$$x, \Phi(x), \Phi^2(x), \ldots \Phi^{t-1}(x).$$
Bowen (15) has shown that
$$\lim_{\varepsilon \to 0} \lim_{t \to \infty} \sup \frac{1}{t} \log N^*(t, \varepsilon) = H(\Phi) \tag{3.2}$$

Hence $H(\Phi)$ is the asymptotic growth rate of the number of finite length orbits as the length goes to infinity. The topological entropy tells us roughly speaking how many orbits Φ has.

By exploiting properties of the shift operator T and the fact that the automata rule R communtes with T, we can characterize more explicitly the randomness of the cellular automaton rule. This randomness will be described by the entropy of the shift operator T on subsets of the configuration space Ω.

First we observe that the topological entropy of the shift operator on the configuration space Ω can be easily computed. We recall that $\Omega = S^Z$, where $S = (1, 2, \ldots m)$.

Write $\quad \alpha_j = [\{x_n\}_{-\infty}^{\infty} \mid x_0 = j]$

then the set $\alpha = (\alpha_1, \alpha_2 \ldots \alpha_m)$ is an open cover for Ω, moreover α is a generator.

Hence $H(T) = h(\alpha, T) = \lim_{n \to \infty} \frac{1}{n} \log N(\bigvee_{i=0}^{n-1} T^{-i}(\alpha)$

$$= \lim_{n \to \infty} \frac{1}{n} \log m^n = \log m \qquad\qquad (3.3)$$

We shall simply write H for the topological entropy $H(T)$.

Now consider the mapping R: $\Omega \to \Omega$ and let $Y_n = R^n(\Omega)$. The sets Y_n are clearly compact subsets of Ω. Consider the shift operator T on Y_n and let H_n denote the topological entropy of the shift T. Since Y_n decreases with n, the entropy H_n also decreases with n. Let $H^* = \lim_{n \to \infty} H_n$. The number H^* measures the degree of organization generated by the rule R. In the case where the operator R maps onto the whole space Ω, the number H^* is precisely the topological entropy of T acting on Ω. In this case we have $H^* = \log m$ as in (3.3).

In general $R(\Omega)$ is a proper subspace of Ω. In this case $H^* < \log m$. The number $\log m - H_n$, which we call the relative topological entropy, increases with n. This increase reflects the tendency towards organization of the cellular automata. This increase in the relative topological entropy for this class of models should be contrasted with the increase in the non-equilibrium entropy described in the models of phenotypic diversity in Section (2).

4. CONCLUSION

The ergodic theory concepts of measure theoretic entropy and topological entropy which have been invoked to characterize diversity in the two classes of evolutionary models, both have their origins in thermodynamic theory. The term

entropy was originally introduced by Clausius to characterize the property of the irreversible flow of heat from a
body of higher temperature to one of lower temperature. The
second law of thermodynamics is a generalization of this
empirical fact. Boltzmann provided a mathematical foundation
to thermodynamic theory by giving a probabilistic characterization of the Clausius entropy. Boltzmann's entropy S is
given by S = k log N where N is the number of microstates
associated with a given macrostate and k is Boltzmann's
constant. The number S thus measures the degree of disorder
of a physical system consisting of many interacting
particles.

The Boltzmann entropy has provided the basis for characterizations of order in other probabilistic contexts.
Shannon observed in his studies of communication systems,
that certain processes whose outcomes are random events have
more uncertainty associated with them than others. Shannon
was able to assign a number to these processes. This number
is called entropy since the measure of uncertainty is
analogous to the Boltzmann concept.

The Shannon entropy describes the uncertainty of processes in which the passage of time does not affect the outcome.
Modern ergodic theory on the other hand deals with mathematical models of a random universe with a mechanism that
describes development in time. A problem of considerable
interest in this theory was to determine techniques for
deciding when two dynamical systems are isomorphic ; iso-
morphism in this case means that there exists a transformation connecting the phase space of the system such that
connected points are always moved by the respected evolution
into connected points. Kolmogorov solved this problem by
introducing the notion of measure theoretic entropy, a
natural generalization of the Shannon concept. The closely
related concept, topological entropy, which is an invariant
of topological conjugacy was introduced in (16).

The mathematical concepts of measure theoretic and topological entropy, which plays such a crucial role in evolutionary theory can be considered as deep generalizations of
the Clausius notion which is indeed the central idea in
thermodynamic theory. These two theories are connected.
Thermodynamic theory in its most general sense is concerned
with understanding changes in the state of matter in so far
as this is determined by changes in temperature. Evolutionary theory can be considered as the study of changes in the
diversity of populations in so far as this diversity arises
from differences in the generation time of the individuals
in the population. The connection between these two theories

have already been rigorously established for the class of models dealing with phenotypic diversity, Demetrius (5). In this class of models, the generation time corresponds to the reciprocal of the temperature, free energy corresponds to growth rate and the measure theoretic entropy which describes the phenotypic diversity corresponds to the Boltzmann entropy. In the models of structural diversity, in which topological entropy is used as the diversity measure, preliminary studies on the connection with thermodynamic theory have been proposed in Wolfram (4) but so far no precise analogues have been established. The relation between measure theoretic entropy and topological entropy together with the methods developed in (5) suggest a basis for rigorously investigating the connection between these two theories in the cellular automata context.

REFERENCES

1. Fisher, R.A. (1930). The Genetical Theory of Natural Selection. Oxford, Clarendon Press, (Reprinted and revised, 1958, New York : Dover).
2. Lotka, A.J. (1925). Elements of Physical Biology. Baltimore: Williams and Watkins. (Reprinted and revised as Elements of Mathematical Biology, 1956, New York: Dover).
3. Von Neumann, J. (1963). Theory of Self-Reproducting Automata, edited by A.W. Burks (University of Illinois, Urbana).
4. Wolfram, S. (1983). Rev. Modern Physics 55, 601.
5. Demetrius, L. (1983). Jour. Stat. Phys. 30, 709.
6. Eigen, M. (1971). Naturwissenschaften 58, 465.
7. Eigen, M. and Schuster, P. (1979). The Hypercycle - a principle of natural self-organisation, Heidelberg: Springer-Verlag.
8. Thompson, C.J. and Mc Bride, J.L. (1974). Math. Biosc. 21, 127.
9. Demetrius, L., Schuster, P. and Sigmund, K. Bull. Math. Biology (submitted).
10. Demetrius, L. and Demongeot, J.,J.Math. Biol. (submitted).
11. Demetrius, L. (1974). Proc. Nat. Acad. Sciences. USA. 74, 384.
12. Goldstein, S. and Penrose, O. (1981). Jour. Stat. Phys. 24, 325.
13. Tuljapurkar, S. (1982). J. Math. Biol. 13, 325.
14. Hedlund, G.A. (1969). Math. Syst. Th. 3, 320.
15. Bowen, R. (1971). Trans. Amer. Math. Soc. 153, 401.
16. Adler, R.L. et al. (1965). Trans.Amer. Math.Soc.114,309.

RANDOM AUTOMATA AND RANDOM FIELDS

J. Demongeot

Institut IMAG - TIM 3
BP 68
38 402 S^{t}Martin d'Hères Cédex
France

Abstract

In this paper, we study the relations existing between the
random automata and the Harris contact processes on $\mathbb{Z}^d$. In
both cases, the asymptotic behavior is described by the canonical measure of a random field. This measure keeps certain properties of the generating process "in mind" : for
example, the spatial Markovian character has the same range
as the transition function defining the underlying automaton.
Finally, we discuss the interest of the introduced tools in
modeling of spatial epidemic processes and in image processing.

1. INTRODUCTION

In a first part, we define random automata and random automata networks ; we give after the definition of a random
field and some properties about Gibbs measures. In a second
part, we study the relation between Harris contact processes
and stochastic automata, by using a concept close to the
Gibbs sampler due to S. and D. Geman (20). Finally, in the
last part, we propose two possible fields of application :
spatial epidemiology and image processing. In both examples,
by introducing "microscopic" invariance or symmetry properties (at the level of the transition function of the random
automaton) and also by defining the range of interaction
between neighbouring sites, we determine asymptotic properties of the transition process. For example, the spatial

DYNAMICAL SYSTEMS
AND CELLULAR AUTOMATA

99

Markovian character of the asymptotic Gibbs measure has same
range as the interaction between sites. Thus, we can predict
by knowing the transition process, the asymptotic behavior
and also, by observing an attractor of the dynamics, infer
some information about the interaction.

2. RANDOM AUTOMATA (after $(5,6,7,8,9,11,17,18,32)$))

Definition 1
We can define a random automaton A as a triplet of sets (Q,
Y,Z) with a pair of functions (F,G) :
$$A = ((Q,Y,Z),(F,G)), \text{ where}$$
- Q is the finite state space
- Y is the finite set of inputs
- Z is the finite set of outputs
and $F : YxQ^2 \rightarrow [0,1]$ verifies :
$$\forall y \epsilon Y, \forall q \epsilon Q, \quad \underset{q' \epsilon Q}{\Sigma} F(y,q,q') = 1$$

and $G : QxZ \rightarrow [0,1]$ verifies :
$$\forall q \epsilon Q, \underset{z \epsilon Z}{\Sigma} G(q,z) = 1$$

In the definition above, $F(y,q,q')$ denote the probability
to go from state q to state q', when the input is y; $G(q,z)$
denotes the probability to have z as output, in the state q.
Definition 2
A random automata network S is defined as a set of cells
(or sites) X with a pair of maps (R,C) :
$$S = (X, (R,C)),$$
where $R : X \rightarrow A$, A denoting the set of all automata A
defined on (Q,Y,Z)
and $C : XxZ^X \rightarrow Y$.
In the definition above, R(x) is the automaton associated
to the site x of X :
$$R(x) = (\ (Q, Y(x), Z), F_x, G_x).$$
C_x is the connexion map of the network N : for every f in
Z^X and x_o in X, $C(x_o,f) = y$ is the input received in x_o,
when each cell x of X gives the output f(x).

In pratice, we choose in this paper :
$$Q = Z = \{0,1\} \ ; \ X \subset \mathbb{Z}^d, \ |X| < +\infty \ , \ 0 \epsilon X$$
and $Y(x) = \{0,1\}^{N(x)}$, where N(x) represents the neighbourhood of a point x of X : $N(x) = (N + x) \cap X$, where $N = N(0)$
is a subset of X containing 0, and where $N + x = \{y+x; y \epsilon N\}$
The map R is then defined by :
$$\forall x \epsilon X, \quad R(x) = ((Q,Y(x),S),F_x,G),$$
Where F_x will be defined below and G verifies :

$$\forall q,z \ \epsilon \{0,1\}, \ G(q,z) = 1, \text{ if } q = z \ (= 0, \text{ if } q \neq z)$$

The map C is defined by :
$$\forall x \in X, \ \forall f \in Z^X, \ C(x,f) = (f(x_{i_1}), \ldots, f(x_{i_k}))$$

where $\{x_{i_j}\}$ is the set of the points of N(x), whose indices correspond to an ordering defined on X. This ordering on X allows us also to determine the sequential dynamics of the network S as follows :
1) if the state of the network S at the step i is defined by the element f_i of Z^X,
2) then, the state at the step i + 1 is defined by

$$\begin{cases} f_{i+1}(x_j) = f_i(x_j), \text{if } j \neq i+1 \bmod (|X|) \\ f_{i+1}(x_j) \text{ is obtained by applying the transition probabi-} \end{cases}$$

lity $F_{x_j}(C(x_j, f_i), f_i(x_j), \cdot)$, if j= i+1 mod($|X|$)

Therefore, the ordering chosen on X permits a walk from x_1 to $x_{|X|}$, each step of this walk corresponding to the

change of one and only one state in the point of X reached at this step.

The main features of this particular random boolean network S are :
- the boolean character
- the range of the connexion map C corresponding to the set N
- the sequential character
- the identity : output ≡ state

Definition 3
A random field on X is a set of random variables $\{T_x\}_{x \in X}$, such that there exists a probability space (Ω, A, P) with :
$$(T_x)_{x \in X} : (\Omega, A, P) \rightarrow (\{0,1\}^X, B(X), P_T),$$
where B(X) denotes the borelian σ-algebra on $\{0,1\}^X$ generated by the cylinders and P_T the canonical measure of the random vector (T_x).

If X is a finite set of $\mathbb{Z}^d$, B(X) is exactly the set of all subsets of $\{0,1\}^X$. Afterwards, we will identify $\{0,1\}^X$ with the set P(X) of all subsets of X.

Definition 4
A probability measure μ on P(X) = $\{0,1\}^X$ is called Gibbs measure associated to the potential
$$U : P(X) \rightarrow \mathbb{R} \text{ if we have} :$$
$$\forall D \subset X, \ \mu([D]) = Z^{-1} e^{-U(D)},$$
where $*Z = \sum_{D \subset X} e^{-U(D)}$ is a normalization constant

$$*[D] = \{E \subset X \ ; \ E \supset D\}$$

A Gibbs measure has the following properties :
1) J.M. Hammersley and P. Clifford proved that every proba-

bility measure μ on $P(X)$ was a Gibbs measure for a certain potential (X being finite), if and only if μ was strictly positive on $B(X)$ (23).

2) U can be considered as defined from an interaction potential V by :

$$\forall D \subset X, \quad U(D) = \sum_{E \subset D} V(E),$$

where V is a map from $P(X)$ to $\mathbb{R}$ verifying $V(\emptyset) = 0$.

The correspondance between U and V is one-to-one (13,14). V has a "range" N, if we have :

$$\forall D \subset X, \quad (\not\exists x \in X ; D + x \subset N) \quad \Rightarrow \quad V(D) = 0$$

In this case, μ is N-Markovian, i.e. :

$$\forall x \in X, \quad \mu(\llbracket \{x\} \rrbracket \mid \llbracket D \rrbracket) = \mu(\llbracket \{x\} \rrbracket \mid \llbracket D \cap N(x) \setminus \{x\} \rrbracket)$$

If $\mu(\llbracket D \rrbracket)$ represents the probability of the occupation by the state 1 for the sites of D, this condition signifies that the occupation of x depends only on the occupation of D in $N(x)$.

3) V is an equilibrated interaction potential, if we have :

$V(D) \neq 0 \Rightarrow D$ is equilibrated,

i.e. D verifies :

$$\forall x, y, z \in D, \quad (x \neq y \text{ and } y \neq z) \Leftrightarrow d(x,y) = d(x,z)$$

If $X \subset \mathbb{Z}^d$, D is a singleton, a pair of sites or an equilateral triplet.

We can prove (3) that the associated Gibbs measure verifies a spatial renewal property : if $\mu(\llbracket D \rrbracket)$ represents the probability of the occupation by the state 1 for the sites of D, the probability $\mu(\llbracket \{x\} \rrbracket \mid \llbracket D \rrbracket)$, where $x \notin D$, is the probability of occupation by 1 for x, knowing the occupation of D. Then we have :

$$\mu(\llbracket \{x\} \rrbracket \mid \llbracket D \rrbracket) = \mu(\llbracket \{x\} \rrbracket \mid \llbracket D_x \rrbracket),$$

where $D_x = \{y \in D ; d(y,x) = d(x,D)\}$.

This property is the same as the temporal renewal property for a binary process : the occupation depends only on the nearest occupied sites.

4) There exists a classical method to extend a Gibbs measure from $P(X)$ to $P(\mathbb{Z}^d)$, if X is a finite subset of $\mathbb{Z}^d$, by using the thermodynamic limit (13,14).

3. HARRIS CONTACT PROCESSES

Definition 5 (after (15,16))

A Markov process on $P(X)$ is called an Harris contact process if its transition semi-group $\{P_t\}$ verifies :

$$\forall D, E \subset X, \quad P_t(D,E) = e^{t \, G(D,E)},$$

where the generator $G(D,E)$ is such that :

$$G(D, D \cup \{x\}) = \beta(x,D) \text{ , if } x \notin D$$
$$G(D \cup \{x\}, D) = \delta(x,D) \text{ , if } x \notin D$$
$$G(D,E) = 0 \text{ , if } |D \Delta E| > 1$$
$$G(D,D) = - \sum_{E \neq D} G(D,E)$$

We can interpret this process as an epidemic process corresponding for example to the contamination of the site x by
the subset D, $\beta(x,D).dt$ being the probability of this infec-
tion between t and t + dt, and $\delta(x,D).dt$ being on the
contrary the probability of the restoration of the healthy
state, knowing that the subset of the infected sites is
equal to D at time t.

The process has a "range" N, if we have :

$$\beta(x,D) = \beta(x, D \cap N(x))$$
$$\delta(x,D) = \delta(x, D \cap N(x))$$

The relation between Harris contact processes and random
automata networks can be described as follows :
- in general, there exists for the Harris process a unique
invariant measure μ, which corresponds to the asymptotic
behavior of the process
- the process corresponding to the successive steps of the
random automata network S is a Markov process on P(X) : the
sub-process corresponding to the times $0, |X|, \ldots, k|X|, \ldots$
is an homogeneous Markov chain and it has in general a
unique invariant measure ν, which corresponds to the asymptotic behavior of S.

We say that S realizes the Harris contact process, if :
$\mu = \nu$.
In this case, we have explained the continuous contagion
process by a discrete sequential automata network.

Let us consider now the network S defined by the following transition function F_x :
if we identify the input space $Y = \{0,1\}^{N(x)}$ of a site x
with the set P(N(x)) of all subsets of N(x), we can associate
to the input $D \subset N(x)$ the probability to go from state q to
state q' for site x defined by the potential U on P(X) as
follows :

$$\forall q \in \{0,1\}, \quad F_x(D, q, q') = \frac{e^{U(D \cup q'x)}}{e^{U(D \cup q'x)} + e^{U(D \cup (1-q')x)}}$$

where $D \cup q'x = D \cup \{x\}$, if q' = 1
$\qquad\qquad = D \setminus \{x\}$, if q' = 0

Proposition 1
The unique invariant measure ν of S is the Gibbs measure
on P(X) associated to the potential U.

Proof
The transition matrix M of the Markov chain on P(X) asso-
ciated to S corresponding to the times $0, |X|, \ldots, k|X|, \ldots$

is such that :
$$M = \prod_{j=1}^{|X|} M_{x_j},$$
where M_{x_j} corresponds to the transition between the steps j and $j+1$. Suppose that the elements of $P(X)$ are ordered by the lexicographic order derived from the ordering $\{x_1,\ldots, x_j, \ldots, x_{|X|}\}$ chosen on X. If $(M_{x_j})_{D,E}$ denotes the general coefficient of the matrix M_{x_j}, that is the probability to go from the state D at the step j to the state E at the step $j+1$, we have :

$$(M_{x_j})_{D,E} = 0 \ , \ \text{if} \ |D \Delta E| > 1$$
$$= F_{x_j}(D \cap N(x_j), \pi_D(x_j), \pi_E(x_j)), \ \text{elsewhere}.$$

We have, by definition of F_x :

$$(M_{x_j})_{D,D \cup \{x_j\}} = \mu([\{x_j\}] \mid [D \cap N(x_j) \setminus \{x_j\}])$$
$$(M_{x_j})_{D,D \setminus \{x_j\}} = 1 - (M_{x_j})_{D,D \cup \{x_j\}}$$

It is very easy to verify that μ is an invariant measure for M_{x_j}, thus an invariant measure for M. Because of M is irreducible and primitive (its coefficients being strictly positive (25)), μ is unique and corresponds to the asymptotic behavior of the associated Markov chain∎

The general coefficient $(M)_{D,E}$ of M is given by :

$$(M)_{D,E} = \prod_{j=1}^{|X|} (M_{x_j})_{D_j,D_{j+1}},$$

where $D_j = D \setminus ((D \setminus E) \cap \{x_1, \ldots, x_{j-1}\}) \cup ((E \setminus D) \cap \{x_1,\ldots,x_{j-1}\})$

$$D_{j+1} = D_j \ , \ \text{if} \ x_j \in D \cap E$$
$$= D_j \setminus \{x_j\} \ , \ \text{if} \ x_j \in D \setminus E$$
$$= D_j \cup \{x_j\} \ , \ \text{if} \ x_j \in E \setminus D$$

Then the Kolmogorov-Sinaï entropy H of the Markov chain associated to M is given by :

$$H = - \sum_{D \subset X} \mu(\{D\}) \sum_{E \subset X} (M)_{D,E} \log ((M)_{D,E}).$$

Proposition 2

Let μ_o be an initial probability measure on $P(X)$ and μ_k the measure obtained on $P(X)^k$ at the step $k|X|$ by multiplying μ_o by M^k :

$$\mu_k = \mu_o M^k,$$

where μ_o is considered as a $2^{|X|}$- dimensional transposed vector.

Then, if δ denotes the Kullback distance on the space of probability measures on $P(X)$, there exist $\lambda > 0$ and $k \in N$, such

that :
$$\forall k \geq K, \quad \delta(\mu_k, \mu) < \lambda e^{-Hk}$$

Proof
This result about the convergence rate of μ_k to μ is a direct application of the results obtained in (24) and (26)∎

Proposition 3 (after(14))
If the graph joining the vertices of X is a Grimmett graph, if V is a pair nearest-neighbours interaction potential (i.e. V(D) equals zero, if the diameter of D is strictly greater than 1) and if we have :
$$\forall \, x \notin A, \quad \frac{\beta(x,A)}{\delta(x,A)} = e^{-\sum_{s \in \Psi(x)} V(s) - V(\{x\})}$$
(where $\Psi(x)$ is the set of pairs (x,y) such that : d(x,y) = 1 and y $\in$ A),
then the Gibbs measure μ associated to the potential U (corresponding to V) is the unique invariant measure of the Harris contact process defined by β and δ.

Proposition 4
With the assumptions of the proposition 3, we have : the random automata network S associated to U realizes the Harris contact process defined by β and δ.

Proof
The proposition 4 is a direct application of the propositions 1 and 3 : both invariant measures of S and of the Harris process are equal to the Gibbs measure associated to the potential U∎

The preceding results illustrate the relations existing between random automata networks and Harris contact processes : a rigorous proof of realization is given only in a particular case, but we are thinking that S realizes the corresponding Harris process in many other more complicated cases.We have here an open field of investigations.

4. APPLICATIONS

4.1 Applications in epidemiology (after (1,2))

In the french department "Pyrénées Atlantiques", we can observe an oak disease : the oak withering. By using photo-maps of different parts of oak-plantations, we can observe different configurations of healthy (0) and withering (1) oaks on the vertices of a regular graph (if planting is not regular, we can denote by 0 (resp.1) each barycenter of a regular partition of the plantation having less (resp.more) withering oaks then healthy : majority rule).

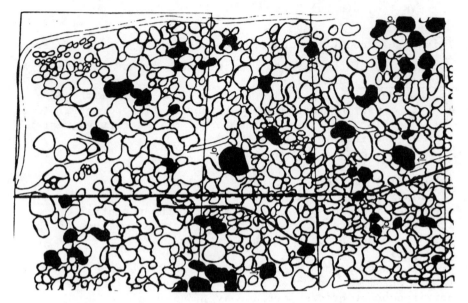

Fig. 1 Oak plantation : withering oaks are black
(the symbol ♀ corresponds to an other variety of
tree)(from a photomap by Dr. D. Larroche, Un. of Pau)

```
                              1        1         1     | 0 0 1 0 0 0 0 1 |
                              1    1                    | 0 0 0 0 0 1 0 0 |
                    1                                   | 0 0 1 0 1 1 1 1 |
               1  1  1                                  | 0 0 0 0 1 0 0 0 |
                                     1          1       | 1 1 |
          1  1
                    1                       1        1
     1
        1                                          1
     1                                  1  1              1     1
  1     1        1 1 1                1
           1        1 1 1        1
                 1     1    1
```

Fig. 2 Configuration corresponding to Fig. 1

If we are assuming that the contagion process has reached its "endemiological" asymptotic behavior, the observation of these configurations coresponds to the observation of a Gibbs measure, whose potential has the same "range" as the potential of the random automata network realizing the contagion process. Hence, we can estimate this range and also the potential U (27) ; after, we can simulate the contagion process with the help of the associated network S (Cf. Fig. 3 and 4, in the case of an equilibrated potential).

Such studies are very important, because they can allow us to detect for example the possibility of percolation (28, 29), that is the possibility of infinite gathering of sites having state 1, when X tends to $\mathbb{Z}^d$. In epidemic phenomena, such results can lead to the decision of a certain percentage of vaccination, in order to modify the parameters of U and suppress the percolation phenomenon.

Finally, we can remark that the properties of symmetry or translation invariance of U still hold for the asymptotic Gibbs measure μ : if $U(0(D)) = U(D)$, for a certain operator 0 on $P(X)$, then $\mu([0(D)]) = \mu([D])$. Then we can infer such properties for U from these observed on μ.

Fig. 3 Initial configuration (equilibrated
potential V)

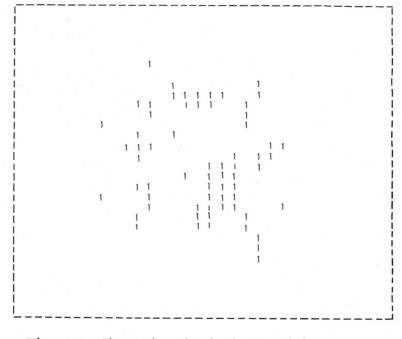

Fig. 4 Configuration obtained after |X| steps.

4.2 <u>Applications in image processing</u> (after(19,20))

In (19) and (20), we propose to use a random automata network to simulate, after a transient behavior (whose length can be estimated from the rate of convergence H, as proposed in section 2. ; see also (10)), a Gibbs measure : this automata network is called a Gibbs sampler by S. and D. Geman (19). An image is considered as defined by an initial "a priori" Gibbs measure on P(X) and we suppose that this a priori measure is observed with a certain noise : then the "a posteriori" measure is still a Gibbs measure. The Gibbs sampler is used to simulate "annealing" of the observation, in order to optimize the estimation procedure ; the existence of an "annealing schedule" (here the random automata network dynamics) guarantees the convergence to a minimum energy state corresponding to the a priori Gibbs measure on P(X). We can note that a good choice of the ordering of X (optimizing the entropy H) can lead to a fast convergence.

5.CONCLUSION

We have shown in this paper that there were close relations between random automata networks and mathematicel tools

often used by the physicists, for instance, the concepts of
entropy and Gibbs measures : in (12), (21), (22) and (30),
several authors are also pointing out these relations and
give other fields of application in physics and biology. The
basic feature of these approaches is the considering of an
hamiltonian-like function U, which serves to define the transition function of the automata. This point is fundamental
for random binary automata. We can note to conclude, that
recent studies (10, 31) show that the considering of such an
hamiltonian is also fruitful for deterministic automata : in
this case, it gives a Lyapunov function which decreases
during the transient. This property is very close to the
following : in the random case, the asymptotic Gibbs measure
corresponds also to the minimum of a certain function ; it
minimizes the entropy among the measures having the same
mean energy (14). We have here an interesting analogy between
random and deterministic automata, which can lead to new
results about their asymptotic behaviors.

REFERENCES

1. Demongeot, J. (1983). Coupling of Markov processes and
   Holley's inequalities for Gibbs measures. *In* "Proceedings
   of the IXth Prague Conference".pp.183-189. Academia,
   Prague.
2. Demongeot, J. (1981). Etude asymptotique d'un processus
   de contagion. *In* "Biométrie et Epidémiologie"(Eds J.M.
   Legay et al.). pp. 143-152.INRA, Paris.
3. Fricot, J. (to appear). Thesis, Grenoble.
4. Dobrushin, R.L. (1978). "Locally interacting systems and
   their application in biology". Springer Verlag, New York.
5. Milgram, M. (1982). Thesis, Compiègne.
6. Paz, A. (1971). "Introduction to probabilistic Automata".
   Academic Press, New York.
7. Rabin, M. O. (1966). Probabilistic Automata, *Inf. and
   Control* 6, 230-248.
8. Arbib, M. (1969). "Theories of Abtract Automata".
   Prentice Hall, New Jersey.
9. Salomaa, A. (1965). On probabilistic automata with one
   input letter, *Ann. Univ. Turku. Ser.* A1,85.
10. Fogelman, F., Goles, E. and Weisbuch G. (1983). Transient
    length in sequential iteration of threshold functions,
    Discrete Applied Maths 6, 95 98.
11. Doberkat, E.E. (1981). "Stochastic automata : stability,
    non determinism and prediction". Springer Verlag, New
    York.
12. Cocozza, C. In this volume.

13. Preston, C.J. (1974). "Gibbs states on countable sets".
    Cambridge Un. Press, Cambridge.
14. Spitzer, F. (1974). Introduction aux processus de Markov
    à paramètres dans $\mathbb{Z}^d$, *Lect. Notes in Maths* **390**, 114-189.
15. Harris, T.E. (1974). Contact interactions on a lattice,
    Ann. of Prob. **2**, 969-988.
16. Liggett, T.M. (1977). The stochastic evolution of infi-
    nite systems of interacting particles, *Lect. Notes in
    Maths* **598**, 188-248.
17. Chmiel, K. (1983). Cycles in the SC-image of a nondeter-
    ministic finite automata, *ICS-PAS Reports Warsaw* **498**.
18. Carlyle, J.W. (1961). Equivalent stochastic sequential
    machines, *Elec. Res. Lab. Ser. Berkeley* **415**.
19. Grenander, U. (1983). Tutorial in pattern theory, pre-
    print Brown University.
20. Geman, S. and Geman, D. (1983). Stochastic relaxation,
    Gibbs distributions and the Bayesian Restoration of
    Images, preprint Brown University.
21. Vichniac, Y.G. (1983). Simulating physics with cellular
    automata, preprint MIT.
22. Wolfram, S. (1983). Universality and complexity in
    cellular automata, preprint Institute for Advanced
    Study Princeton.
23. Besag, J. (1974). Spatial interaction and the statisti-
    cal analysis of lattice systems, *J. Royal Statist. Soc.*
    B36, 192-326.
24. Goldstein, S. (1981). Entropy increase in dynamical
    systems, *Isr. J. of Maths* **38**, 241-256.
25. Gantmacher, F.R. (1959). "Applications of the theory of
    matrices". Interscience Publishers, New York.
26. Tuljapurkar, S.D. (1982). Why use population entropy?
    It determines the rate of convergence, *J. Math. Biol.* **13**,
    325-345.
27. Demongeot, J. (1981). Asymptotic inference for Markov
    random fields on $\mathbb{Z}^d$, *Springer Series in Synergetics* **9**,
    254-267.
28. Smythe, R.T. and Wierman, J.C.(1978). "First-passage
    percolation on the square lattice". Springer Verlag,
    New York.
29. Griffeath, D. (1979). "Additive and cancellative inter-
    acting particle systems". Springer Verlag, New York.
30. Hopfield, J. (1982). Neural networks and physical systems
    with emergent collective computational abilities, *Proc.
    Nat. Acad. Sci. USA* **79**, 2554-2558.
31. Goles E. (to appear). Thesis, Grenoble.
32. Tautu, P. (1975).A stochastic automaton model for inter-
    acting systems. *In* "Perspectives in Probability and Sta-
    tistics" (Ed. J. Gani). pp. 403-415. Applied Prob. Trust,
    London.

COMPARATIVE PLANT MORPHOGENESIS

FOUNDED ON MAP AND STEREOMAP GENERATING SYSTEMS

Jacqueline Lück and Hermann B. Lück

Laboratoire de Botanique analytique et Structuralisme végétal
Faculté des Sciences et Techniques de St-Jérôme
CNRS - ER 161, rue H. Poincaré, 13397 Marseille cedex 13
France

INTRODUCTION

Plant cells enlarge and divide at the border and within the meristematic tissue. The mechanisms which assure a coherent organization of cell arrangement during the development, and by which the initial cell bulk brings forth 'an invariant form, remain still problematic.

Parallel map generating systems (1 to 6) and stereomap systems (5,7) are appropriate to describe the development of the cell wall net of cell layers and 3d-cell groups respectively. We investigate a class of such systems, *double wall map systems*, abbr. dw-map systems (3,4,5,8), and define them exhaustively. From them, systems with an integrated wall alphabet are derived to elucidate morphological aspects of development.

We suppose that new cell walls are inserted at determinate sites. These insertion sites are considered to be identical for all cells or for all pairs of sister cells which, therefore, may behave differentially. As the walls of a cell are heterogeneous in regard to their relative age of inception, we define the position of the new division wall in respect to the position of the last formed wall. The morphogenetic consequences of this hypothesis lead to the definition of developmental archetypes which would be generated if all cell divisions were synchronous and if cellular differentiation were absent. The complexity of the generated patterns is measured by the size of the integrated cell wall alphabet which discriminates inner from environmental cell walls.

DYNAMICAL SYSTEMS
AND CELLULAR AUTOMATA

111

BASIC DOUBLE WALL MAP GENERATING SYSTEMS

A map is a finite set of bounded and non intersecting regions.
In a plane, regions stand for cells and edges for the projection of cell walls. Walls are assumed to be heterogeneous
and are considered as a closed string S of segments separated by corners. Each cell has its own boundary. Two touching
cells are separated by a double wall. In each cell, segments
are labeled clockwise, beginning conventionally with the
youngest wall. A dw-map generating system $G = (\Sigma,"/",P,K,M_0)$
has an alphabet Σ of wall-segment labels w_i (a wall can contain several wall-segments),

$$\Sigma = \bigcup_{i=1}^{p} S_i \quad , \quad i = 1,\ldots,p \text{ and } S_i = w_1 \ldots w_{m_i} .$$

Cell corners are not labeled except division wall insertion
corners specified by slashes "/". The productions of the set
P over the segment labels are of the form $w \to s$, with $w \in \Sigma$
and $s \in \Sigma^*$. The sequence s may contain 1 or 2 slashes, but a
slash should never be placed at an end of s to avoid more
than 4 walls to meet at a cell corner. The rule K defines
the way how a new division wall is drawn within a dividing
region : $K = (s_i,w_i,S_i,s_j,w_j,S_j)$ means that the open strings
s_i and s_j, disconnected by two slashes in the boundary D surrounding a pair of daughter cells (called polycyte boundary),
are completed by the new walls w_i and w_j drawn in parallel
and spanning the slashes, in order to produce two new closed
daughter cell boundaries S_i and S_j, $1 \le i \le j \le p$. If $i = j = 1$,
all cells have identical boundaries S ; we speak about G_m
systems. If $p = 2$, the boundaries of sister cells are distinguished, of length m and n resp., and the corresponding systems are called $G_{m,n}$. M_0 is an initial map, for example a
boundary S embedded, if not specified, in an environment e ,
$e \in \Sigma$. (Cf.(5) for a more explicite description of dw-maps).
    At each iteration, the map is redrawn by replacing each
wall-segment by its produced sequence s and, according to the
rule K, division walls are added (Ex.1 and Fig.1). Thus, maps
grow by introduction of nodes into arcs and edges into regions.

Example 1 : $S_1 = 1\ 2\ 3\ 4\ 5$, $S_2 = 6\ 7\ 8\ 9\ 10$, $m = n = 5$, $p = 2$;
         $G_{5,5} = (\Sigma,"/",P,K,M_0)$ with $\Sigma = \{1,\ldots,10,e\}$,

$$P = P_1 \cup P_2 \cup \{e \to e\}; \quad P_1 = \begin{cases} 1 \to 4\ 5/7 \\ 2 \to 8 \\ 3 \to 9 \\ 4 \to 10/2 \\ 5 \to 3 \end{cases} \quad P_2 = \begin{cases} 6 \to 5/7\ 8 \\ 7 \to 9 \\ 8 \to 10/2 \\ 9 \to 3 \\ 10 \to 4 \end{cases}$$

$K = (2\ 3\ 4\ 5,\ 1,\ S_1,7\ 8\ 9\ 10,\ 6,\ S_2)$; $M_0 = S_1$.

Fig. 1 Sequence of generated topological maps

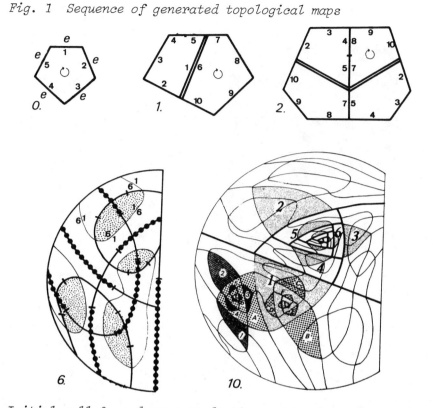

Initial cell 0. and maps at developmental steps 1., 2., 6., and incomplete map 10. Half map 6.: tiny lines for last division walls labeled 1 and 6; shadowed regions : 3-sided apical cells; dotted lines : walls with finite growth; slash : virtual corner. Half inc. map 10.: heavy lines: clockwise spiral segmentation for the first apical cell; large numbers and grey tones : its derivatives (one or more cells) which give rise to apices of order 1; letters and dots : counter-clockwise segmentation of first order apices (given here only for the first order apex N° 1); small numbers and black tones : derivatives of second order apices (given for apex A); they segmentate again clockwise. A derived apex results from the division of a lateral cell. An apex and the directly subordered ones are separated by walls with finite growth (see map 6.).

The sequence of maps in figure 1 leads to a cell pattern similar to that followed during the apical segmentation of *Psilotum* (Fig.2), discussed by Hagemann (9). It is obvious that a strict positioning device for division walls may describe those biological phenomena.

Fig. 2 A biological realization

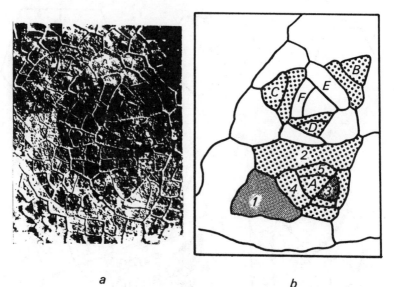

a   b

*Surface view of an apex of the fern Psilotum triquetrum. Fig.
2a : from a photomicrograph, by courtesy from W. Hagemann. Fig.
2b : its schematic analysis showing the   counter-clockwise
segmentation (from A to F) with the formation of first order
apices and their clockwise segmentation (from 1 to 5) with
second order apices. Fig. 1 represents a similar structure.*

GEOMETRIC INTERPRETATION OF THE MAPS

The quantification of the wall lengths given by the number of
segments they are composed of, transforms topologic maps into
geometric ones. In most patterns, the geometric constraints
lead to non planar surfaces (Fig.3). Conventionally, all cur-
vatures of the walls are turned towards the same side of the
map and the figures are drawn at the best in order to get regions interpreted like natural cells. Plant like structures
are thus incepted (Fig.7); they are described by the superfi-
cial form taken by their protodermal envelope.
     The geometric interpretation of *3d-stereomaps*, defined by
corners and edges surrounding polyhedral-like regions, needs
some supplementary conditions. Particularly, wall segments
need variable lengths to assure the executability of drawings.

*Fig. 3 Geometric map interpretations concerning the first de-
velopmental steps of example 1 and Fig. 1 . Slash : virtual*

*corner; dotted line : future division wall. Quantified wall
lengths involve geometric constraints which impel the develop-*

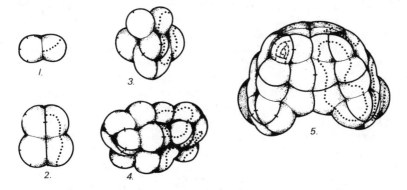

*ing cell layer to take a cap-like shape, evocating so the do-
me-shaped protodermal layer of shoot apices of higher plants.*

AN EXHAUSTIVE DEFINITION OF A CLASS OF SYSTEMS

Our system-dial (Fig.4) procures the entire class of systems
for a given number p of cell boundaries. The set P of wall-
segment productions is specified by the m or n segments in the
mother cell boundary S_i (inner ring) which procure the left
hand side terms of the productions. These segments produce
partition segments in the polycyte boundary D (outer ring)
which represent the right hand side terms of the productions.
Wall insertion slashes are forbidden to separate partition
segments of D. Each position of S_i and D rings defines a different partition of D, i.e. a different system G.

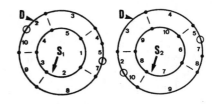

*Fig. 4 System-dial (in posi-
tion corr. to ex. 1):* • *: corner;* ∅ *: division wall insertion corner;* S_1, S_2 *: wall-
segment sequences of sister
cells; D : polycyte wall surrounding a pair of sister
cells. By turning the rings,
a segment* α *of* S_1 *or* S_2 *faces a partition segment* β *of D;*
α → β *: a production of* $P = P_1 \cup P_2$ *;* $P_1 = \{α→β|\ α \in S_1,\ β \subset D\}$ *;*
$P_2 = \{α→β\ |\ α \in S_2,\ β \subset D\}$ *.*

DERIVED DOUBLE WALL MAP GENERATING SYSTEMS

The relationship between the huge number of basic systems (e.

g. 28 824 for m = n = 6 (3)) is revealed by derived systems
which generate the same maps, but also inform about their patterns. A derived system H treats double walls as integrities
composed of two adjacent and inversely running sequences of
wall-segments. Such double wall labels issue from the transition graph of H. This graph can be generated by system G if
its initial map M_0 represents a double division wall (Fig.5a).
Whereas all basic systems G, in a class defined by p wall sequences, have an identic wall-segment alphabet Σ, the double
wall alphabet Δ of the corresponding derived systems H becomes variable in size and component labels.

A system $H = (\Delta, "/", Q, K, M_0)$ has labels like
$\dfrac{\alpha}{\text{inv } \beta}$, and productions of the form $\dfrac{\alpha}{\text{inv } \beta} \rightarrow \dfrac{\gamma}{\text{inv } \delta}$,

for α, β, γ, and $\delta \in \Sigma^*$. In Δ, two disjoint subalphabets, Δ_I
and Δ_E, resp. concern walls inside the tissue and walls fa-
cing the environment of the whole tissue. In Δ_E one wall label is e . The set of the productions is $Q = Q_I \cup Q_E$. The r.
h.s. term of a production is itself a double wall label if it
is free of slashes. Otherwise, it is split up at the places
of slashes.

Example 1 (continuation) : $H_{5,5} = (\Delta, "/", Q, K, M_0)$;

$$\Delta = \Delta_I \cup \Delta_E ; \qquad Q = Q_I \cup Q_E ; \qquad C = 22 \ (8,14) ;$$

$$\Delta_I = \left\{ \frac{1}{6}, \frac{2}{10}, \frac{3}{9}, \frac{4}{8}, \frac{5}{7}, \frac{2\ 3}{10\ 9}, \frac{3\ 4}{9\ 8}, \frac{4\ 5}{8\ 7} \right\} ; \quad Q_I = \left\{ \frac{1}{6} \rightarrow \frac{4\ 5/7}{8\ 7/5}; \frac{4\ 5}{8\ 7} \right.$$

$$\rightarrow \frac{10/2\ 3}{2/10\ 9}, \frac{7}{5} \rightarrow \frac{9}{3} \rightarrow \frac{3}{9}, \frac{2}{10} \rightarrow \frac{8}{4} \rightarrow \frac{10/2}{2/10}, \frac{2\ 3}{10\ 9} \rightarrow \frac{8\ 9}{4\ 3} \rightarrow \left. \frac{10/2\ 3}{2/10\ 9} \right\} ;$$

$$\Delta_E = \left\{ \frac{1\ 2\ 3\ 4\ 5}{e}, \frac{6\ 7\ 8\ 9\ 10}{e}, \frac{2\ 3\ 4\ 5}{e}, \frac{7\ 8\ 9\ 10}{e}, \frac{2\ 3\ 4}{e}, \right.$$

$$\left. \frac{8\ 9\ 10}{e}, \frac{2\ 3}{e}, \frac{3\ 4}{e}, \frac{8\ 9}{e}, \frac{9\ 10}{e}, \frac{2}{e}, \frac{4}{e}, \frac{8}{e}, \frac{10}{e} \right\} ; \text{ omitting } e,$$

$$Q_E = \begin{cases} 2 \rightarrow 8 & , & 2\ 3 \rightarrow 8\ 9 & , & 2\ 3\ 4 & \rightarrow 8\ 9\ 10/2 \\ 4 \rightarrow 10/2, & 3\ 4 \rightarrow 9\ 10/2, & 8\ 9\ 10 & \rightarrow 10/2\ 3\ 4 \\ 8 \rightarrow 10/2, & 8\ 9 \rightarrow 10/2\ 3, & 2\ 3\ 4\ 5 \rightarrow 8\ 9\ 10/2\ 3 \\ 10 \rightarrow 4 & , & 9\ 10 \rightarrow 3\ 4 & , & 7\ 8\ 9\ 10 \rightarrow 9\ 10/2\ 3\ 4 \\ & S_1 \rightarrow D & , & S_2 \qquad \rightarrow D \end{cases}$$

KINETICS OF DERIVED SYSTEMS

An analysis of the structure of the transition graph of H
(Fig.5) gives an insight into the produced morphologies. Specific substructures of the graph can be related to specific
morphologic features, such as apical growth of filaments,

lateral ramification (5), leaf tips, etc., and also to features such as X, Y, and Z in figure 5. A *developmental arche-*
type is conditioned by a specific combination of substructures of the graph. Those which condition the "spiraled segmentation of 3-sided apical cells" of example 1 is

$$\frac{\alpha}{\text{inv } \beta} \rightarrow \frac{\gamma \; / \; \delta}{\text{inv } \zeta / \text{inv } \eta} \; ,$$

$$\frac{\gamma}{\text{inv } \zeta} \rightarrow \frac{\cdot \; / \; \cdot}{\cdot \; / \; \cdot} \; , \qquad (\cdot \; :\text{non specified sequence})$$

$$\frac{\delta}{\text{inv } \eta} \rightarrow \frac{\cdot}{\cdot} \; , \qquad \alpha, \beta, \gamma, \delta, \zeta, \eta, \text{ and } \cdot \in \Sigma^* \; .$$

This signifies that a cell wall divides into two walls. One
of them divides immediately again, and the other divides never. Formerly, we listed seven morphogenetic archetypes (3,4,
5) we got by means of G_m systems. $G_{m,n}$ systems procure a
further archetype representing leaves (non published results)
and presently a new one, an "apex with spiraled segmentation"
(example 1).

Fig. 5 Double-wall transition graph.Fig.5a : Graph concerning
example 1. Fig.5b :Graph concerning an equivalent $G_{m,n}$ system

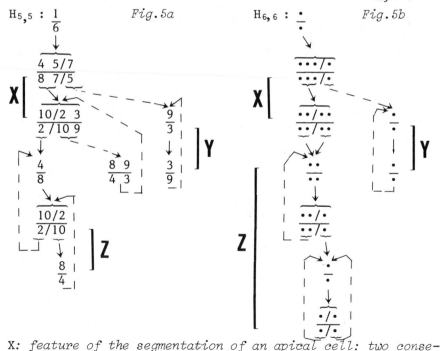

X: *feature of the segmentation of an apical cell: two conse-*

cutive wall divisions. Y: finite wall growth. Z: crossed tet-
rads of cells: the walls divide each second step. (X,Y,Z) :
archetype: 3-sided apical cell with spiraled segmentation.

FILIATION DIAGRAM OF DEVELOPMENTAL ARCHETYPES

For systems with parity, i.e. both walls in a dw-label have
the same length, the alphabet Δ is finite. The vector of the
sizes of its subalphabets leads to the definition of an alphabet size complexity

$$C = |\Delta| \ (|\Delta_I|, |\Delta_E|) \ .$$

It indicates the complexity of archetypal developments. As the
two subalphabets are disjoint, the increment of their sizes
by successive integers may be taken as a criterion by which a
filiation diagram of archetypes or their various forms can be
established (Fig.6). We use, as construction device of such a

Fig. 6 Filiation diagram of developmental archetypes.

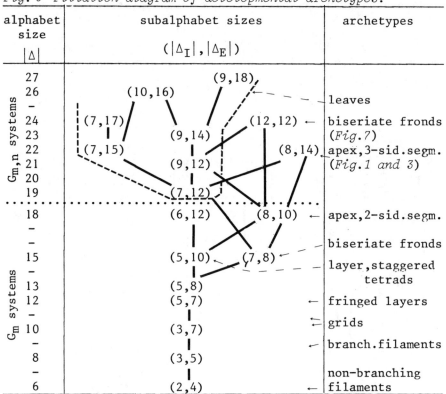

alphabet size $\|\Delta\|$	subalphabet sizes $(\|\Delta_I\|, \|\Delta_E\|)$		archetypes
27		(9,18)	
26	(10,16)		leaves
–			
24	(7,17)	(12,12) ←	biseriate fronds
23		(9,14)	*(Fig.7)*
22	(7,15)	(8,14)	apex,3-sid.segm.
21		(9,12)	*(Fig.1 and 3)*
20			
19		(7,12)	
18		(6,12) (8,10) ←	apex,2-sid.segm.
–			
–			biseriate fronds
15		(5,10) (7,8) ←	layer,staggered
–			tetrads
13		(5,8)	
12		(5,7) ←	fringed layers
–			
10		(3,7) ⇆	grids
–			branch.filaments
8		(3,5)	
–			non-branching
6		(2,4) ←	filaments

($G_{m,n}$ systems — upper group; G_m systems — lower group)

For further explications, cf. text.

diagram P. Delattre's method (*). It becomes possible to quantify a "morphogenetic distance" between archetypes, and this
on the base of the difference ($|\Delta|_1 - |\Delta|_2$) between their com-
plete dw-alphabets.

DEPTH FOR CREATION OF MORPHOLOGIC ITEMS

The depth of a transition graph (Fig.5) is sufficient to procure all *wall types* of the alphabet Δ. But the determination
of the morphologic features asks for a reinforced knowledge of
the derivation depth of maps. It is given by the filiation
graph of all possible *cell types* used to built up the pattern. These types differ by the number of neighbor cells and
types of complementary walls .For example in step 4 of figure
7, the cell a with 6 neighbor walls of type 7,e,11,e,9, and
e, divides in step 5 into two daughter cells from which one
is again a cell a with exactly the same neighborhood than the
mother cell, and the other, a cell b with 4 neighbor walls
labeled 1,e,3 4 5, and e. In this example, the depth is 12;
there are 52 cell types from which 38 touch the outer environment (e) by one or several walls. The remaining 14 types
are entirely embedded in a cellular environment. The former
cells are used to build up the procumbent part of the plant
body, whereas the latter ones appear uniquely in the erected
tubular part (Fig.7).

COMPARATIVE BEHAVIOR OF SYSTEMS

Systems which concern a same archetype are found as well within than between subclasses G_m and $G_{m,n}$, based on 1 or 2 different division wall insertions. The structure of their dw-tran-
sition graphs indicates if the expected patterns are equivalent, symmetric, or similar.
  Equivalent systems lead to identic maps, i.e. identic topologies and wall lengths. Disregarding wall labels, such systems have identic transition graph structures.
  Symmetric systems generate patterns with mirror symmetry.
They are revealed by mirror symmetric graph structures.
  Systems are said *similar* if they generate patterns with
identic topologies but different wall lengths. Their dw-tran-
sition graphs show the same sequence of dividing and not dividing walls (see $H_{5,5}$ and $H_{6,6}$ in Fig.5). The geometry of similar maps displays more or less curled surfaces, but con-
serves the neighborhood relationships.

(*) P. Delattre,"Sur la recherche des filiations en phylo-
genèse", conf. on the 2ème Séminaire de l'École de Biologie
théorique, Solignac, 1982.

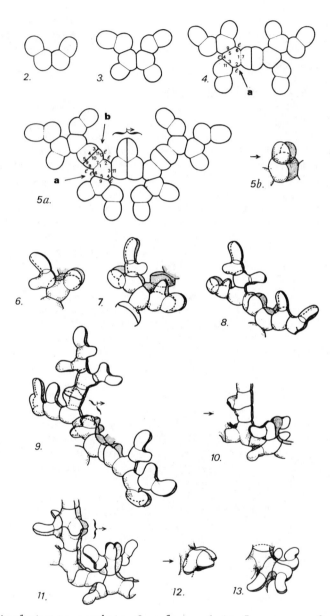

Fig. 7 Archetype: seriate fronds .−1 to 5a: maps of creeping branching filaments.−5b to 13: geometric interpreted maps: bi-seriate alternately branching fronds rise up.−5b: detail of 5a. −10: detail of 9; outgrowing 4-seriate fronds branching bila-terally.−12: detail of 11.− $G_{6,6}$: {1 → 3,2 → 4,3 → 5,4 → 6/8,5 → 9 10 11,6 → 12/2,7 → 11,8 → 12/2,9 → 3 4 5,10 → 6/8,11 → 9,12 → 10} .

CONCLUSIONS

Growing cell layers described by maps can also be generated by stereomaps which treat cells as polyhedrals. For bulky cell patterns, stereomaps are obtained by verifying wall parity in the boundaries of double wall faces.

Recent biological works (10) point to the localization of microtubule organizing sites in defined regions of the cell membrane, and which may control the placement of the cross wall (Hepler). Further, septum wall formation and primary wall growth are possibly involved by incorporation of vesicle materials (Kiermayer). Map systems can help to elucidate a positional order in organizing sites and localized wall growth, and this for widespread series of developing cell patterns.

ACKNOWLEDGEMENT

Eliane Isnard contributed efficaciously to the graphical work.

REFERENCES

1. Carlyle, J.W., Greibach, S.A., and Paz, A. (1974).A two-dimensional generating system modeling growth by binary cell division.Proc.15th Ann.Symp.Switch.Autom.Th.: 1-12.
2. Lindenmayer, A., and Rozenberg, G. (1979). Parallel generation of maps: developmental systems for cell layers.*In* Lect. Notes in Computer Sci. 73:301-316. Springer,Berlin.
3. Lück, J., and Lück, H.B. (1981). Proposition d'une typologie de l'organisation cellulaire des tissus végétaux. *In* Actes I Sém. Ec. Biol. théor.CNRS: 335-371. ENSTA, Paris.
4. Lück, J., and Lück, H.B. (1982). Sur la structure de l'organisation tissulaire et son incidence sur la morphogenèse. *In* Actes II Sém. Ec.Biol.théor.CNRS: 385-397.Univ.,Rouen.
5. Lück, J., and Lück, H.B. (1983).Generation of 3-dimensional plant bodies by double wall map and stereomap systems.*In* Lect. Notes in Comp.Sci.153: 219-231,Springer, Berlin.
6. de Does, M, and Lindenmayer, A. (1983). Algorithms for the generation and drawing of maps representing cell clones. *In* Lect.Notes in Comp.Sci. 153: 39-57. Springer, Berlin.
7. Mayoh, B.H. (1974).Multidimensional Lindenmayer organisms. *In* Lect. Notes in Comp.Sci. 15: 302-326. Springer,Berlin.
8. Lück, J., Lindenmayer, A., and Lück, H.B. (1984). Cell neighbor and descent relationships in meristematic cell layers and tetrad analysis. (To be published)
9. Hagemann, W. (1980).Über den Verzweigungsvorgang bei *Psilotum* u. *Selaginella*. Plant System. & Evol.133: 181-197.
10. Cytomorphogenesis in plants (ed. O. Kiermayer),(1981). Cell Biol. Monogr. 8 . Springer, Wien.

COMPLEXITY OF GROWING PATTERNS IN
CELLULAR AUTOMATA

Norman H. Packard[*]

Institute for Advanced Study
Princeton, NJ 08540, USA

1. INTRODUCTION

Cellular automata have seen a wide variety of applications since their invention in the late 1940's (*cf.* the conference proceedings (1,2) and the extensive bibliography of Wolfram (3)), but the motivations of this paper hark back to those of the inventor of cellular automata, J. von Neumann, whose perspective we will briefly review.[1] He observed that although physics has had amazing success in recent years providing fundamental explanations for many phenomena, there remains a wide class of phenomena that have been unapproachable with the conventional tools in a physicist's toolbox. The most intriguing examples are found in biology, where there seems to be a sense in which Nature makes something from nothing under the right conditions, as in the genesis of life, biological evolution, or in any organism's learning process. All these examples involve, in a naive sense, something proceeding from a simple beginning through states of increasing complexity (whether toward some final state or not is arguable; *cf.* section 5). Many people have suggested that there should be physical laws governing such phenomena; these laws are the unfulfilled goal motivating this work as they motivated von Neumann. He felt that the real biological world was far too complicated to model directly, and developed cellular automata to serve as simpler systems that display some of the essential dynamical features that might form the basis for such laws (analogous to the way that certain simple statistical mechanical models might be in the same universality class as real thermodynamic systems).

Despite the similarity in motivation, the approach used here is significantly different from that of von Neumann. He chose some complicated natural phenomenon and then to construct a cellular automaton that would manifest this phenomenon. The phenomenon he chose, motivated by the goal mentioned

---

[*]Much of this work was accomplished at the Institut des Hautes Etudes Scientifiques under the auspices of a NATO fellowship. It was also supported in part by the U. S. Office of Naval Research under contract number N00014-80-C0657.

[1] Actually, the origin of cellular automata can be traced to the joint efforts of J. von Neumann and S. Ulam, though the first published material on the subject seems to be von Neumann's book (4). The above impressions were gleaned from some very illuminating lectures contained in this book.

above, was the self- reproduction of an entity which could function as a universal Turing machine. Though von Neumann's proof of the existence of such a cellular automaton is a formidable accomplishment, there are some problems associated with his approach. For example, though von Neumann's invention mimics an aspect of reality, it cannot necessarily be considered a "fundamental model" because there is no identification between the microscopic mechanisms in both the model and reality that end up producing similar macroscopic phenomena. Another problem is that von Neumann's construction is very special in the sense that small perturbations in either the rule governing the evolution of states or in the initial condition will destroy the desired behavior (*i.e.* self-reproduction). Our approach is to try to understand the laws that govern complex phenomena by studying the evolution of typical initial conditions under the action of typical cellular automata models. Making rigorous the idea of "typical" is nontrivial, and will be discussed below, but the physicality of such a requirement for a model has its roots in statistical mechanics, where physically possible behavior is deemed unobservable if it has zero probability of occurring.[2] S. Kauffman (5) used this idea of searching for typical behavior in his study of automata on random lattices (to model the genome), and more recently S. Wolfram (6) has attempted a scheme to classify observed qualitative dynamics of simple cellular automata on the basis of the evolution of initial conditions taken at random from the set of all possible configurations on a finite lattice.

We must, with von Neumann, acknowledge certain difficulties that make ours a rather unusual scientific endeavor. Above, we gave examples of the type of phenomena we hope eventually to model, characterizing them by evolution from simple to increasingly complex states. A very great difficulty is that we have no good definition of complexity. We cannot yet develop a theory of complex evolution because we have no good ideas for what experimentalists should measure to test the theory. Thus, our endeavor must be both theoretical and experimental, with the experiments being performed, as von Neumann envisioned, on the very simple "toy universes" provided by cellular automata. The large task which we merely begin here is to find an appropriate definition of complexity by expanding upon the currently available notions of complexity as well as by observing simple systems that display it.

Several beginnings of definitions of complexity are presently available. Von Neumann (6) offered the admittedly heuristic idea that "complication," as he called it, should measure the ability of a system to do difficult and involved purposive operations, but he made no progress toward formalizing this idea. Most progress in formalizing notions of complexity has come from the study of randomness in various contexts. Kolmogorov (7) and Chaitin (8) introduced the idea that complexity (*vis.* "randomness") of a string of numbers should be

---

[2] J. Conway has pointed out that in a cellular automaton that is capable of supporting self-reproducing structures, zero probability configurations may "take over" the asymptotic behavior with probability one, thus overturning the conventional wisdom discussed above. A complete analysis of this question must address thorny questions of time scale which are beyond the scope of this work. We merely note that paleontologists tell us that life appeared relatively quickly after it was chemically possible (17, essay 19), supporting the idea that it was "likely" in the sense it having a positive measure of initial conditions that would produce it.

equated with the size of the minimal program needed to reproduce it. This idea turns out to be equivalent, in many cases, to the idea of entropy in dynamical systems theory (Kolmogorov (9)) and in communications theory (Shannon (10)). Entropy measures the exponential growth rate of the number of distinct orbits (messages, in communication theory) as a function of time, where two orbits are distinct if they fall in different partition elements of some appropriate partition. Entropy is empirically measurable for simple chaotic dynamical systems using symbolic dynamics (11,12,13). The spectrum of Lyapunov characteristic exponents is a set of closely related quantities which measure the time averaged local asymptotic divergence rate, and which are numerically computable given the equations of motion (14,15). The averaged sum of all the positive Lyapunov characteristic exponents is always less than or equal to the metric entropy; equality is conjectured, and in many cases (*e.g.* for all diffeomorphisms) can be proven. All of these quantities are essentially measures of randomness, a particular kind of complexity. The complexity of the biological systems mentioned above, however, seems to have some combination of randomness and intricate macroscopic structure which these measures of randomness cannot discern.

As our point of departure for the development of more appropriate measures of complexity, we will begin with the computation of Lyapunov characteristic exponents for cellular automata. In section three we will present their numerical computation for a set of example systems. One reason they seem an appropriate beginning is that there is a powerful analogy between the diversity of a process like evolution and exponential divergence of trajectories in some state space. Evolution is unquestionably punctuated by various fits and starts (*e.g.* the precambrian explosion; see Gould (16,17,18) for a description of this and other instances), but there is also some observational evidence for exponential increase in complexity over an evolutionary time scale (*e.g.* the increase of the ratio of brain mass to body mass as a function of time). Evolution theorists from T. de Chardin to G. Bateson have eloquently characterized biological evolution as a "divergent process," and Shaw (19) has suggested that some generalization of Lyapunov characteristic exponents might quantify such divergence.

There is, however, a problem with the analogy: Lyapunov characteristic exponents measure the average *local* (infinitesimal) divergence of trajectories. In all cases where they can be computed, trajectories remain confined to a compact subset of the state space, so there can be no *global* divergence. In contrast, processes like evolution seem to display constant global macroscopic divergence properties. In section four, below, I will argue that cellular automata can provide models for such global macroscopic divergence. An unconventional metric enables the measurement of the divergence of finite growing patterns, thus providing one step in the direction of new measures of complexity.

2. Mathematical Preliminaries

We will consider one-dimensional cellular automata as dynamical systems whose state space consists of configurations of a set of k symbols S over a one dimensional integer spatial lattice:

$$\Sigma = S^Z.$$

The dynamic is given by a map $f:\Sigma\rightarrow\Sigma$ that is local in the sense that for $x\,\epsilon\,\Sigma$

$$(f(x))_i = f_{loc}(x_{i-r}, \ldots, x_{i+r}),$$

where $f_{loc}:S^{2r+1}\rightarrow S$. The local map $f_{loc}:S^{2r+1}\rightarrow S$ is then repeated at every spatial lattice site to obtain the image of x under f.

We will be interested in the asymptotic behavior of f, which is to say, what happens after many iterations of f. It is easy to see that $f\Sigma\subseteq\Sigma$, and the limit set is

$$\Lambda = \lim_{n\rightarrow\infty} f^n\Sigma.\qquad\qquad 1$$

Of interest is the structure of the limit set, and the nature of the dynamics on it. In our numerical computations, we will compute average quantities using time averages, which will be constant if f is ergodic, and which may take on a range of values if Λ turns out to be a collection of non-transitive pieces. The numerical computations will be done on a finite lattice, and the connection between the computations on a finite lattice and the behavior of the infinite system is, as usual, not well understood (12,20), but we will use a range of lattice sizes to try to obtain some idea of what happens in the infinite limit.

We give S the discrete topology, and Σ the product topology, which makes it a compact, metrizable space homeomorphic to the Cantor set, and f is continuous with respect to this topology. The open sets of Σ are n-cylinders: $c^n = \{x\,|\,x_i=c_i, \ldots, x_{n-1}=c_{n-1}\}$. The usual metric on Σ is obtained by taking a metric on S which we denote $|x_i-y_i|$ for $x_i,y_i,\epsilon S$, and forming the power series:

$$|x - y| = \sum_{-\infty}^{\infty} \frac{|x_i - y_i|}{k^{|i|}},\qquad\qquad 2$$

using subtraction mod k as the metric on S.

The Cantor set may be visualized by forming coordinates by mapping some configuration x into the unit interval. Two coordinates used by Grassberger (21) are

$$X = \sum_{i=0}^{\infty} \frac{x_i}{q^i} \quad\text{and}\quad Y = \sum_{i=0}^{\infty} \frac{x_{-i-1}}{q^{i+1}}\qquad\qquad 3$$

Note that if we define the difference coordinates

$$\Delta X = \sum_{i=0}^{\infty} \frac{x_i-y_i}{q^i} \quad\text{and}\quad \Delta Y = \sum_{i=0}^{\infty} \frac{x_{-i-1}-y_{-i-1}}{q^{i+1}},\qquad\qquad 4$$

then $|x - y| = \Delta X + \Delta Y$. Arbitrarily many coordinates may be constructed this way by taking different subsets of the spatial lattice to form the power series:

$$X_j = \sum_{i(j)} \frac{x_{i(j)}}{q^{|i(j)|}},\qquad\qquad 5$$

where $i(j)$ is some monotonically increasing or decreasing subset of $\mathbf{Z}$ corresponding to the jth coordinate. These coordinates add a geometrical flavor to the Cantor set, as will be seen below in the computation of Lyapunov characteristic exponents.

The limit sets are known for relatively few cellular automata. Many automata are easily seen to have as limit sets fixed points (Wolfram's class I) or periodic orbits (Wolfram's class II; see Wolfram (8), for examples). Others seem to behave chaotically (in the sense that they seem to have positive entropy; Wolfram's class III) on some Cantor set. Wolfram has also identified a fourth class of qualitative behavior that displays various complicated moving structures. This classification scheme is so far empirical, and it is not known how sharp the boundaries are between them, especially between class IV and either class II or class III. One goal is to find computable quantities that can distinguish between these qualitative classes.

The limit sets for some cellular automata are known. Those one dimensional automata that are one-to-one with respect to site values at either end of the neighborhood of f_{loc} are onto, meaning that the limit set is all of Σ, and the product measure[4] is invariant with respect to all such rules. The entropy of these automata is easily computed to be $2r \, log(k)$. A subset of these are the so called linear rules with k prime, i.e. those rules that can be represented in the form

$$(f(x))_i = \sum_{j=-r}^{r} \alpha_j z_{i+j}$$

for some $\{\alpha_j \epsilon S\}$, where all arithmetic is done mod k. Grassberger (22) identified another class of limiting behavior that seems to be asymptotic to linear rules in an average sense. He noted that certain automata f have disjoint sets $\Sigma_1, \Sigma_2 \subset \Sigma$ that are invariant under f^2, and that f restricted to these subsets act as a linear rule. The examples investigated by Grassberger (elementary rules 18, 122, 126, 146, and 182) had as Σ_1 the set of all configurations with zeros on even sites, and as Σ_2 the set of all configurations with zeros on odd sites. A random (with respect to the product measure) initial condition will be a mixture of these two "phases," with some pieces of the lattice belonging to phase one, and some pieces to phase two, separated by phase boundaries, or "kinks". Under the iteration of f, the kinks appear to execute a random walk, annihilating when they meet, so that the density of kinks goes like $\rho(t) \sim t^{1/2}$. Thus the regions of the lattice containing pure phase one and pure phase two are growing with time, but any point on the lattice will eventually be passed by a kink, and change from one phase to the other, with the passage of kinks becoming increasingly infrequent with time.[5] More complicated rules are seen to display multi-phase

---

[3] We will conform to Wolfram's labeling of rules (5,8). For example, the "rule number" is given by $R_f = \sum f_{loc}(z_{i-r}, \ldots, z_{i+r}) k^{\sum k^{r-j} z_{i+j}}$, where the first sum ranges over all configurations of sites in the neighborhood, and the second sum goes from -r to r. This yield a number between 0 and 255 for nearest neighbor binary state rules. "Totalistic rules" are a special case where f_{loc} depends only on the sum (mod k) of the neighborhood sites, and are labeled with a numerical code $C_f = \sum d^n f_{loc}(n)$, with n running over all possible values of the sum, from 0 to $(2r+1)(k-1)$. "Elementary rules" refer to binary nearest neighbor rules ($k=2$, $r=1$).

[4] By product measure we mean the measure that assigns equal probability to all n-cylinders of a fixed length.

[5] Wolfram (this volume) displays a good picture of Grassberger's multi-phase phenomenon.

behavior with more than two phases, *e.g.* totalistic rule 30 has four phase behavior.

It is easy to construct rules that combine various kinds of behavior, having for instance, type I behavior when the dynamics is restricted to some subset of Σ, but which dies and leave type III behavior for almost every initial condition.[6] Likewise, it is easy to find examples of type III transients that die off to leave a fixed point. Type IV behavior embedded in a matrix of type III behavior has been observed, and Grassberger's multi-phase phenomenon can be seen as different phases of type III behavior coexisting.

3. LYAPUNOV CHARACTERISTIC EXPONENTS

For smooth dynamical systems, Lyapunov characteristic exponents measure the average rates of expansion and contraction along an orbit. They may be thought of as the expansion and contraction rates of the semi-major and -minor axes of an ellipse growing from an infinitesimal ball around x (22). Numerical algorithms for computing all the Lyapunov characteristic exponents have been developed and applied to a variety of systems (*e.g.* (15,22,23)), and our algorithm uses the same ideas.

For cellular automata we may use the metric mentioned above to measure how small subsets of Σ expand under the action of some automata rule $f : \Sigma \rightarrow \Sigma$. Note that small balls containing x are long n-cylinders, and as these sets grow (if they grow), the n-cylinders become shorter. Figure 1 shows some examples of the difference between two different configurations as a function of time, when they begin matching on a large portion of the lattice. Note that for linear rules the expansion is uniform, and for multi-phase rules like rule 18 it is very nearly uniform, with the only nonuniformities coming from the boundaries between phases, and since the density of the phase boundaries goes to zero with time, the expansion rate goes to a constant. Rule 86 shows uniform expansion on one side because of the linear dependence on the site at that end of the neighborhood of f_{loc}, but nonuniform expansion on the other side, where f_{loc} acts nonlinearly. Essentially nonlinear rules like rule 22 or rule 193 are seen to have nonuniform expansion on both sides. Rule 22 is the only symmetric nearest neighbor binary state automaton that displays nonuniform expansion, which is undoubtedly related to the nonuniformity of the probability distribution noticed by Grassberger (21). This situation is analogous to the case of one dimensional maps, where iterated maps with constant linear expansion have a flat probability distribution (Lebesgue measure is invariant), and nonlinear maps (or even piecewise linear maps with nonconstant slope) yield nonuniform probability distributions. For iterated maps there is an effective algorithm for constructing the maps' probability distributions: the Frobenius-Perron algorithm. The analogous technique for cellular automata has not been discovered.

Rule 193, as noted above, is asymmetric, and is observed to have different expansion rates on either side of the n-cylinder. This means that f is expanding different amounts in the X and Y coordinate "directions", so it is tempting to identify these two different expansion rates with the values of two unequal

---

[6] Unless otherwise specified, "almost every" will be with respect to the product measure.

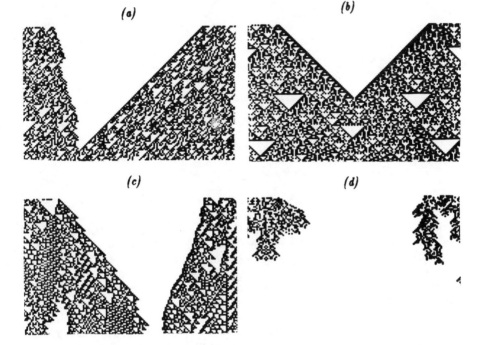

Figure 1. Difference patterns for elementary rules 18 (a), 22 (b), 86 (c), 90 (d), 193(e), and totalistic (r=2, k=2) rule 20 (f). These were made by initially taking two nearby patterns (nearby means matching on a large piece of the lattice), letting each of them evolve, and displaying the difference between the two. The horizontal direction is the spatial extent of the lattice, and the vertical direction is time. Constant slope of the difference pattern implies uniform expansion, nonconstant slope implies that the expansion rate varies along the orbit. The patchy effect in (a) is due to overlapping of regions of opposite phase.

Lyapunov characteristic exponents. The relationship between the values of these quantities to exponents computed using arbitrarily many different generalized coordinates defined above is unclear, however; the space of difference patterns does not have quite the same nice properties as the tangent space of a smooth dynamical system. The pullback algorithm described below is more involved for the computation of expansion rates in more than one direction, and will be implemented in the future. Of course, for symmetric rules, average expansion in the X and Y directions must be the same, and if consider the spreading in each coordinate separately, the sum of these two spreading rates would equal the entropy for the case of linear maps, where the entropy is known.

Lyapunov characteristic exponents	
rule 90	1.0
rule 18	.99
rule 193	.5
rule 86	.98
rule 22	.82

Table 1. Lyapunov characteristic exponents for a variety of elementary cellular automata. All the measurements have errors $\sim \pm$ *.03.*

Table one contains some numerical computations of Lyapunov characteristic exponents. The computations were done on a lattice with 1001 sites using periodic boundary conditions. We used the pullback technique of Bennetin *etal.* (15), beginning with close initial conditions (two configurations from a 100-cylinder), regarding one as a fiducial trajectory, and the other as the diverging trajectory. Each was evolved for 20 time steps, then the expansion (into a smaller n-cylinder) was measured and accumulated in an ongoing average. The diverging trajectory was then "pulled back" to a configuration in the same 100-cylinder.[7] The average was allowed to converge for 10^4 time steps. All computations were done for 100 initial conditions chosen at random to test the ergodicity of the cellular automata; all tries converged to the same result, asymptotically, giving numerical evidence that the these cellular automata are ergodic, so that the Lyapunov characteristic exponents are almost everywhere constant.

All of the elementary rules listed in table one are class III; in fact, positive Lyapunov characteristic exponents (or positive entropy) might be a good definition of class III rules. Clearly classes I and II rules will have zero Lyapunov characteristic exponent, as patterns never propagate. Class IV rules do have patterns that propagate, but numerical evidence based on the examination of several examples is that small perturbations from any given trajectory typically

---

[7] When the trajectory is pulled back, the same separation vector is always added to the fiducial trajectory, in contrast to the usual pullback algorithm in which the separation vector is allowed to change, and it aligns itself with the direction of maximal expansion. Thus, our measurements can only be regarded as a lower bound on the maximal Lyapunov characteristic exponent.

die out (after some time that depends on the rule), causing the Lyapunov characteristic exponent to converge to zero. Thus, the Lyapunov characteristic exponent is inadequate to distinguish between the qualitatively different behavior of class IV and classes I and II.

4. GROWING PATTERNS

Though we can't claim to model any real system at this point, there are many examples where an analysis of pattern growth might be useful. One example is spread of disease; *e.g.* automata models for tree blight (24). Pattern growth in cellular automata is also remiscent of a model for prebiotic evolution due to Rössler (25): he considered the prebiotic chemical soup as modeled by a gigantic set of coupled reaction diffusion equations, with the concentration of most of the variables (all those corresponding to large organic molecules, for example) set to zero at time zero. Cellular automata would represent a vast discretization of the concentration variables, and a very regular coupling of such a system of equations.

Any time a cellular automaton has a fixed point configuration, *i.e.* if $f(x_0)=x_0$ for some x_0, one may ask what happens to finite pattern on the background x_0, *i.e.* what happens to some y that is equal to x_0 except on a finite number of lattice sites. For the special case that x_0 consists of the same symbol s_0 at every site, the condition for x_0 to be a fixed point to f is $f_{loc}(s_0, \ldots, s_0)=s_0$. This is the case we will be concerned with, and some typical growth patterns are shown in figure 2, where the initial state is a finite string of length 10 chosen at random.

Growth is clearly bounded by a maximum velocity, sometimes analogous to the speed of light, $v_{max}=r$ *sites per unit time*. A sufficient condition for maximal growth velocity (to the left) in a background of s_0 is that $(s_0, \ldots, s_0, s) \epsilon f_{loc}^{-1}(s)$ for $s \neq s_0$. For $k=2$ this is also a necessary condition; for $k>2$ there is an analogous necessary condition to accommodate the possibility that the pattern's boundary sites change periodically (which they must do with $1 \leq period < k$). Pattern growth rate depends on the background symbol s_0, and there may be more than one possible background symbol. For example, the $k=2$, $r=2$ rule number 117 has pattern growth velocity $v_{max}=2$ in a background of ones, and velocity one in a background of zeros. A sufficient condition for sub maximal growth velocity is that

$$\{(s_0,..,s_0,s,s') \mid s' \text{ takes on all values in } S\} \subseteq f_{loc}^{-1}(s) .$$

Another interesting possibility is that patterns can grow with a constantly decreasing velocity. Growth from a single nonzero site under the composition of the elementary rule 225 composed with the left shift (rule 170) displays this behavior,[8] growing with a velocity $v(t) \sim t^{-1/2}$.

For patterns that grow at a constant velocity, figure 2 shows two distinct types of behavior: patterns that grow eventually into a fractal pattern (26), and patterns that don't. The eventually fractal patterns appear to mirror the fractal pattern obtained from the growth of a single nonzero initial site, with a random

---

[8] This was pointed out to me by J. Franks and R. Williams.

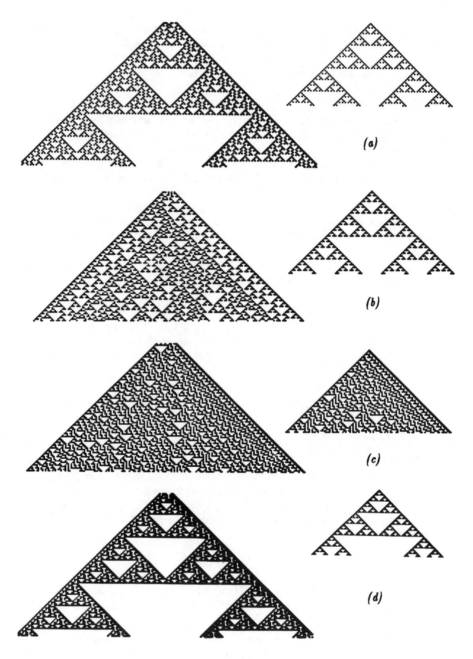

Figure 2. Growing patterns from a finite initial condition chosen at random for elementary rules 18 (a), 22 (b), 86 (c) and 90 (d). For reference, the pattern growing from a single nonzero initial site is placed on a smaller scale to the right.

smearing occurring on the length scale of the length of the initial condition. The
length scale of the spatial extent of the initial condition also introduces a characteristic time scale that marks the emergence of the fractal "order." Rule 22
demonstrates that fractal growth from a single nonzero initial condition is not,
however, a sufficient condition to ensure an eventually fractal pattern. In both
cases, the growing patterns have a periodic structure seen by looking along a
diagonal parallel to the pattern boundary. In other words, f composed with the
shift has a periodic temporal sequence at any given lattice site. In general, the
period grows with the distance q away from the pattern boundary; the eventually fractal patterns have maximal growth of the period as a function of q:
$P(q) \sim q$, and the nonfractal patterns appear to have slower growth of $P(q)$,
with the periodicity emerging only after some time proportional to q.

In modeling pattern growth in real systems, there is some motivation to
regard a random initial condition on finite piece of lattice as a "microscopic
seed" growing to a macroscopic structure. One analogy in this regard is the
growth of a snowflake from a presumably random microscopic seed crystal. The
metric of eq. 2 is inadequate for this analogy, with this metric the high order bits
of a pattern's "size"[9] come from sites near the origin of the lattice, and as the
pattern grows, its "size" goes to a constant. We may alleviate this difficulty by
introducing a new metric

$$|| x{-}y || = \sum_{-\infty}^{\infty} \frac{| x_i - y_i |}{k^{Q-|i|}} \, , \qquad\qquad 6$$

again using subtraction mod k as the metric on S, and introducing an "inversion" length scale Q. This metric has the desired property that as two finite
patterns grow, their distance from each other increases. Figure 3 shows some
examples of 2^9 finite initial conditions of length 11 (the boundaries are fixed to
be one) for a few elementary automata. These pictures have used a coordinate
map from sequences to the unit square analogous to eq. 3, but with the same
modification as the metric:

$$X' = \sum_{i=0}^{\infty} \frac{x_i}{q^{Q-i}} \quad \text{and} \quad Y' = \sum_{i=0}^{\infty} \frac{x_{-i-1}}{q^{Q-i-1}} \, . \qquad\qquad 7$$

For all the pictures, all 2^9 trajectories initially begin in a small square region
$(2^{-11}$ of the size of each plot) spread homogeneously. As the pattern grows, they
spread over an increasingly large region of the unit square, and become unevenly
distributed, except in the case of the linear rule 90.

The coordinate representation of the limit
$f^t\{x \mid x$ is a finite pattern of extent $Q\}$ with $t \to \infty$, $Q \to \infty$, and $q \to \infty$, with
$Q/q \to 0$, appears to yield a Cantor set for the nonlinear symmetric elementary
class III rules 18 and 22. In general, this limiting Cantor set should not be the
same as not the limit set Λ of eq. 1, since it should depend on the background
configuration. The nonsymmetric rule 86 collapses differently in the different
coordinate directions; it is unclear whether the vertical lines display cantor set

---

[9] Size, in this context, means the distance between a given pattern x and the background pat-
tern x_0 using the metric $|x{-}x_0|$, not the spatial extent of the pattern as above.

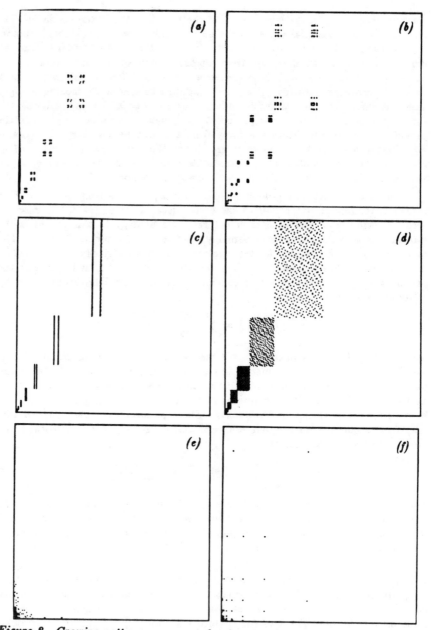

Figure 3. Growing patterns represented using the metric of eq. 6 for elementary rules 18 (a), 22 (b), 86 (c), 90 (d), and class IV totalistic rules 20 (e) and 52 (f). Both X and Y coordinates are very small to begin with, and they grow with time. Plotted are the trajectories of all 2^9 finite patterns with spatial extent 11, for 10 time steps.

gaps at the current resolution. For all the class III elementary rules, since there is a one on the boundary of the growing pattern, and all the initial conditions grow at the same velocity, the images of the ensemble of growing patterns occupy disjoint regions of the unit square at each time step, each further from the origin with each step.

For the two class IV rules, we see a markedly different limiting behavior. The fact that most of the initial conditions die out or go to localized periodic structures after initial transients is reflected in the concentration of points around the origin (obviously the regions occcupied at successive time steps are not disjoint). The few initial states that continue to grow are represented by the few scattered points in the center of the unit square.

5. CONCLUSIONS

We have viewed the states of a cellular automaton, *i.e.* configurations of symbols over a lattice, as patterns that evolve under the action of a local, spatially homogeneous cellular automaton rule. On the space of patterns we have considered two different kinds of metrics that enable the measurement of distance between patterns. Using one metric the Lyapunov exponent has been computed for several examples, measuring the average asymptotic exponential separation of initially nearby patterns. A possible definition for Wolfram's class III automata might be a positive Lyapunov characteristic exponent for almost all initial conditions.

A second metric allows one to regard finite initial patterns as microscopic seeds growing under the action of an automaton rule. Linear rules are seen to have all finite initial patterns of a given spatial extent spread evenly over the possible final states after they grow for a time t. Nonlinear rules have an uneven distribution, reflecting the structure of the asymptotic patterns. The distribution has been represented using coordinates constructed from the patterns, with the property that patterns with small spatial extent correspond to small coordinate values.

We have as yet performed no statistical measurements such as dimension, spatial correlations, etc., on this representation of the growing patterns, but a few qualitative conclusions may be drawn. From the examples observed so far, there seems to be a clear distinction between the distributions of class III automata and those of class IV automata (and of course those of classes I and II, which might be defined by the property of not having growing patterns for most initial conditions). Class III automata seem to have patterns that grow into disjoint regions of the constructed space, and class IV automata have an extremely uneven distribution that remains clustered about the origin, with a small fraction of trajectories growing indefinitely. The detailed structure of these distributions, and their exact relationship to the growing patterns themselves remains to be clarified.

ACKNOWLEDGEMENTS

My work on cellular automata began during a stimulating summer at the Center for Nonlinear Studies in Los Alamos, NM. Since then I have benefited from many helpful conversations with J. Conway, J. Crutchfield, J. Franks, J.

Milnor, O. Rssler, and S. Wolfram.

REFERENCES

1. *Physics of Computation* conference proceedings (1982). E. Fredkin, R. Landauer, T. Toffoli, eds. in Int. J. Theo. Phys. **21**, nos. 3/4, 6/7, 12.

2. *Cellular Automata: Proceedings of an Interdisciplinary Workshop (Los Alamos, March 7-11, 1983)*, North Holland.

3. S. Wolfram (1983). "Statistical Mechanics of Cellular Automata," Rev. Mod. Phys. **55**, 601-644.

4. J. von Neumann (1966). *Theory of Self-Reproducing Automata* A. W. Burks, ed., University of Illinois Press, Urbana.

5. S. A. Kauffman (1969). "Metabolic Stability and Epigenesis in Randomly constructed genetic nets," J. Theor. Bio. **22**, 437. Also see S. A. Kauffman, this volume.

6. S. Wolfram (1983). "Universality and Complexity in Cellular Automata," to be published in Physica D and *Cellular Automata: Proceedings of an Interdisciplinary Workshop (Los Alamos, March 7-11, 1983)*, North Holland.

7. A. N. Kolmogorov (1965). "Three Aproaches to the Quantitative Definition of Information", Prob. Info. Trans. **1**, 1.

8. G. Chaitin (1966). "On the Length of Programs for Computing Binary Sequences", J. ACM **13**, 547.

9. A. N. Kolmogorov (1958). Dolk. Akad. Nauk. **119**, 754.

10. C. Shannon (1948). Bell Tech. Jour., **27**, 379.

11. J. Crutchfield and N. Packard (1982a). "Symbolic Dynamics: Entropies, Finite Precision, and Noise", Int. J. Th. Phys. **21**, 493.

12. J. Crutchfield and N. Packard (1982b). Symbolic Dynamics of Noisy Chaos", in *Order in Chaos*, (D. Campbell, H. Rose, eds.) North-Holland.

13. P. Grassberger and I. Procaccia (1983). "Estimation of the Kolmogorov Entropy from a Chaotic Signal", Phys. Rev. **A28**, 2591.

14. G. Bennetin, L. Galgani, and J. Strelcyn, (1976). Phys. Rev. **A14**, 2338; also G. Bennetin, L. Galgani, A. Giorgilli, and J. Strelcyn, (1980). "Lyapunov Characteristic Exponents for Smooth Dynamical Systems and for Hamiltonian Systems: A Method for Computing All of Them", Meccanica, **15**, 9.

15. R. Shaw (1980). "Strange Atractors, Chaotic Behavior, and Information Flow", Z. Naturforsch. **36a**, 81.

16. S. J. Gould (1973). *Ever Since Darwin,* W. W. Norton, New York.

17. S. J. Gould (1982). *The Panda's Thumb,* W. W. Norton, New York.

18. N. Eldridge, S. J. Gould (1972). in *Models in Paleobilogy* (T. J. M. Schoph, ed.) Freeman, San Fransisco.

19. R. Shaw (1980). PhD. thesis (appendix II), University of California, Santa Cruz.

20. O. Martin, A. M. Odlyzko, S. Wolfram (1983). "Algebraic Properties of Cellular Automata," to appear in Comm. Math. Phys.

21. P. Grassberger (1938). "Chaos and Diffusion in Deterministic Cellular Automata," to be published in Physica D and *Cellular Automata: Proceedings of an Interdisciplinary Workshop (Los Alamos, March 7-11, 1983),"* North Holland.

22. J. D. Farmer (1982). "Chaotic Atractors of an Infinite-Dimensional Dynamical System", Physica **4D**, 366-393.

23. J. P. Crutchfield (1981). Undergraduate thesis, University of California, Santa Cruz, CA.

24. J. Demongeot, this volume.

25. O. Rössler (1983). "Deductive Prebiology" in *Molecular Evolution and the Prebiological Paradigm* (K. Matsumo, K. Dose, K. Harada, K. L. Rolfing, eds.), Plenum Press.

26. B. Mandelbrot (1983). "The Fractal Geometry of Nature", Freeman.

DISCRETE SIMULATION OF PHYLLOTAXIS

N.RIVIER[*+],R.OCCELLI[+],J.PANTALONI[+],A.LISSOWSKI[*$]

+ Laboratoire de Dynamique et Thermophysique des Fluides,Université de Provence,Rue H.POINCARRE 13397 MARSEILLE Cedex 13 ,FRANCE

* Blackett Laboratory, Imperial College , London SW7 2BZ, United Kingdom.

$ Departement of Psychology ,Polish Academy of Science, Warsaw,POLAND.

I-INTRODUCTION

This paper is concerned with crystallography in two spatial dimensions in the presence of cylindrical symmetry.A manifestation of cylindrical symmetry can be found in the structure of daisies,sunflowers,... called phyllotaxis (or leaf-arrangement) which constitutes a space filling problem solved by some algorithm or code governing the successive generation of cells or florets from the stem,and whose manifestation is through a sequence of numbers of opposite spirals(the Fibonacci sequence/1/2/).We have searched for crystalline structure filling the space available ,with as much homogeneity, and made of as similar ,isotropic cells, as compatible with the boundary conditions.
  In this paper we present only the main results, and the reader should find a fairly complete record of the study in ref/3/.(There is another field of application in the case of cellular patterns in Bénard convection whereby a fluid heated from below exhibits convective motion above a certain temperature threshold.Cylindrical symmetry is imposed by the shape of the container and by the earth's rotation/3//4/.)

II THE ALGORITHM

Symmetry suggests polar coordinates for the cell centers.

$$r(1) = a\sqrt{1}$$
$$\theta(1) = 2\pi\lambda \quad 1 \qquad\qquad (1)$$

In this relation $1=1,...,N$ labels the individual cells proceeding outwards from the centre of symmetry (a is the typical cell's linear dimension). λ is proportionnal to 1 and r increases as $\sqrt{1}$ for uniformity and homogeneity in cell densities . .The algorithm (1) introduces cells regularly on a generative spiral in the reverse order of successive leaves budding from the stem of a plant. Once the centers are specified the cells are constructed by VORONOI partition of space around each center/5/The only structural information which is encoded is the angle λ between florets, and the fact that successive florets appear at regular time intervals. The shape of each floret (hexagon, pentagon,...) is not coded. It is simply a consequence of filling an area.

III) CHOICE OF THE STRUCTURAL PARAMETER λ

a) For definition $0 < \lambda < 1$. The area can be regarded as partitioned into concentric quasi-circular shells containing an integer number of cells. Thus λ cannot be rational (ex: if $\lambda = A/B$; A,B integers ,the resulting cellular structure resembles a spider's web (Fig.1))The shells correspond to successive rational approximations of λ , or a division of 2π by successive integers, the number of cells per shells.

b)Theory of number yield these successive rational approximations of λ by representation as a continued fraction /6/

$$\lambda = \cfrac{1}{q_1 + \cfrac{1}{q_2 + \cfrac{1}{q_3 + \cdots}}}$$

with q_i integers. This decomposition is unique, thus $\lambda = \{q_i\}$ is given by the set of integers which constitute a code. If λ is irrational the continued fraction is infinite. Successive truncations of the continued fraction yields convergents which are rational approximants to .

$$\lambda \simeq 1/(q_1 + 1/(q_2 + ...1/q_m)) = A_m/B_m$$

where A_m and B_m are (relatively prime) positive integers monotically increasing with m.

<u>Fig.1:</u>   The spider web constructed with $\lambda = 13/21$ rational

These convergents yield the successives shells:B_m is the
number of cells in the shell described by the mth convergent,
and A_m the number of turns of the generative spiral necessary
to fill the shell.
c)λ should not be <u>transcendant</u> , either because the continued
fraction(i.e. the <u>sequence</u> $\{\overline{q_i}\}$) would not be periodic;
and the code $\{q_i\}$ would be infinite.Alternatively, <u>radial</u>
<u>homogeneity</u>, or <u>self-similarity</u> requires a periodic continued
fraction.Thus λ is quadratic irrational,i.e. a solution
of the quadratic equation
          $a\lambda^2 + b\lambda + c = 0$; a,b,c integers.

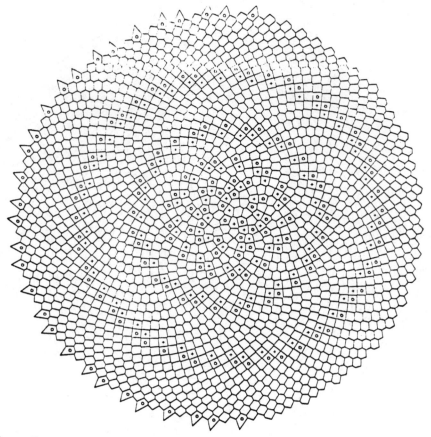

Figure 2 :
Structure generated with $\lambda = 1/\tau$ .Symbol: o=pentagon
+=heptagon,o+=dislocation;none=hexagon.

d) Homogeneity of the structure requires that,after a few
arbitrary q_i , the continued fraction must merge into the
inverse of the Golden section,

$1/\tau = \frac{1}{2}(\sqrt{5}-1) = 1/(1+1/(1+...)) = 0.618...$ Thus /7/,

homogeneity $\Rightarrow$ $\lambda = 1/(q_1+1/(q_2+....+1/\tau))$;such λ have been
called Noble numbers by Percival/6/.Thus $1/\tau \Rightarrow \{q_i\}=1$,the
simpest code possible(solution of the equation $\tau^2-\tau-1 = 0$)
In this case the numerator and the denominator of the
convergents of $(\{ A_m \},\{ B_m \})$ constitute the Fibonacci
series : $\{B_m\}=1,1,2,3,5,8,13,21,34,55,89,...$; familiar to
rabbit-breeders.

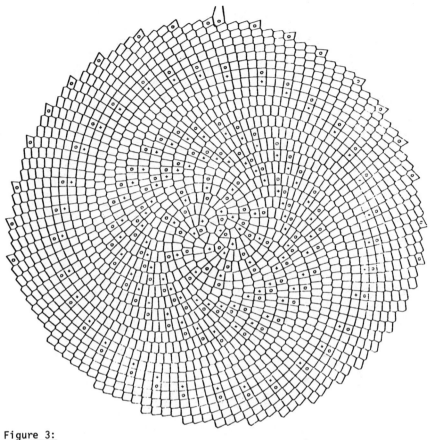

Figure 3:
False phyllotaxis constructed with $\lambda = 1/(3+1/(3+...3))$
$\lambda = (\sqrt{13} - 3)/2$

Figure 2 shows the generated structure with $\lambda = 1/\tau$ (the daisy).The agreement with botanical structures is manifest (we can observe circular grains which are conspicuous in sunflowers).The structure is self-similar in that every grain or grain boundary is similar to every other.
The simplest code possible $\{q_m\} = 1$ is required to generate the entire structure.Structure $\{q_m\} = 2$ or 3 ,which do not have the same desirable properties of self similarity as $\{q_m\} = 1$ (fig 3) require more bits to encode.
Homoqeneity only requires $\{q_i\} = 1$ after a few stages,i.e. with the first q_i arbitrary. This advantage has been taken to generate structures which are   most similar to a closed packed crystal.Figure 4 shows the structure generated with $\lambda = 2/(11 + 5) = \{6,\underline{1}\}$; ending up with $1/\tau$. /8/

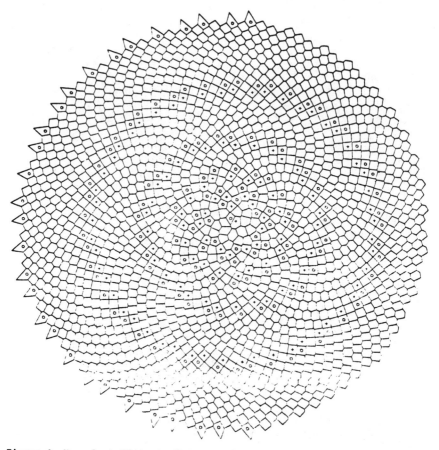

Figure 4: Normal phyllotaxis (the daisy) generated with
$\lambda = 2/(11+\sqrt{5}) = \{6, \underline{1}\}$;ending up with $1/\tau$.

IV) CONCLUSION

We have shown that 2-dimensional cellular structures with
cylindrical symmetry constitute a genuine crystallography.
The requirements of homogeneity and self-similarity are
sufficient to reduce drastically the number of possible
structures. The simplest possible code, $\{q_m\} = 1$ yields the
whole 2D crystallography in cylindrical or axial symmetry;
which is parametrized by Noble numbers. It is therefore
no accident that the Fibonacci integers are so pervasive
in phyllotaxis. We have related the requirements of homogeneity
and self-similarity with the numerical concept of code,

which is obviously relevant in biology,but also in physics
via the entropy.The open question is how does the plant
transfer digital ($\{q_i\}$) into geometrical information,i.e.
compute the continued fraction.

REFERENCES

/1/ D'Arcy W. Thompson,On Growth and Form;Cambridge UP,
2nd edition,1952,ch 14

/2/R. Jean,Phytomathématique,Presse Univ. du Québec,1978

/3/ N.Rivier,R.Occelli,J.Pantaloni,A.Lissowski;J.de Physique
45 (1984),January.

/4/ J.Pantaloni,P.Cerisier,R.Bailleux,C.Gerbaud;J.de Phys.
Let. 42 (1981) L147.

/5/ See R.Occelli,Thèse Université de Provence 1984,or
W.Brostow,J.P.Dussault and B.L.Fox,J.Comp.Phys. 29(1978)81,
or J.L. Finney J.Comp.Phys. 32 (1979) 137; for computer
programmes generating VORONOI polygons.

/6/ I.C.Percival,Physica D 6D(1982) 67.

/7/ Proof:Cell 1 has a reference frame made of its nearest
neighbours, cell $1 \pm B_m$,$1 \pm B_{m-1}$ and $1 \pm B_{m-2}$. Homogeneity
requires this relationship to be valid for any 1. One can
reach neighbour $1+B_m$ from 1 either directly,or through
cell $1+B_{m-1}$ by using the B_{m-1} and B_{m-2} spirals,thus,
  $1+B_m = (1+B_{m-1}) + B_{m-2}$ $\forall 1$,
or $B_m = B_{m-1}+B_{m-2}$,which is the Fibonacci recursion relation
associated $\{q_i\} = 1$ in the continued fraction(2-3).

/8/ In this example of Fig.4,$\{B_m\} = 1,6,7,13,20,33,53,86,139...$
still satisfy the Fibonacci recursion relation/7/(homogeneity
condition).

STOCHASTIC MODELS OF CLUSTER GROWTH

J. Vannimenus[*], J.P. Nadal[*] and B. Derrida[+]

[*]*Groupe de Physique des Solides de l'E.N.S.*
24 rue Lhomond, 75231 Paris Cedex 05 (France)

[+]*Service de Physique Théorique, CEN-Saclay*
91191 Gif-sur-Yvette (France)

1. INTRODUCTION

Like many theoretical physicists we have been working on cellular automata without knowing it - after all, the most studied model of statistical mechanics, the Ising model, is just
a particular probabilistic cellular automaton ! Conversely
many problems in our domain, like percolation, the gelation
of polymers..., may interest people working primarily on
automata. For reasons of space limitation we shall concentrate
our contribution on some recent developments in the study of
cluster growth, hoping this may be a good starting point for
interactions.

The *equilibrium* statistics of lattice clusters are now
fairly well known for a number of cases of physical interest :
percolation, linear polymers (i.e. self-avoiding walks),
branched polymers (i.e. lattice animals). Few exact results
have been obtained, but a combination of numerical studies
and renormalization-group arguments has provided a coherent
picture. The main general result is that there exists a few
"universality classes" for a broad range of problems : for
instance, the average gyration radius of a self-avoiding
walk of N steps behaves asymptotically as $N^{0.75}$ in two
dimensions, whatever the lattice.

However, in the real systems these models are meant to
describe, *kinetic* effects often play a major role. Recently
studied situations include :
- the formation of metallic aggregates, soot, smoke...(1),
- two-phase flow in porous media (2) ("invasion percolation"),
- the pattern of electric discharge in solids (3).

The first questions to be answered are : do the kinetic effects modify the "universal" equilibrium properties ? If yes, can one find new classes with a weaker universality ? In the following, we concentrate on the problem of aggregate formation, which is presently very active (4-6).

2. MODELS OF CLUSTER FORMATION

It is experimentally observed (7) that clusters obtained by agglomeration of many small particles do not behave at all like ordinary matter : their density goes to zero when their size increases. The remarkable fact is that the relation between the number N of constituent particles (equivalently, the total mass) and the size R of the cluster is often given by a simple scaling law of the form :

$$N \sim A R^D .$$ (1)

D is the fractal dimension of the cluster and if $D < d$, the dimension of space, the average density vanishes for large N.

The direct theoretical study of a realistic situation is very difficult, so it is useful to consider first simplified models and see whether they predict values of D in agreement with the observations. In these models, "particles" are represented by discrete variables n_i on a cubic lattice : $n_i = 1$ if site i is occupied, = 0 otherwise. Deterministic rules of growth, such as for Convay's game of "Life", seem too restrictive to give the desired fractal patterns (but see Wolfram's contribution to this colloquium) and various stochastic rules have been studied in detail :

- *Lattice animals* (8) : all connected configurations of N sites on the lattice are taken into account with *equal* weight to calculate the average value of R. This neglects all kinetic effects.

- *Eden model* (8) (or "growing animals") : starting from an initial cluster (seed) one adds new particles at random on its empty boundary sites, with the same probability for all sites.

- *Diffusion-limited aggregation* (1) (DLA) : the particles are allowed to diffuse randomly before joining the growing cluster or leaving to infinity. This takes into account the fact that growth is more likely to occur at exposed external sites (tips), and the structures so obtained are very ramified, with $D < d$ (1,4-6).

- *Clustering of clusters* (9-10) : the diffusing particles form clusters of various sizes, which are themselves allowed to move and aggregate into larger entities.

Results obtained up to now, mostly by numerical simulations, are summarized in Table I. Experimental results for D lie in range $1.5 - 1.8$ for $d = 2$ (11), $1.7 - 1.9$ for $d = 3$ (7).

TABLE I

Fractal dimension D for various models

d	Lattice animals	Eden Model	DLA	Clustering
2	1.56	~ 2	1.7 ± 0.07	$1.4 - 1.5$
3	2	~ 3	2.4 ± 0.2	
4	12/5		3.3 ± 0.2	

These results clearly show that kinetic effects modify the fractal dimension, but other classes of models may still be discovered in the future and there is no general framework at the present stage. For many problems in statistical physics, the situation is simpler at high space dimensions, because mean-field theory becomes valid and provides a starting point for more elaborate theories. Here such a useful first-order approximation seems to be still lacking (12, 13).

3.   DIRECTED CLUSTER

An interesting extension of the previous models consists in studying situations where the diffusion of particles is biased in a particular direction : this may be due in practice to gravity, electric fields, imposed flows... The clusters then grow preferentially in that direction and their average width W and length L scale in general as different powers of the number of particles N :

$$L \sim a \, N^{\nu_{||}} \quad , \quad W \sim b \, N^{\nu_{\perp}} \quad . \tag{2}$$

To study the importance of kinetic effects in that case, we have considered two models :
− *Directed animals* (14,15) : any site of the cluster can be reached from the root via an orientated path on the lattice. The averages in Eq.(2) are performed with the same weight for all possible configurations, neglecting kinetic constraints. Fig.(1) shows typical directed animals, obtained by

a Monte Carlo method (16) on a square lattice oriented along the diagonal.

  - *Directed aggregation* (16,17) : the particles drift at a constant speed in the preferred direction and have a diffusive motion in the others. Fig.(2) shows directed aggregates obtained by numerical simulation.

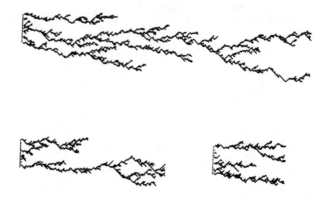

Fig.1 Directed animals drawn by a Monte Carlo method

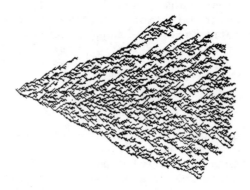

Fig.2 Directed aggregate

These two types of clusters are clearly very different
objects. The animals are very elongated, with $\nu_\perp = 1/2$ (exact
result (14,15)) and $\nu_{//} \sim 0.82$ (14). The aggregates are compact, with $\nu_\perp \sim \nu_{//} \sim 0.5$ (16,17). Heuristic arguments (16)
can be given that $\nu_\perp = \nu_{//} = 1/2$ but a rigorous derivation
would be valuable.

4. CONCLUSION

The realization that rather simple probabilistic rules for
cluster growth can give rise to fractal objects has led to
a wave of interest for these problems. A lot of numerical
work has already been done but few analytic results have
been obtained and a classification of the various rules
would be highly desirable. The connection between the growth
models and cellular automata is direct enough, and interesting ideas might emerge from a confrontation of both areas :
in particular, one may hope to devise deterministic models
(18) which contain some of the most striking features observed numerically.

REFERENCES

1. Witten T.A. and Sander L.M. (1981) Phys. Rev. Lett. 47,
    1400-1403.
2. Wilkinson D. and Willemsen J.F. (1983). J. Physics A 16,
    3365-3376.
3. Sawada Y. Ohta S., Yamazaki M. and Honjo H. (1982). Phys.
    Rev. A 26, 3557-3563.
4. Meakin P. (1983). Phys. Rev. A 27, 1495-1507.
5. Witten T.A. and Sander L.M. (1983). Phys. Rev. B 27,
    5686-5697.
6. Gould H., Family F. and Stanley H.E. (1983). Phys. Rev.
    Lett. 50, 686-689.
    Stanley H.E., Family F. and Gould H. (1983). Kinetics
    of aggregation and gelation, to be published in J. Poly.
    Sci., and references therein.
7. Forest S.R. and Witten T.A. (1979). J. Physics A 12,
    L109-117.
8. Peters H.P., Stauffer D., Hölters H.P. and Loewenich K.
    (1979). Z. Physik B 34, 399-408.
9. Meakin P., (1983). Phys. Rev. Lett. 51, 1119-1123.
10. Kolb M., Botet R. and Jullien R. (1983). Phys. Rev. Lett.
    51, 1123-1126.
11. Allain C. and Jouhier B. (1983). J. Physique Lettres 44,
    L.421-428.
12. Ball R., Nauenberg M. and Witten T.A. (1983). Diffusion-

controlled aggregation in the continuum approximation
(to be published).

13. Vannimenus J., Nickel B. and Hakim V. (1983). Models of
    cluster growth on the Cayley tree (to be published).

14. Nadal J.P., Derrida B. and Vannimenus J. (1982). J.
    Physique (Paris) $\underline{43}$, 1561-1574.
    Redner S., Yang Z.R. (1982). J. Physics A $\underline{15}$, L177-L187.
    Day A.R., Lubensky T.C. (1982). J. Physics A $\underline{15}$, L285.
    Dhar D., Phani M.K., Barma M. (1982). J. Physics A $\underline{15}$,
    L279.

15. Dhar D. (1982). Phys. Rev. Lett. $\underline{49}$, 959-962.
    Dhar D. (1983), Phys. Rev. Lett. $\underline{51}$, 853-856.

16. Nadal J.P., Derrida B. and Vannimenus J. (1983). Directed
    diffusion-controlled aggregation versus directed animals,
    (to be published).

17. Jullien R., Kolb M. and Botet R. (1983). Diffusion-
    limited aggregation with directed and anisotropic
    diffusion (to be published).

18. Vicsek T. (1983). Fractal models for diffusion controlled
    aggregation (to be published).

SOME RECENT RESULTS AND QUESTIONS
ABOUT CELLULAR AUTOMATA

Stephen Wolfram[*]

The Institute for Advanced Study, Princeton NJ 08540, USA.

1. INTRODUCTION

It is common to find that systems consisting of many identical components, each obeying simple rules, exhibit highly complex overall behaviour. Cellular automata are simple examples of mathematical systems with this property. From their study it may ultimately be possible to abstract a general mathematical theory for such behaviour, which would be of importance in many fields. This paper gives an informal sketch of some of the results about cellular automata that have been obtained so far, and mentions some of the questions that have arisen. Most of the material is covered in greater detail, and with many further references, in refs. (1,2,3,4).

A simple example of a cellular automaton consists of a line of sites, with the sites having values 0 or 1. In a series of discrete time steps, the value of each site is updated according to a rule which involves the previous value of the site itself and its two neighbouring sites. One simple example of such a rule takes the value $a_i^{(t)}$ of site i at time step t to be the sum modulo two of the values of its neighbours on the previous time step:

$$a_i^{(t)} = \left(a_{i-1}^{(t-1)} + a_{i+1}^{(t-1)} \right) \bmod 2 \ . \tag{1}$$

In a more general case, the sites of a one-dimensional cellular automaton take on, say, k possible values, and evolve according to a rule of the form

$$a_i^{(t)} = \mathbf{F}\left[a_{i-r}^{(t-1)}, a_{i-r+1}^{(t-1)}, .., a_i^{(t-1)}, .., a_{i+r}^{(t-1)} \right] \ , \tag{2}$$

in which the value of a site depends on the values of a neighbourhood $2r+1$ sites on the previous time step.

Cellular automata such as eqn. (2) may be viewed as discrete dynamical systems, and their properties analysed using the methods of dynamical systems and ergodic theory (e.g. (5)). When the line of sites is infinite in extent, the possible configurations of the cellular automaton may be considered to form a Cantor set. The cellular automaton time evolution is then an iterated continuous

---

[*]Work supported in part by the U. S. Office of Naval Research under contract number N00014-80-C0657.

mapping of this Cantor set to itself, similar to the iterated mappings of intervals of the real line studied extensively in dynamical systems theory.

Cellular automata may alternatively be viewed as digital computers. Their initial configurations may be considered to represent programs and initial data, "processed" by the cellular automaton time evolution, to yield configurations representing results of computations. Methods from the theory of computation (e.g. (6)) may then be used to analyse the general behaviour of cellular automata.

Cellular automata with finite numbers of sites (with, for example, periodic boundary conditions) are similar to feedback shift registers. Following this third analogy, techniques from discrete mathematics and number theory, used in the analysis of shift register sequences (e.g. (7)), may thus also be applied to cellular automata.

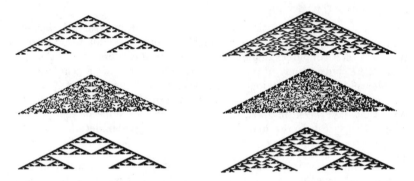

Fig. 1 Patterns generated by evolution according to three typical cellular automaton rules with $k=2$, $r=2$, from initial states containing one or a few nonzero sites. Sites with value 1 are represented by black squares. Successive time steps are shown on successive lines. Fractal and chaotic patterns are seen.

2. QUALITATIVE CHARACTERISTICS OF CELLULAR AUTOMATA

Figure 1 shows some patterns "grown" by the evolution of a variety of cellular automata from initial states with all sites zero, except for a small "seed" consisting of one or a few nonzero sites. In a few cases (less for larger k and r), all sites quickly attain value zero, or a pattern of fixed size is generated. The majority of rules, however, yield growing patterns, with boundaries expanding at a fixed speed. Patterns of this kind are found to fall into three basic classes: uniform, fractal and chaotic. Uniform patterns are regular and spatially periodic. Fractal patterns have a self-similar structure; the total average number of nonzero sites they contain after time t grows as t^d, where d is their fractal dimension. Chaotic patterns have a complicated form, in which the density of nonzero sites tends to a nonzero limiting value. No simple criterion is known to determine which type of pattern is generated by any particular cellular automaton rule. However, if in the rule $\mathbf{F}[a,0,0,\cdots,0]\neq0$ (and $\mathbf{F}[0,0,\cdots,0]=0$), then some finite fraction of patterns should grow at a speed of r sites (on each side) per time step. Many expanding patterns do grow at this rate, but some grow at speeds as low as $1/k$ or even lower.

Self-similarity provides one distinguishing feature of patterns generated with some cellular automaton rules. Self-similar patterns with simple appearance tend to be much more common than more complicated ones. The pattern generated by the first and third rules shown in fig. 1, with fractal dimension $\log_2 3$, is especially common. For any particular cellular automaton rule, it is straightforward to establish by direct construction whether a fractal pattern is formed, and to determine its fractal dimension. A necessary but very insufficient condition for the generation of fractal patterns is the presence of "growth inhibition": adding more nonzero sites in the neighbourhood must sometimes cause the value of the function F to turn from nonzero to zero. A general sufficient condition is not known. However, for certain additive rules, it is possible to show that fractal patterns are always generated (cf. (8)). Additive rules are ones for which F is linear in the site values modulo k, and time evolution may be represented by multiplication of characteristic polynomials corresponding to configurations by fixed polynomials (2). For prime k and simple time evolution polynomials, a simple formula for the fractal dimension has been found. A general formula in terms of algebraic properties of the time evolution polynomial should be derived.

For large k and r, an increasing fraction of cellular automaton rules generate chaotic patterns. No simple characterization of these patterns is known. Even though the patterns may be specified by simple rules and simple initial states, they appear to have a complex and random form. Statistical properties are typically almost indistinguishable from those of uncorrelated site value sequences.

The second column in fig. 1 shows examples of patterns grown from extended seeds containing several nonzero sites, rather than just a single nonzero site. For some cellular automaton rules, the patterns obtained in these two cases are qualitatively similar (e.g. the second and third rules of fig. 1); for other rules, however, very different patterns may be generated. If some simple superposition principle is operative, as in the additive rules mentioned above, then a seed consisting of s sites should generate a pattern similar to that from a single site seed on length scales larger than about s sites. In practice, it is often found that a cellular automaton rule which with small seeds yields regular fractal patterns yields chaotic patterns when grown from larger seeds, even on arbitrarily large scales.

Figure 2 shows examples of some cellular automaton rules evolving from "disordered" initial states, in which each site is chosen independently, with equal probabilities for each possible value. These pictures suggest the general observation, supported by an extensive empirical study, that the patterns generated by all (one-dimensional) cellular automaton rules evolving from disordered initial states fall into four basic qualitative classes:

1.   All sites ultimately attain the same value.

2.   Simple stable or periodic separated structures are formed.

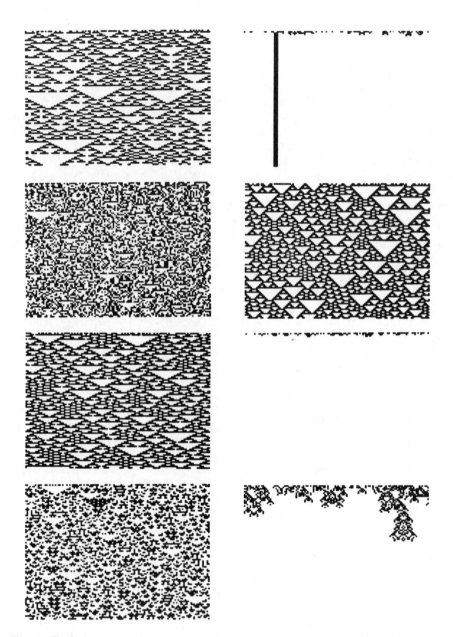

Fig. 2 Evolution according to a typical selection of cellular automaton rules with
$k=2$ and $r=2$ from disordered initial states. Four qualitative classes of
behaviour are seen.

3. Chaotic patterns are formed.

4. Complex localized structures are formed.

Quantitative characterizations of these classes and their properties are required.

"Green functions" may provide one distinguishing feature. Green functions measure the change in the pattern generated by evolution of the cellular automaton resulting from small changes in the initial state (for example, a change in a single site value), and give the average probability for sites at a particular distance to be changed after a certain time. Class 1 cellular automata have unique final states, unaffected by any change in initial conditions. For class 2 cellular automata, a small change in the initial state affects only a fixed finite region of sites. The Green function thus vanishes for all distances beyond some critical range. In class 3 cellular automata, almost all changes in initial conditions affect a finite fraction of the sites in a region of linearly increasing size. The Green function in this case is thus nonzero throughout a triangular spacetime region ("light cone"). The behaviour of class 4 cellular automata is so irregular that the averaging necessary to define the Green function cannot meaningfully be done. Some perturbations on some initial states appear to propagate for ever, while others die out after a finite time.

The patterns of growth from simple seeds illustrated in figure 1 show the effects of a few nonzero sites on a background of zero sites. Green functions measure the average effect of changing a few site values on a more general background. Cellular automata are usually found to be of class 3 or 4, as indicated by their Green functions, when they grow expanding patterns from simple seeds, but no precise connection is known.

3. GLOBAL ANALYSIS OF CELLULAR AUTOMATA

The global properties of a cellular automaton are determined by the evolution of the probabilities for the ensemble of its possible configurations. An important feature of most cellular automaton rules is their irreversibility: a single configuration may have many predecessors, so that the global cellular automaton mapping is not one-to-one. In addition, in many cases, there exist configurations with no predecessors, so that the global cellular automaton mapping is not onto, but is instead contractive. Configurations with no predecessors under cellular automaton evolution cannot be generated by the evolution, and can occur only as initial states.

The contractive nature of most cellular automata implies that even starting from all possible initial configurations, only a subset of configurations may occur at large times. Such behaviour is quite opposite to that found in reversible systems, for which coarse-grained entropy tends to a maximum, and all coarse-grained states ultimately appear with equal probability. (A detailed analysis of this point requires consideration of coarse graining for cellular automata, and will not be discussed here.) The contraction to a subset of configurations under cellular automaton evolution allows for "self-organizing" behaviour, in which an initially disordered system evolves to be dominated by a limit set consisting of a special subset of its possible states. The properties of this limit set determine the nature of the self-organization which occurs. These properties differ considerably between the four classes of cellular automata mentioned above. For class 1 cellular automata, the limit set consists of a unique configuration. For class 2 cellular automata, it consists of a few special stable or periodic configurations. For class 3 cellular automata, it consists of a complicated set of aperiodic configurations.

Consideration of the nature of the limit set in cellular automaton evolution provides a connection with the theory of continuous dynamical systems and ordinary (non-linear) differential equations: the first three classes of cellular automata yield subsets analogous to limit points, limit cycles and chaotic (or "strange") attractors.

The sets of configurations generated by evolution from disordered initial states in class 3 cellular automata may be characterized in part by their limiting entropies or dimensions. The dimension of a set of configurations may be computed as

$$d = \lim_{X \to \infty} \log_k [N(X)]/X \quad , \tag{3}$$

where $N(X)$ is the number of distinct site value sequences (out of a possible total of k^X) which occur in blocks of X sites in configurations in the set. (In ergodic theory terminology, this dimension is the topological entropy of the shift mapping on the set of configurations.) If all possible configurations occur, as for globally onto cellular automaton mappings, then $d=1$. In most cases, however, $d<1$, so that the set of configurations generated forms a fractal subspace of the space of all possible configurations, just as for chaotic attractors in continuous dynamical systems (as deduced from their symbolic dynamics).

Exact computations of the dimension (3) are possible in only a few cases. Any rule for which $\mathbf{F}[a_{i-r}, \cdots, a_{i+r}]$ is a one-to-one function of a_{i-r} or a_{i+r} yields $d=1$. Examples are most additive rules. A finite algorithm to determine whether $d=1$ for any cellular automaton rule is known (see sect. 5), and a few rules with $d=1$ not satisfying the sufficient condition just mentioned are known. Direct numerical computations of d are essentially impossible because of the loss of statistical accuracy associated with the $X \to \infty$ limit, and the absence of known lower bounds on d. It is however possible to determine the value of d attained after a small finite number of time steps, using a method outlined in sect. 5. However, there are strong indications that the set of configurations generated by class 3 and 4 cellular automaton evolution usually continues to contract for many time steps, and possibly for ever, so that the final value of d cannot be computed by this method. It is in fact not even known whether $d>0$ for all class 3 cellular automata. Computations of d for some specific class 3 cellular automata would be considerable interest.

The overall structure of the limit set for cellular automaton evolution is essentially unknown, except in a very few rather simple cases. For example, the orbit or trajectory representing evolution from one configuration in the limit set may in the course of time visit all configurations in the limit set (or, more precisely, may come arbitrarily close to any point in the limit set). Alternatively, the limit set may consist of several disjoint pieces, perhaps with even qualitatively different properties. There may in general be several "invariant sets" of configurations which are closed under the cellular automaton mapping. Some of these invariant sets may be attractors to which many initial configurations evolve; others may be repellors reached from a set of initial configurations of measure zero. Specific examples of such phenomena would be valuable.

The possible sequences of values attained by individual sites as a function of time represent another global aspect of cellular automaton evolution. A dimension may be associated with the set of such temporal sequences which occur, just as the dimension (3) is associated with the set of spatial sequences. This temporal dimension vanishes whenever the patterns generated by cellular automata have finite periods, as for class 1 and 2 rules. The ratio of temporal to spatial dimensions is bounded by the maximum propagation rate in the cellular automaton, as measured by the Green function.

The standard concepts of ergodic theory may be applied to cellular automata. The continually increasing effect of small changes in initial states found in class 3 cellular automata is akin to the expansiveness property in ergodic theory. It presumably implies that these cellular automaton mappings exhibit some form of mixing and ergodicity. Detailed proofs of these properties are not yet available.

In ergodic theory, two systems are considered equivalent if their configurations can be related by an arbitrary invertible mapping. Such mappings may be represented by one-to-one cellular automata. Many different cellular automaton rules related in this way undoubtedly exist, but no non-trivial examples are yet known. Two rules may be equivalent only if their "invariant entropies" are equal. These entropies are given by a formula analogous to (3), but for space-time patches, with the limit of infinite time extent taken first. In general, the limit of infinite spatial extent must then also be taken, but expansiveness properties may render consideration of spatially finite patches sufficient. Even so, direct numerical computation of invariant entropies again appears essentially impossible. The most promising approach to their computation appears to be the identification of "elementary excitations" in cellular automaton rules analogous to the elementary excitations considered in condensed matter theory. Any initial configuration would then be decomposed in terms of these elementary excitations, and the evolution of the configuration would be represented by a series of simple interactions between the elementary excitations.

4. BASES FOR UNIVERSALITY

Empirical evidence suggests considerable universality in the overall behaviour of cellular automata. Many different cellular automaton rules appear to lead to qualitatively similar behaviour. It is important to find a mathematical explanation for this empirical observation. A method loosely related to renormalization group as applied to critical phenomena may possibly provide such an explanation.

The basis of this method is the fact that special configurations in a cellular automaton with one rule may evolve like configurations in a cellular automaton with another rule. The configurations in the two cellular automata may be related by "blocking transformations". An example of such a transformation replaces every pair of sites in cellular automaton A by a single site in cellular automaton B according to some definite blocking rule. For example, every 00 block in A could be replaced by a 0 site in B, and every 11 by a 1. In this case, cellular automaton A may be considered to "simulate" B under the blocking transformation. For any configuration of B, a configuration of A may be constructed which reproduces the behaviour of B after the blocking transformation. In general a particular cellular automaton rule A may simulate several rules B through different blocking transformations. Some possible blocking transformations may however lead to systems corresponding to cellular automata with larger values of k or r.

Within a specific set of cellular automata, one may develop a "simulation network" in the space of possible cellular automaton rules. Each node of the network represents a rule, and the nodes are joined by directed arcs representing possible blocking transformations up to some block length. (Ref. (4) gives an example of such a network.) Note that if rule A simulates B, and B simulates C, then A must simulate C directly, but usually only through a blocking transformation with a larger block length.

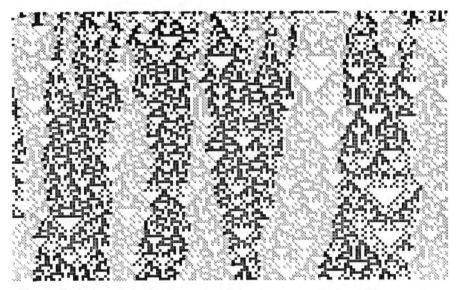

Fig. 3 Evolution of the $k=2$, $r=1$ cellular automaton "rule 18" from a disordered initial state, shown with a factor of two blocking in space and time. Pairs of site values are combined and represented by four grey levels. Alternate time steps are shown. Two "phases", separated by domain walls, are evident.

The simulation of rule B by rule A implies that there exists a special class of configurations which evolve under rule A just as they would under rule B. A remarkable phenomenon observed in several cases is that starting from all possible initial configurations, a cellular automaton A may evolve to generate only configurations in the special class for which it simulates some other cellular automaton B. An example is provided by the $k=2$, $r=1$ rule in which 100 and 001 neighbourhoods yield value 1, and all other neighbourhoods give value 0 ("rule

18") (cf. (9)). This rule simulates the additive rule of eqn. (1) ("rule 90") under the blocking transformations 00→0 and 01→1 or 10→1. Any initial configuration may be decomposed into a sequence of "domains", in which either all even-numbered sites have value 0, or all odd-numbered sites have value 0, separated by "domain walls". The configurations within each domain have the special forms required by the blocking transformations, and evolve as they would under rule 90. The domain walls do not evolve according to rule 90, but instead execute approximately random walks, annihilating in pairs, so that their density decreases as $t^{-1/2}$. After a sufficiently long time, arbitrarily large domains thus exist, and the behaviour of the whole system becomes arbitrarily close to that for rule 90. Rule 18 may thus be considered to exhibit an "attractive" simulation path to rule 90.

The occurrence of multiple "phases" separated by domain walls which execute approximately random motion, as in rule 18, appears quite common among cellular automaton rules. When $k > 2$, these phases may be evident even without blocking transformations.

Some cellular automaton rules are invariant under certain blocking transformations. Rule 90 and some other additive rules, including the identity rule and the null rule (which transforms all neighbourhoods to zero) are examples. These rules may be considered as fixed points of the blocking transformation, and exhibit a form of scale invariance. Essentially all the cellular automaton rules in the limited sets whose properties under blocking transformations have so far been investigated have been found to follow attractive simulation paths to one of these fixed point rules. One may therefore speculate that at large times, and on large spatial scales, the behaviour of all cellular automaton rules may tend to these fixed points. The small number of fixed point rules may possibly account for the universality observed in the qualitative behaviour of cellular automata. All class 1 rules may tend to the null rule, all class 2 rules to the identity rule, and class 3 rules to fixed point class 3 rules. All the fixed point class 3 rules identified so far have been found to be onto. This property may be a general feature of fixed point rules, necessary to yield time-independent behaviour. If all cellular automata tend at large scales to fixed point rules, then the statistical properties of cellular automaton attractors may in large part be determined just from the blocking transformations required to reach the fixed point rules.

5. COMPUTATIONAL CHARACTERISTICS OF CELLULAR AUTOMATA

The analyses sketched above treat cellular automata as dynamical systems, and use techniques from ergodic theory and statistical mechanics. Cellular automata may also be viewed as computers, and analysed using methods from the theory of computation.

One application of this approach is to the characterizations of the limit sets generated by cellular automaton evolution. Entropy and dimension give coarse measures of the statistical properties of these sets. Formal language theory appears to provide the basis for more complete characterizations. It may perhaps also suggest a general approach for the generalization of thermodynamics to irreversible and self-organizing systems.

Formal languages consist of sets of sequences (usually called "words") composed of symbols from a finite alphabet, and constructed by a "grammar" consisting of a finite set of grammatical rules. Sets of configurations generated by cellular automata may be considered as formal languages with alphabets $0,1,..,k-1$.

The sets of sequences found in the symbolic dynamics of most systems which have so far been investigated correspond to "regular" or even simpler languages. Words in regular grammars are defined to be generated or recognized by finite state machines, specified by finite state transition graphs. A symbol is associated with each arc in the graph, and the possible words in the language correspond to the possible paths through the graph. The entropy or dimension of the language is thus for example determined by the logarithm of the largest eigenvalue of the graph incidence matrix.

Regular languages may be specified by regular expressions. The regular expression $(0\backslash/1)*$ represents 0 or 1 repeated any number of times, and corresponds to the set of all possible sequences of zeroes and ones. $((0\backslash/1)0)*$ represents all sequences with no adjacent pairs of ones.

It can be shown that the set of configurations generated by any finite number of time steps of (one-dimensional) cellular automaton evolution corresponds to a regular language. Hence, for example, starting from all possible configurations, specified by the regular expression $(0\backslash/1)*$, one time step of evolution according to "rule 18" generates only the configurations $(0(0*)1(0*)1)*$, as generated by a three state machine. The dimension of this set of configurations is given by $\log_2\kappa\simeq0.81$, where κ is the largest root of the equation z^3-2z^2+z-1 (cf. (2,10)).

For class 1 and 2 cellular automata it appears that this correspondence survives even after an infinite number of time steps. For class 3 cellular automata, however, the size of the minimal finite state machine required to describe the sets generated increases very rapidly with time, and the limiting set is presumably not described by a regular grammar.

The number of nodes in the state transition graph for the (deterministic) finite state machine that generates the set of configurations obtained after t time steps of cellular automaton evolution is at most $2^{t^{2rt}}$. The existence of this bound allows the structure of the finite state machine to be found with a finite algorithm. (The algorithm first constructs a non-deterministic finite state machine in which each state represents a sequence of $2rt$ site values, and then applies the subset construction to obtain an equivalent deterministic finite state machine.) From this structure the entropy or dimension of the set of configurations may be computed, and the global surjectiveness of the cellular automaton mapping determined.

A further algorithm yields the minimal finite state machine (unique up to state relabellings) that generates any particular regular language. If, for example, the same minimal finite state machine is found at two successive time steps in cellular automaton evolution, then the limiting set of configurations must have been reached, and is represented by this machine. The number of states in the minimal finite state machine may be considered as measure of the "complexity" of the set.

It appears that the minimal finite state machines that represent the configurations of class 2 cellular automata tend to a fixed form after a few time steps. (In some cases, however, terms such as 10^t1 appear.) The limit sets for class 2 cellular automata thus appear to correspond to regular languages. For class 3 cellular automata, however, the number of states in the minimal finite state machine typically grows rapidly with time, and the resulting limit sets presumably do not usually correspond to regular languages.

Context-free languages form a class more general than regular ones, and perhaps adequate to describe the limit sets for class 3 cellular automata. Words in context-free languages correspond to the sequences of leaves (terminal nodes) in trees constructed of vertices chosen from a finite set which specify the grammar. (The basic grammars for most modern practical computer languages are supposed to be context-free.) The recognition (or "parsing") of context-free languages requires machines with a potentially infinite memory, but the memory may be arranged as a last-in first-out stack, with only one element accessible at any given time. No specific example of the generation of context-free sets by cellular automata has yet been found. (Such sets may, however, be generated even after a finite number of time steps by two and higher dimensional cellular automata.)

As mentioned above, cellular automata may be viewed as computers, "processing" initial configurations by their time evolution. The nature of the sets of configurations generated is presumably related to the computational power of the cellular automaton.

Figure 4 shows some examples of the evolution of a typical class 4 cellular automaton. Very complicated behaviour is evident. Periodic and propagating structures are sometimes formed. It seems likely that this and other class 4 cellular automata are capable of "universal computation", implying that with suitable initial configurations, the evolution of these cellular automata could implement any finite algorithm. (Such a capability has been proved for the two-dimensional "Game of Life" cellular automaton (e.g. (11).) With appropriate "programming" and encoding of input and output, a system capable of universal computation can simulate the behaviour of any other computational system. If the necessary encodings are simple enough, this suggests that any computationally-universal cellular automaton rule would be connected to all other cellular automaton rules in the simulation networks discussed above, albeit perhaps by indirection through many intermediate nodes, and therefore by blocking transformations with large block lengths.

The behaviour of a universal computer is fundamentally unpredictable. The only way to predict the behaviour of any system is by a computation on a universal computer. This computation in general reduces to a direct simulation if the system whose behaviour is to be predicted is itself capable of universal computation. But then the speed of this simulation differs only by roughly a constant factor from the speed at which the simulated system itself evolves. Thus questions concerning the infinite time behaviour of a system capable of universal computation may in general be unanswerable in any finite time, and are thus formally undecidable. An example of such a question is the "halting problem", roughly equivalent, for a cellular automaton, to the question of whether some

Fig. 4 Two examples of the evolution of a class 4 cellular automaton with $k=2$ and $r=2$ from disordered initial states. Persistent periodic and propagating structures are seen. The system is conjectured to be capable of universal computation. (The rule illustrated takes a site to have value one if and only if a total of two or four of the sites in the five-site neighbourhood at the previous time step have value one.)

finite initial configuration will eventually evolve to the null configuration, or will continue to generate configurations containing nonzero sites for ever. Notice that while the capability of a system for universal computation implies the existence of undecidable questions about that system, it is possible for some questions about a system to be undecidable without the system being capable of universal computation.

A system may be proved capable of universal computation by showing that it is equivalent to some system already known to be computationally universal. It may be possible to identify in class 4 cellular automata structures corresponding to the wires, gates, clocks and other components of a standard digital electronic computer. In a perhaps more promising approach, persistent structures in a class 4 cellular automaton could be identified as "symbols", and the interactions of periodic and propagating structures could then be placed in correspondence with the state transitions of a Turing machine head, or the productions of a string manipulation system.

In an alternative approach one could consider a cellular automaton simulation network, and show that as encodings of successively greater lengths were included, the number of distinct cellular automaton rules simulated by the candidate universal rule always increased. With arbitrarily long encodings, the rule could then simulate arbitrarily many others.

If a simple class 4 cellular automaton could be proved to be computationally universal, then it would probably represent the simplest known example of a universal computer.

A universal computer can generate in a finite time any formal language with a "phrase structure" grammar, which consists of a finite number of arbitrary transformation rules between symbols in the language. Such languages are much more general than the context-free and regular ones mentioned above. The sets of configurations generated by class 4 cellular automata are expected to be described in general by phrase structure grammars. However, the sets generated by such cellular automata after an infinite time are probably still more complicated, and may not be recursive.

If class 4 cellular automata are indeed universal computers, then they should in some sense be capable of arbitrarily complicated behaviour. Thus, for example, there should exist initial site value sequences which generate self-reproducing structures. Such structures, once generated, would replicate, and their progeny would eventually dominate the statistical behaviour of the cellular automaton. The generation of a self-reproducing structure may well require a specific very long sequence of initial site values. The probability for such a sequence to occur at a particular point in a disordered initial configuration may be infinitesimal; however, it must eventually occur in any arbitrarily long disordered initial configuration, and ultimately dominate the behaviour of the cellular automaton. Such a cellular automaton thus has no ordinary "infinite volume" or "thermodynamic" limit. Statistical averages based on successively larger numbers of sites do not converge smoothly to a limit. This phenomenon is related to the potentially non-recursive nature of the sets generated by such systems at arbitrarily large times.

For a particular class 4 cellular automaton, many initial configurations may exhibit very simple behaviour. Only a small fraction of initial configurations may yield potentially undecidable behaviour. In general, it is in fact formally undecidable whether a particular cellular automaton rule is capable of universal computation, since an arbitrarily small fraction of its configurations may be responsible. In fact, it remains possible that with very special initial configurations some class 3 cellular automata may formally be capable of universal computation. If this is the case, it seems likely that the necessary initial configurations form a set of measure zero.

6. DISCUSSION

This paper has outlined some features of cellular automaton behaviour. It is clear that despite their simple construction, cellular automata may behave in very complex ways. They exhibit on the one hand analogues of chaotic attractors in (continuous) dynamical systems, and on the other hand show complex computational capabilities. Nevertheless, there are several empirical and other indications that the overall behaviour of cellular automata may be governed by a few rather general laws. These laws probably involve elements from both dynamical systems theory and computation theory. Their form may provide a paradigm for still more general laws, applicable to a very wide range of complex natural and other systems.

ACKNOWLEDGEMENTS

I am grateful to many people for discussions about cellular automata. Points raised in this paper were discussed particularly with J. Conway, D. Lind, J. Milnor, N. Packard and L. Priese.

REFERENCES

1.   S. Wolfram, "Statistical mechanics of cellular automata", Rev. Mod. Phys. 55 (1983) 601.

2.   O. Martin, A. M. Odlyzko and S. Wolfram, "Algebraic properties of cellular automata", Comm. Math. Phys., to be published.

3.   S. Wolfram, "Universality and complexity in cellular automata", Physica D, to be published.

4.   S. Wolfram, "Cellular automata", Los Alamos Science, to be published.

5.   P. Walters, "An introduction to ergodic theory", Springer-Verlag (1982).

6.   J. E. Hopcroft and J. D. Ullman, "Introduction to automata theory, languages, and computation", Addison-Wesley (1979).

7.   S. W. Golomb, "Shift register sequences", Holden-Day (1967).

8.   S. Willson, "Growth rates and fractional dimensions of cellular automata", Physica D, to be published.

9.   P. Grassberger, "A new mechanism for deterministic diffusion", Phys. Rev. A, to be published; "Chaos and diffusion in deterministic cellular automata", Physica D, to be published.

10.   D. Lind, "Applications of ergodic theory and sofic systems to cellular automata", Physica D, to be published.

11. E. R. Berlekamp, J. H. Conway and R. K. Guy, "Winning ways, for your mathematical plays", vol. 2, Academic press (1982), chap. 25.

Models of Complex Interacting Systems

H. Atlan

Two instances of Self Organization in Probabilistic
Automata Networks: Epigenesis of cellular networks and
"self"-generated criteria for pattern discrimination

Department of Medical Biophysics, Hadassah University
Hospital, Ein Karem, Jerusalem, Israel.

Introduction

I will present two examples of the use of random automata
networks, probabilistic in two different senses, to provide
very simplified simulations of highly integrated biological
phenomena taking place at the level of the organism, with
interactions between billions of cells, namely the embryonic
development and non-directed learning in cellular networks.
The purpose of the simulation is to suggest mechanisms of
global behavior leading to new concepts and new ways of
thinking in the tradition of the kind of lose interactions
which existed at the beginning of molecular biology between
biological problems and cybernetics. These interactions led
to metaphors of heuristic value such as genetic information,
genetic program, regulation of gene expression etc. Today,
several of these metaphors must be replaced or at least
modified because they have lost their creative value and are
being used more like metaphysical explanatory principles.
(I am thinking in particular of the so-called genetic program which is supposed to account for everything but cannot
be localized in an organism, at least in the form of a computer program written in some computer-like language). I
want to show that automata theory and especially the analysis
of networks of probabilistic automata is of help in this task
because it allows for testing several ways of mixing deterministic and probabilistic procedures. And such mixing has
been recognized already (1,2) as the necessary recipe for
producing self-organizing properties of the kind observed in
embryonic development and non-directed learning.

EPIGENESIS OF A CELLULAR NETWORK

My first example is taken from a work in collaboration
with Maurice Milgram (3), which was aimed at simulating some
features of the embryonic development of a network of inter-

connected cells. The question we asked ourselves was the
following: Assuming a given network of cells, interconnected
in some specific way (i.e. by synapses, as in the central
nervous system, or by recognition sites, as in the immune
cellular network), and assuming that this network is the
product of one single initial cell, after a large number of
divisions and establishments of connections, what kind of
program must have been contained in this initial cell from'
which the whole network was generated?

As a simple case, let us assume that we want a network in
the form of a chain to be generated, i.e. starting from one
initial cell, a chain made of a given number of cells connected one after the other. And we want to define a finite
automaton to simulate the behavior of the initial cell and
to produce the chain. By means of a deterministic automaton,
this can be done only if the automaton contains a kind of
device like a counter, i.e. something able to count the
number of cells as they are generated by division and connected. This means that if the chain length k if very large
the initial automaton must contain at least k states to be
used in its counter.

However, if we use a probabilistic automaton it is still
possible to obtain a chain, i.e. we are not going to obtain
necessarily a mess with any recognizable form; and even more
interesting is that we obtain a chain with a number of states
drastically reduced in the initial automaton. In fact, as
will be seen, a chain of length k with k arbitrarily large
can be produced from a probabilistic automaton with only 5
states by a proper adjustment of the transition probabilities.

The 5 states are defined as initial q_o, s for source, g
for goal, d for division and r for rest. Fig. 1 shows the
transition graph of the automaton upon receiving two different inputs α and β.The following connection function relates
the input to be considered with the state of connection of
the automaton: if the automaton has received a connection
its transition graph is the one for input β; as long as it
has not received a connection from another automaton it is
assumed to receive input α (i.e. its transition graph remains
the one figured for input α). In other words, when an automaton is created - by division of a previous one - it is not
connected to start with, and it is assumed to receive input
α as long as it does not receive a connection from another
cell.

Therefore the first automaton goes from its initial state
q_o to state s, where it would send a connection to a cell if
it would exist (but it cannot do it yet) until it passes the

state g (where it would receive a connection from a source
if, again, another cell would exist), and then reaches d
where it divides. At the next time interval the second cell
is in q_o while the first one is in s; then the second one
goes to s while the first one goes to g: at that time a connection can be established from the second cell in s to the
first one in g.

Figure 1

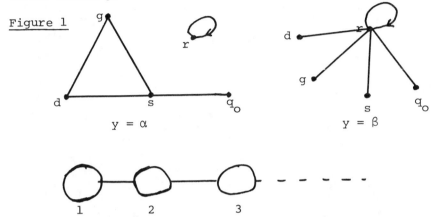

$$y = \alpha \qquad\qquad y = \beta$$

As soon as this first cell has received a connection, its
input becomes β, and it will behave according to the corresponding transition graph (Fig. 1), i.e. it will go to rest
and stay there. Then the second cell will divide and produce
a third cell which will send a connection to it and so on.

Now if this automaton is deterministic, i.e. all the transitions occur with certainty, this process will go on for
ever and never stop; unless a counter will count the cells
and stop the process when the number of cells reaches k.

However a similar result can be obtained without such a
counter by allowing the automaton to be probabilistic in the
sense that the transition from s to g is not certain. While
all other transitions are maintained with a 1 probability
the s to g is given a probability 1-ε with a non-zero probability ε of going from s to rest. When that happens, of
course, the process is stopped.

The expectation for a chain length to be achieved, i.e.
its mean value is of course a function of the transition
probability ε. It can easily be shown to be equal to $1/\varepsilon-1$;
i.e. if a wanted value for k is one thousand it can be pro-
duced as an average, with a $\varepsilon=10^{-3}$.

In other words, obviously, the probabilistic automaton
will not produce a given chain length k with certainty. Most
actual chains will have a different length, but as an average
their mean value will be k. Thus, as could be expected, we

have lost some accuracy in using probabilistic automata instead of deterministic; and this loss, as the dispersion
around the mean value, can be measured by the standard deviation of the length, or else by some distance with a probability value between possible different realizations of the
chain. However, we have gained a drastic reduction in the
number of necessary states; theoretically, an infinite reduction, since the number of states is now independent from
the length.

We can obtain a slightly modified automaton, by permuting
the states s and g. As a result, every time a new cell is
created it is connected as a goal - with the first cell being always the source. Thus, instead of a chain, a network
in the form of a wheel is produced, with a given number of
radii instead of a given length.

Figure 2

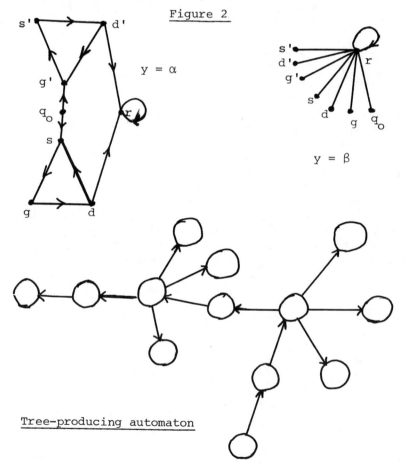

Tree-producing automaton

Now, by combining a chain producing and a wheel producing automaton one can made a probabilistic automaton of 8 states (Fig. 2) which will produce branching, tree-like networks. A newly created cell in state q_o will go into the subset of states leading to chain formation (i.e. s,d,g) or to wheel formation (i.e. s',d',g') with probabilities a and b respectively, and a probability ε of going to resting state. As a result, elements of chains of variable length and elements of wheels, i.e. branching points with variable number of branches will be formed.

At this point, when we end up with networks in the shape of trees, the question of complexity and specificity of the resulting network begins to be non-trivial.

And what is interesting is to analyze more precisely what is gained and what is lost by making use of probabilistic automata instead of deterministic and which kind of compromise can be achieved to retain specificity in the end product.

First of all, two features of a probabilistic automaton can account for the accuracy of the resulting network: its more or less chaotic feature and the effects of different inputs. The first one can be measured from the transition probability matrix of the automaton by the value of its entropy function

$$H(A) = -\sum_{ij} a_{ij} \log a_{ij}$$

which reduces to zero in a deterministic automaton (where the a_{ij} transition probabilities from state i to state j are 0 or 1) and is maximum in the case of a chaotic automaton with equal transition probabilities.

The second feature is the effect of the different inputs on the form of different transition matrices. It can be measured by a quantity called the capacity of an automaton as a generalization of Shannon's capacity of communication channels. It measures an upper bound of the difference betweeen the entropy of an average matrix

$$H\left[\sum_i p(y_i) \; A(y_i)\right]$$

and the average of the entropies $\sum_i p(y_i) \; H\left[A(y_i)\right]$

where the averaging is made over the different transition matrices $A(y_i)$ produced by different inputs y_i.*

---

* More precisely the capacity of an automaton A is defined as $C(A) = \text{Sup}_p \; (H\left[\sum p(y_i)A(y_i)\right] - \sum p(y_i)H\left[A(y_i)\right])$ where the upper bound is taken over all the probability distributions p of the input set Y(4).

However, in addition to these two features, the reduction
in accuracy produced by a probabilistic automaton depends
also upon the structure of the network to be produced at the
end. As I have mentioned before, using probabilistic automata instead of deterministic leads to a reduction in the
number of necessary states and the price is paid by a loss
in the accuracy of the end product.

Figure 3

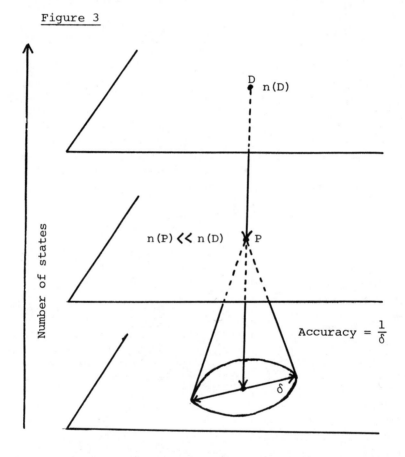

The schematic drawing of Fig. 3 shows the kind of compromise
which can be expected. To each probabilistic automaton P we
can always associate a deterministic one D which would do
the same thing with infinite accuracy but with a much larger
number of states represented here in the ordinate.

If the accuracy is defined as the reciprocal of a dispersion around a target to be reached, or, simply a dispersion of the resulting networks produced by the automaton, the
maximum accuracy with a zero dispersion is obtained by this
deterministic automaton. The number of necessary states is
reduced by the use of a probabilistic one, while the dispersion is increased. Now, it is interesting to try to analyze
what can influence this kind of compromise, what can make it
more or less worthwhile, because what is gained and what is
lost - and this is the important point - are not going to be
the same for any kind of network to be produced at the end.

In particular, the more redundant this network, the
smaller the difference in the accuracy of the process between a deterministic and a probabilistic automaton, whereas
the number of necessary states will still be reduced significantly for the probabilistic one. Conversely, if the
complexity of the resulting net is high, the price being
paid in accuracy by the probabilistic automaton is high for
the same reduction in the number of necesssary states.

Thus, it appears that the notion of specificity commonly
used in biology, is in fact a superposition of two different
although interrelated notions: that of accuracy as the reciprocal of a dispersion, and that of complexity defined and
measured as classically, by a minimum number of states (or
the minimum length of an algorithm) necessary to produce
deterministically a given structure or to perform a given
task. Therefore, we can understand that if natural networks
are built by probabilistic automata, redundant networks will
be produced more easily and more accurately than complex
ones, and this would appear as a first step in the building
of biological specificity: relatively accurate building of
redundancy.Only in a second step,more complexity,whereby more
specificity, will be produced by reduction of the initial
redundancy. This can be produced also randomly to some extent,
as I suggested to be the case in my previous work on self-
organizing systems (2,5,6).

In fact, such initial redundancy has been observed in
several instances with a subsequent reduction in the course
of development; namely, in the examples of initial poly-
innervation reduced subsequently in the development of the
Central Nervous System and of the initial degeneracy of the
immunological network, which is progressively reduced as the
immune response becomes more and more specific to a given
organism due to its encounters - partially random - of new
antigens.

At the time of the main discoveries of molecular biology
in the sixties, if automata theory and artificial intelligence would have been as advanced as today it is not sure

that the metaphor of the genetic program would have been
accepted so enthusiastically. Rather, we would be talking
today about genetic heuristics or genetic probabilistic
automata networks.

Along these lines, I would like to suggest an alternative
metaphor to account for the observed genetic determinations
and their localization in the DNA molecules.

The physicochemistry of the living cell can be viewed as
a network of probabilistic automata since chemical interactions follow statistical laws with a non-negligible amount
of indeterminacy and noise, especially at the relatively
high temperature of the organisms. The same network is found
with few variations in all known species since biochemical
reactions are more or less the same in all species, as well
as the spatial and temporal organization of the cell. Thus,
the specificity of this network, although it is not nil, is
not very high. What makes such a network behave in a specific way, characteristic of a given species and different from
the others, is obviously the genetic material in the form
of the DNA molecules. As mentioned above, one of the results
of the work I am presenting is that the specificity of a
given network producing automaton is largely increased when
the effects of different inputs are different. Thus, it
seems to me more satisfactory to look at the effects of DNA
as those of data or inputs to be processed by biochemical
automata networks rather than to imagine a computer program
written in the DNA molecules. The only program which is
necessary is that written in the whole physico-chemistry of
the metabolic network, which realizes a relatively non-
specific automaton, common to all living species. What makes
the difference is the specificity of the DNA molecules given
as inputs to this automaton. More or less in the same manner
as what makes the specificity of an algorithm to perform an
addition for example, is not so much the algorithm itself
which can produce any number as a result but the input
numbers added which will determine more specifically the
result.

STRING RECOGNITION BY SELF GENERATED CRITERION

My second example is taken from a work in collaboration
with Esther Ben Ezra, Gerard Weisbuch and Francoise
Fogelman. It is a report that a random network of boolean
automata can behave like a recognizer, where the sequences
to be recognized are defined by a randomly built structure
of the network. Thus algorithms are created randomly to
distinguish between classes of patterns, and the criterion
for this distinction is nothing else than the algorithm

itself which cannot be defined a priori, before the building of the network.

In particular, pseudo-random sequences can be recognized as not random because they belong to a class which is defined by the specific structure of such a network able to recognize it.

The process goes as follows: we start from a random network of boolean automata of the kind which was studied first by Stuart Kauffman (7,8). I will remind you quickly with well known facts about such networks before showing you these new and preliminary results on sequence recognition. Every element in the network is an automaton which can be in either one of two states. Its state is the result of a boolean function of two input binary variables. There are 16 different possible kinds of such elements according to the 16 boolean functions. Each element, defined by one of these functions, receives two inputs, each of them from one of its neighbors. It sends two outputs equal to its state, one at each of its two other neighbors. In a typical experiment, a network of such elements is constructed as a 16x16 element matrix, closed on itself on a torus in such a way that the last element in a row is connected to the first element of the row, and the same for a column. The different possible boolean functions are randomly distributed on the elements of the network and random initial conditions are set up.

Well established results (7,8) have shown that these networks exhibit an asymptotic behavior which is relatively stable to variations in initial conditions.

After a relatively small number of iterations (no more than one or two hundred, run in parallel) the network enters a limit cycle with a relatively short periodicity or cycle length.

In addition a structuration by subnets appears in the network. It is characterized by the fact that all the elements in a subnet are oscillating and they are separated by other subnets where all the elements are stable. The cycle length of the whole network is the smallest common multiple of the cycle lengths of the individual oscillating subnets.

For a given network, this structuration in subnets is relatively stable to changes in the initial conditions. Some elements are always oscillating and some others ("wishy-washy") are either stable or oscillating depending on the set of initial conditions.

In Fig. 4 the asymptotic behavior of a network for 150 different sets of initial conditions is represented by showing the number of times a given element has been found to be oscillating.

Figure 4

0	0	49	49	49	150	0	0	148	148	148	150	150	150	150	150
146	0	49	49	150	150	0	0	149	148	148	150	150	150	150	150
146	150	150	49	150	150	64	150	148	148	128	150	150	150	150	150
146	150	150	150	150	150	150	150	149	149	138	149	149	150	150	146
20	56	78	82	150	150	150	150	150	149	138	134	149	150	132	132
132	71	85	74	58	150	150	150	150	150	138	150	150	122	133	131
133	17	84	84	84	150	150	71	71	41	33	150	150	118	133	131
14	16	97	84	81	0	0	70	70	69	33	142	142	130	130	130
11	68	93	91	78	69	0	0	70	69	70	94	141	125	128	9
60	58	71	85	61	63	132	17	70	69	140	140	51	53	130	62
64	8	93	91	83	69	132	132	0	0	140	140	132	144	5	62
63	8	97	84	81	60	132	134	0	0	142	142	144	144	67	59
150	150	73	91	70	81	148	148	145	150	150	0	144	144	61	60
150	150	78	57	75	77	148	148	145	150	150	0	116	60	60	150
49	49	0	0	0	0	148	148	148	150	150	150	131	150	150	150
0	49	49	49	0	0	0	0	148	148	148	150	150	150	150	0

This figure shows the number of times a particular element (i,j) in a 16x16 random boolean network was found to be oscillating in the limit cycle for 150 sets of initial conditions. For example, module (1,3) was found to be oscillating in 49 out of the 150 limit cycles that were observed for this network.

Thus the elements labeled with zero constitute the stable core and the ones appearing with 150 constitute the oscillating core, while the others are the wishy washy.

To understand the mechanisms of this structuration the influence of the different kinds of boolean functions has been studied. The existence of forcing loops and forcing domains explains more or less the structuration in subnets relatively stable to changes in initial conditions. Francoise Fogelman will talk about that more extensively in her presentation.

What I want to concentrate on now is the study we have initiated recently on the effects of permanent perturbations

or noise injected ~~from~~ outside on one or more elements of
the network.

By permanent perturbations, or noise, I mean pseudorandom sequences of 0 and 1 inserted for a time long enough
to cover at least two cycle lengths and maintained for
another two cycle lengths during which the new asymptotic
behavior of the network is compared with the one of the
original noiseless networks.

Of course this new asymptotic behavior cannot be a true
limit cycle with a stationary structure since at least the
elements where the noise is inserted are varying by construction in a non-periodic fashion. Therefore the analysis
of the behavior was limited to the network without the elements where the noise is inserted and without their four
first neighbors. This is what we called the pseudostructure
of these noisy networks.

Under this condition, very interesting preliminary
results have been obtained.

Figure 5a shows the behavior of a network for 15 different sets of initial conditions, where the zeros are good
approximations of the stable core, the 15 represent the
oscillating elements and the others the wishy washy.

Then, for one of the initial conditions permanent perturbations were inserted in the 3 elements marked by a star;
disregarding these elements and their first neighbors the
rest of the network showed an asymptotic behavior where the
zeros represent stable elements and the 1's oscillating
ones (Fig. 5b).

Now, we can see that some elements of the stable core in
the noiseless network (i.e. the zeros) become oscillating;
and surprisingly enough, one element (1,16) which was always
oscillating in the noiseless network appears to be stabilized
by the noise.

In fact this last result was already obtained by inserting
noise in one element only (2,1) indicated by an inward arrow
and the stabilization of element (1,16) indicated by the
outward arrow was the only change observed in the basic
structure of the noiseless network.

At this point it is interesting to try to understand what
is going on and to look for possible features of the pseudorandom sequences which produce this stabilization since one
can interpret this phenomenon as a kind of recognition process: the element where the noise is inserted receives the
input sequences to be recognized; when the other element
(1,16) which is normally oscillating, is stabilized by a
given sequence it means that the sequence has been
'recognized".

Figure 5

5a) Stable and oscillating cores are determined by limit
cycles of 15 sets of initial conditions:

```
 8  15 15  0   0   0   0  15 15 15  9   9  15 15 15 15
 8   8 15 15   0   0   0   0 15 15  9  15 15  6  15 15
 0   8 15 15   0   0   0   0 15 15  0  15 15  8   8  0
 0  15 15  0   0   0   0   0 15 15  9  15  8   8  0
15  15  0  0   0   0   0  15 15 15 15  5   5   8  15 15
15   0  0  0   0   0   0  15 15 10 10 12  12  12 15 15
15   0  0  0   0   0  15  15 15 12 10 12  12  12 12 15
10   0  0  0   0   0   0  11 11 12 12  0   0   0   0 10
10   0 12 12  12   0   0  11 12 12 12  0   0   0  10 10
 0   0 12 12  12  12   0  12 12 10 10 10  0   0  15 15
14   6 12 12   6   6  15  15 12 12 10 10 15  15 14 14
12   6 12 12   6  15  15  15  0 10 10  9 15  15 14 14
12  12 12 12  12  15  15  15  0 10  5 15  9  15  9 14
12  12 12 12  15  15  15  15 15 15  5 15 15  15  0  0
12  12 15 15  15  15  15  15 15  0  0 15 15  15 15  0
 8  15 15 15   0   0   0  15 15 15  0 15 15  15 15  0
```

5b) Structure of limit cycle after noise was inserted at 3
simultaneous elements located at points (1,4), (2,1),
(5,2). A value of 0 means that it was stable throughout
the pseudo-limit cycle, a value of 1 means that it was
oscillating, a "*" means that it was not considered
since it was a point of noise insert, and "-" means that
it was not considered since it was a neighbor of an element
where noise was inserted.

```
  -  1  -  *  -  0  0  1  1  1  1  1  1  1  1 [0]→
→ *  -  1  - [0][1] 0  0  1  1  1  1  1  1  1
  -  1  1  1  0  0  0  0  1  1  0  1  1  1  1  0
  0  -  1  0  0  0  0  0  0  1  1  1  1  1  1  0
  -  *  -  0  0  0  0  1  1  1  1  1  1  1  1  1
  1  -  0  0  0  0  0  1  1  1  1  1  1  1  1  1
  1  0  0  0  0  0  1  1  1  1  1  1  1  1  1  1
  1  0  0  0  0  0  1  1  1  1  0  0  0  0  1
  1  0  1  1  1  0  0  1  1  1  1  0  0  0  1  1
  0  0  1  1  1  1  0  1  1  1  1  1  0  0  1  1
  1  1  1  1  1  1  1  1  1  1  1  1  1  1  1  1
  1  1  1  1  1  1  1  1  0  1  1  1  1  1  1  1
  1  1  1  1  1  1  1  1  0  1  1  1  1  1  1  1
  1  1  1  1  1  1  1  1  1  1  1  1  1  0  0
  1  1  1  1  1  1  1  1  1  0  0  1  1  1  1  0
  0  1  1  - [1] 0  0  1  1  1  0  1  1  1  1  0
```

A detailed analysis of this particular case will help to understand what is going on. For this purpose it is enough to consider the elements connected from the input element to the output and their subsequent states knowing their actual boolean functions and their connections.

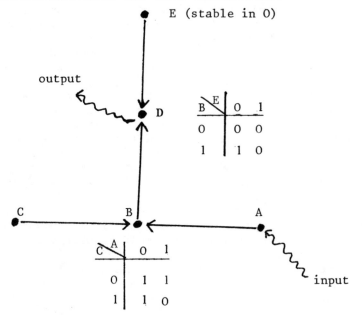

C is oscillating with cycle length 8 and following sequence of states in the cycle:

$$0\ 1\ 0\ 1\ 0\ 1\ 1\ 0$$

In the input sequence a 0 is necessary to match a 1 in C : . 0 . 0 . 00 . The   other input digits are indifferent.

Figure 6

This is represented in Figure 6 where the input element (2,1) in Fig. 5b (where the noise is inserted) is called A and the output element (1,16) (i.e. the stabilized one in Fig. 5b) is called D. Connected elements B, C, and E are responsible for the state of D.

In addition, it is important to consider the boolean
functions of elements B and D figurated here: In the limit
cycle of the noiseless,network C is an oscillating element
and E is stable in state 0. Now, the stabilization of D
under insertion of permanent noise in A, is observed in
state 1. This means that, due to its boolean function, in
order for D to be in state 1, since it receives a zero input
from E, it must receive a 1 input from B.

In order for this to be the case element B must always be
in state 1. Now, due to the NAND function of B, in order for
it to be always in 1, the only thing which is necessary is
that its two inputs should never be 1 at the same time; i.e.
whenever C - which is oscillating - is in state 1, the input
element A must be in 0. In all other circumstances, i.e. when
C is in state zero, the state of the input element A is indifferent and this is what is important for our purpose.

C is oscillating with a very short periodicity of 8 time
intervals and it goes through the following states:

                    01010110 repetitively.

For an input sequence to stabilize D it is enough that its
2nd, 4th, 6th and 7th digits should be 0 in order to match
the 1 (going from element C to B), and fulfil the condition
for 2 to stay always in state 1. The other 4 digits in this
sequence of 8 are indifferent. The overall period of the
whole network was 96 in this example, which means that the
inputs were sequences of 192 digits out of which half of them
are arbitrary for this recognition problem.

Therefore it is possible to build a long sequence by
repeating these 8 digits with the only constraint of the
digits 2,4,6,7 being zero. Since the others are arbitrary,
there is a large class of such pseudorandom sequences with
this hidden periodic structure in it which is going to be
recognized by our device.

Applying any pseudorandom sequence to the input element A
one can expect that it will be recognized if, and only if,
it belongs to this class.

Thus a specific pathway in the network between a given
element serving as input and another one serving as output
defines a class of sequences to be recognized.

Sometimes the same input element can be used to produce
such recognition responses by more than one output element.
This is easily recognized by injecting to the input element
a permanent inversion perturbation i.e. by inverting its

state systematically and looking for changes in the status of other elements, different from the neighbors. Then, if two separate elements have been modified in their asymptotic behavior as compared with the unperturbed network, it means that two different pathways can be used, defining two different classes of sequences.

The particular sequence made by the inversion of the state of the input element is an intersection of these two different classes.

One of the questions raised by this kind of phenomenon is whether or not such random boolean networks used to recognize pseudorandom sequences are more efficient than classical autocorrelation functions which are used for the same purpose.

It can be speculated that recognition procedures by this system should be faster because they are performed in parallel; and, maybe, that at least for relatively short sequences this system would succeed in finding out the hidden pseudoperiodic structure of a sequence while the auto-correlation test would fail.

Anyway, this model shows a mechanism by which a set of messages is divided into those which are recognized and those which are not, while the criterion for this demarcation, which is akin for example to making sense and not making sense to a cognitive system, is nothing else than a given inner structure which has no other meaning than being able to produce this demarcation, and, itself may have come about randomly.

CONCLUSION

The two models presented in this paper are examples of how the use of randomness in building algorithms may help to overcome difficulties in problem solving due to the complexity of the system to be analyzed.

It is as if complexity which appears as an apparent non-reducible randomness can be removed by means of a kind of orderliness which did not come about as a result of planning but as a result itself of indeterminacy and randomness. This, to my opinion, is the consequence of the close relationship between complexity and disorder in natural systems not planned and ordered by man, the only difference being the existence of an apparent meaning or function in the former to the eyes of the observer (9).

REFERENCES

(1)   Atlan, H. (1978). Sources of Information in Biological
      Systems in "Information and Systems" (Ed. B. Dubuisson)
      Pergamon Press, N.Y. pp. 177-184.
(2)   Atlan, H. (1979). Entre le Cristal et la Fumee, Seuil
      Paris.
(3)   Milgram, M. & Atlan, H. (1983). Probabilistic Automata
      as a Model for Epigenesis of Cellular Networks, J.
      Theoret. Biol., 103, 523-547.
(4)   Milgram, M. (1981). Barycentric Entropy of Markov
      Systems, 12, 141-178. Cybernetics and systems.
(5)   Atlan, H. (1974). On a Formal Definition of Organization,
      J. Theoret. Biol. 45, 295-304.
(6)   Atlan, H. (1981). Hierarchical Self Organization in
      living systems. In "Autopoiesis: a theory of living
      organization" (Ed. M. Zeleny) pp. 185-208, North
      Holland Publ. N.Y.
(7)   Kauffman, S. (1970). Behavior of randomly constructed
      genetic nets: Binary element nets. In "Towards a
      Theoretical Biology" (Ed. C.H. Waddington) Vol. 3,
      pp. 18-37. Edinburgh University Press.
(8)   Atlan, H., Fogelman-Soulie, F., Salomon, J., Weisbuch,
      G (1981). Random Boolean Networks, Cybernetics and
      Systems, 12, 103-121.
(9)   Atlan, H. (1981). Disorder, Complexity and Meaning.
      In "Symposium on Disorder and Order". Stanford Univ.
      Press. In Press.

DYNAMICS OF SELF-ORGANISATION IN COMPLEX ADAPTIVE NETWORKS

D. d'Humières* and B.A. Huberman+

* *Groupe de Physique des Solides de l'E.N.S.*
24 rue Lhomond, 75231 Paris Cedex 05, France
+ *Xerox Palo Alto Research Center, Palo Alto, CA 94304, USA*

1. INTRODUCTION

Complex automata are structures situated in between the systems with few degrees of freedom and the simplifying disorder encountered in many-body systems like gases. As such they pose a special challenge when trying to understand their dynamical properties as a function of given parameters and inputs. And yet, even partial knowledge about self-organization of complex networks is essential to understand the higher brain functions, such as learning and associative memory, which is one of the central problems of neurobiology.

Although statistical methods used in many body systems were applied to cellular automata (1), we believe that layered adaptive networks lend themselves to techniques developed in the study of non-linear systems with few degrees of freedom (2). And rather than optimizing an existing network or trying to closely model neurobiological systems, we have concentrated our study on the general dynamical properties of adaptive automata.

Thus we first present a stroboscopic technique to analyze the time evolution of a layered adaptive network during unsupervised learning of periodic sequences of training sets (Section 2). The method is then applied to the experimental study of a particular network with distributed memory described in Section 3. We show that with general input pattern sequences, changing the relative levels of excitation versus inhibition leads to fixed points, oscillatory states and chaotic wanderings during the adaptive process (Section 4a). Moreover, using the self-organized network after the adaptive process has taken place, we study the filtering properties of the network as a function of excitation and inhibition (Section 4b). A conclusion summarizes our findings and leaves some questions open for future research.

DYNAMICAL SYSTEMS
AND CELLULAR AUTOMATA

187

2. STROBOSCOPIC ANALYSIS OF LAYERED NETWORKS

Let us consider a layered synchronous network having p layers, each layer containing n cells. Each layer ℓ of the net is caracterized at step k by vectors $I_\ell(k)$ and $O_\ell(k)$, made up of the inputs and outputs of the layer's cells, respectively, and an array $S_\ell(k)$ which stores the information about the internal state of the network. Each cell i is a device with n afferent inputs stemming from the n cell's outputs of the preceding layer, i.e. : $I_\ell(k) = O_{\ell-1}(k)$, which in turn produces an output $o_{\ell,i}(k)$ whose value depends on the internal state $s_{\ell,i}(k)$, the input vector and the propagation rules given by a transformation f relating the cell's output to its input vector and its state, i.e. :
$o_{\ell,i}(k) = f\big(s_{\ell,i}(k), I_\ell(k)\big)$. At each step k, the input vector of the first layer is given externally, and the information transmitted from layer to layer, down to the last layer, which we call the output layer.

The adaptive process considered here, consists of a feedback mechanism through which the output of each layer acts back on its own internal state. The adaptive process for the layer ℓ is then given by a transformation G_ℓ, relating its state at step k+1 to its state and its output at step k, i.e. : $S_\ell(k+1) = G_\ell\big(S_\ell(k), O_\ell(k)\big)$.

In principle, the time evolution of the network is completely determined by the transformations f and G_ℓ, and the input sequence. However, our knowledge about the dynamics of systems with many degrees of freedom does not provide many useful insights, and we have to reduce the dimensionality of the space in which we study the problem.

A first reduction comes from the layered and hierarchical structure of the network, which allows us to consider as equivalent all states producing the same output at the last layer for a given input vector. Thus, we only need to focus on the last output vector, thereby reducing the number of variables from p.n.(1+m.n) to n (if the dimensionality of S_ℓ is m.n.n.). This simplification reduces in our network the number of independant variables by almost three order of magnitude.

A second simplification arises if the input vectors are presented to the network in a periodic sequence, i.e. :
$I_1(k+K) = I_1(k)$ for k = 0 to K-1. In this case we can use a method similar to the mappings at a period used in classical mechanics (2), and which consists in sampling the state of the network at each period of the input sequence. Therefore, analysing the periodically sampled data, it is easier to draw conclusions about the system's behavior. In particular, one point in the sampled hyperspace will indicate a fixed point

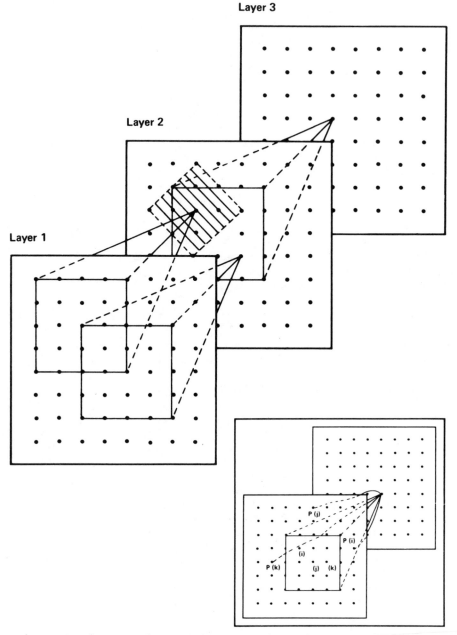

Fig.1 Implementation of the network. All the cells inside the squares act as inputs for the cell at the top of the cones, the insert shows the random set of connections. The shaded diamond gives the range of comparison between outputs.

or a cycle with the input frequency, several points a cycle
with a fundamental period which is a harmonic of the input
pattern period, a closed trajectory a quasiperiodic motion
with a frequency incommensurate with that of the input
sequence, and a cloud of points the presence of chaotic
behavior.

As we are interested in the existence of fixed point for
the system, but as their positions are unknown, a last reduction is achieved by measuring the distance between two successive outputs produced by the same input pattern. In what
follows, the distance between two vectors V and W will be
defined by :

$$<V,W> \;=\; 1 - V.W/(\|V\|.\|W\|)$$

where V.W is their inner product and $\|V\|$ and $\|W\|$ their
Euclidian measures.

Thus, to describe the network, we will measure for each
period K of the input sequence, the maximum and the minimum
values of the quantities :

$$<O_p(k+\kappa.K)\;,\;O_p(k+(\kappa+1).K)> \tag{1}$$

evaluated at the last layer for $k = 0'$ to K-1 as a function
of time κ, with the time unit equal to the input period.

3.   EXPERIMENTAL ADAPTIVE MACHINE

In what follows we will describe a particular adaptive network which we used in order to test the theoretical ideas
presented in section II. Among the very many cell structures,
we chose the one introduced by Fukushima (3). The details of
this particular network will be found in Refs. 3 and 4 and
here, we will only present its basic features.

The automaton was simulated in the configuration shown
Fig.1. The network is made up of three layers (p = 3) and
the input and output vectors are folded into 12 by 12
matrices (n = 144). Each cell was connected to its 25 nearest
neighbors in the previous layer (the square shown Fig.1) and
to 25 cells deduced from the preceding ones by a random permutation within the layer (see the insert in Fig.1). The output of each cell was given (in a condensed form) by :

$$o \;=\; \phi\big((P-M)/(1+M)\big)$$

with $\phi(x) = x$ if $x > 0$
and $\phi(x) = 0$ if $x \leqslant 0$,

the excitation P given by :

$$P = \Sigma s_j i_j$$

and the inhibition M by :

$$M = b \Sigma i_j$$

where the i_j are the input values, the s_j and b the excitatory and inhibitory strengthes respectively, and the summations are taken over all the effective connections as defined above.

Then, during the adaptive process, the states of the cell producing a maximum output in a 5 by 5 diamond area within the same layer (as shown by the shaded area in Fig.1) are updated according to :

$$s_j(k+1) = s_j(k) + Q_0 \alpha i_j$$

and

$$b(k+1) = b(k) + \alpha \Sigma i_j^2 / \Sigma i_j$$

where α defines the time constant of the adaptive process, Q_0 measures the ratio excitation to inhibition and the summations are defined above.

4. RESULTS OF ADAPTIVE EXPERIMENTS

(a) Dynamical properties

We studied the dynamics of the network for two training sets of input patterns, as a function of the ratio of excitation to inhibition Q_0. The first training set consisted of all the horizontal and vertical lines which could be arranged into the square input matrix, whereas the second was conposed of the 26 capital letters A to Z plus the ten digits 0 to 9, arranged in a 9 by 11 matrix within the 12 by 12 array.

The maximum and the minimum of the quantities defined by Eq.1 were measured over many periods of the input pattern sequence and their decimal logarithm was plotted as a function of time. Typical curves are shown in Fig.2 a to c for the line set and Fig.2 d to f for the alphabetic one. For both sets of patterns, the best convergence properties were found for $Q_0 \simeq 2$. As expected, the time to reach a fixed point was longer for the more complicated set of input patterns. As Q_0 was decreased or increased away from that value, we found out that the convergence of the adaptive

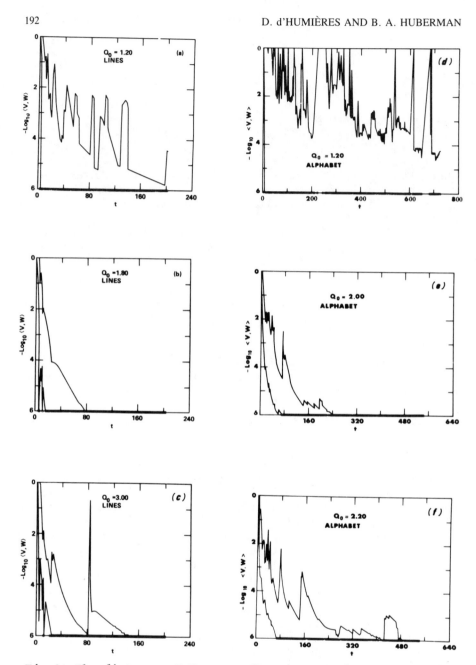

Fig. 2 The distance <V,W> as a function of time for a training set of lines for Q_0 = 1.2 (a), 1.8 (b), 3.0 (c) and an
alphabetic set for Q_0 = 1.2 (d), 2.0 (e), 2.2 (f). The time
unit is defined as the period of the input sequence.

process was altered and even disappeared for $Q_0 = 1.2$ with the alphabetic set (Fig.2d). Another interesting phenomenon is illustrated in Fig.2c. For long times, the network showed mono-tonic convergence towards a self-organized state with a simple fixed point, only to start unraveling itself at later times.

A more surprising feature is illustrated Fig.3. With the input set composed of lines, we found a chaotic behavior (Fig.3b) and a pseudo-periodic one (Fig.3d) entangled in regimes for which the network flowed towards a fixed point (Figs.3a and c). These results are to be contrasted with dynamical systems with few degrees of freedom in which the sequences of attractors one observes are both simpler and more regular (5).

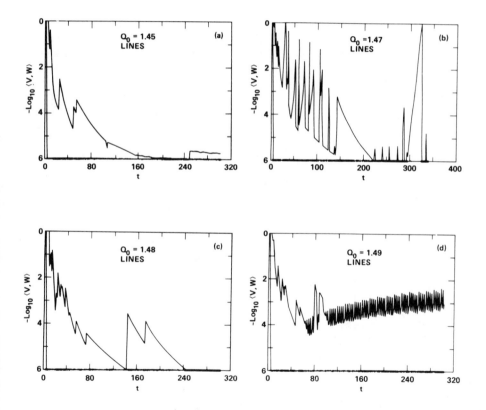

Fig.3 The distance <V,W> as a function of time for a trai-ning set of lines for $Q_0 = 1.45$ (a), 1.47 (b), 1.48 (c) and 1.49 (d).

(b) *Filtering properties*

In order to understand the dynamical properties of the network, we studied its filtering properties after the adaptive
process took place. First, we measured the output power produced at the last layer by each input pattern.

We also measured for each couple of the input patterns
$\{I_1(k), I_1(k')\}$ their distance at the input layer :

$$d_i = <I_1(k), I_1(k')>$$

and the corresponding distance of their output vectors at
the last layer :

$$d_0 = <O_p(k), O_p(k')>$$

The results obtained with the letter S are summarized in
Table 1. One can easily see that the output power and the
number of patterns producing an output close to S increase
as Q_0 increases from 1.2 to 2.2. (In order to define the
word "close", we use two criteria, one based on $d_0 < d_i$ and
the other such that the distance between the last pattern
in a class and the first one excluded is a maximum, see
Table 1 in Ref.4.).

TABLE 1

Filtering properties

Q	Power	Criteria 1	Criteria 2
1.2	0.382	0	0
1.4	0.962	0	0
1.6	1.430	2	3
1.8	2.443	7	8
2.0	6.021	6	11
2.2	7.520	12	11

These results show that as the ratio excitation to inhibition increases the network evolves from a state with
very sharp discrimination between similar patterns, to a
state with broad class aggregation. This effect suggests
the following interpretation of the dynamical behavior. For

the low values of Q_0 one pattern is lost during the training, thus leading to a bootstrapping procedure wich in turn affects deeply the state of the network. On the other hand, for large value of Q_0, one pattern can go from one class to another leading to the phenomena observed in Fig.3 d.

5. CONCLUSION

In this paper, we have shown that stroboscopic methods allowed us to obtain crisp information on the dynamics of a particular adaptive network. In that fashion we showed that, in addition to regimes where asymptotic learning can take place, there exist scenarii characterized by periodic oscillations and chaos. In addition, we concluded that the higher the ratio of excitation to inhibition, the broader the equivalence class into which patterns are lumped together.

Although this is only a tentative conclusion, we believe these effects are likely to be found in any layered automaton obeying local computational rules.

REFERENCES

1. Hinton G.E. and Sepnowski T.J. (1983)."Optimal perceptual inference", Proceeding of the IEEE Conference on Computer Vision and Pattern Recognition, 448-453.
2. See for example, Arnold V.I. (1978). In "Mathematical Methods of Classical Mechanics". Springer, New-York.
3. Fukushima K. (1975). "Self-Organizing Multilayered Neural Network", Systems Computers Controls 6, 15-22.
4. d'Humières D. and Huberman B.A., "Dynamics of Self-Organization in Complex Adaptive Networks" to appear in the February 1984 issue of the Journal of Statistical Physics.
5. Crutchfield J.P., Farmer D. and Huberman B.A., (1982). "Fluctuations and Simple Chaotic Dynamics", Phys. Repts. 92, 45-82.

BOOLEAN AND CONTINUOUS MODELS FOR THE GENERATION OF BIOLOGICAL RHYTHMS

Leon Glass

Department of Physiology, McGill University
3655 Drummond Street, Montreal, Quebec   H3G 1Y6
Canada

1.   INTRODUCTION

Logical or Boolean models have often been proposed to represent the interactions, dynamics and function in biological systems. Such models have been particularly useful in genetics (1-4), neurobiology (5-7), and immunology (8). Since the Boolean models are highly oversimplified, it is often possible to determine the behavior of the model using simple techniques. Consequently, the properties of the model, and hopefully of the biological system which it represents, can be easily appreciated without extensive technical analysis. Although such simplifications have provided insight into biological function, it remains necessary to show that the properties of the Boolean models remain intact in more realistic continuous models.

This note deals with the dynamics in two different models of biological systems. I first consider the generation of stable limit cycle oscillations by complex networks of elements with switchlike interactions. Next I consider the dynamics of "integrate and fire" models which have oscillatory thresholds. The common thread in both analyses is to construct the appropriate Poincaré or return map $f:M \to M$. Since much of this material has appeared in earlier publications, I will only summarize the main results and emphasize aspects which have not been previously published.

2.   OSCILLATIONS IN BIOCHEMICAL AND NEURAL NETWORKS.

In many biological systems one has negative feedback loops. The interactions in a negative feedback loop with two

DYNAMICAL SYSTEMS
AND CELLULAR AUTOMATA

elements x, y are shown schematically in Fig. 1a. Here x
stimulates y and y inhibits x. The variables x and y might
refer to activities of nerve cells, concentrations of
biochemicals, or densities of circulating cells in the
blood. This negative feedback loop can be represented by a
Boolean network whose truth table is

xy(t+1)	xy(t)
11	01
10	11
01	00
00	10

Note that there is a cycle $11 \rightarrow 01 \rightarrow 00 \rightarrow 10 \rightarrow 11 \dots$.
   In 1972, S. Kauffman and I proposed that the
interactions in such networks could be given by the
nonlinear ordinary differential equations

$$\frac{dx}{dt} = \frac{\lambda \theta^n}{\theta^n + y^n} - x, \qquad \frac{dy}{dt} = \frac{\lambda x^n}{\theta^n + x^n} - y, \qquad (1)$$

where λ, θ and n are positive parameters (9). Kauffman and
I were interested in analyzing the changes in the dynamics
that arose in Eq. (1), and related examples, as n varied.
In the examples considered we found numerically only a
limited repertoire of behavior, either stable limit cycle
oscillations or stable steady states over a range of n.
Thus, the qualitative features of the dynamics were not very
sensitive to n.
   To better appreciate the dynamics of Eq. (1) consider the
special case $\lambda = 1$, $\theta = 0.5$, $n \rightarrow \infty$. For this case the equations
are piecewise linear and can be immediately integrated. The
trajectories are piecewise linear, piecewise focussed
towards the corners of square whose vertices are at (0,0)
(1,0) (1,1) (0,1). The equations can be integrated by
drawing straight lines with pencil and ruler (Fig 1b). The
thresholds at x = 0.5, y = 0.5 separate the phase space into
four quadrants which are labelled by Boolean variables (Fig.
1c). Note that the flows across the thresholds are of
unique orientation and consequently the dynamics in the
system can be schematically represented by a state
transition diagram drawn as directed edges on the Boolean
2-cube (Fig. 1d). The cycle in the logical network is thus
also found in the continuous system.
   We now generalize Eq. (1) in the limit $n \rightarrow \infty$ to represent
an N-dimensional network of elements interacting by
switchlike interactions (10). For an N-dimensional system

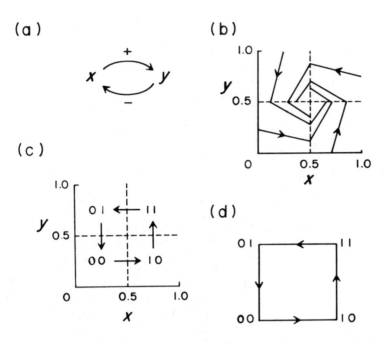

Fig. 1 (a) A simple negative feedback loop. (b) Phase plane for piecewise linear equation (see text). (c) The phase space subdivided into quadrants. (d) Boolean 2-cube showing flows between quadrants.

there are N real, positive variables given by x_1, x_2, ... x_N. To each continuous variable, x_i there is an associated Boolean variable, $\tilde{x}_i$ defined

$$\tilde{x}_i = 0, \quad \text{if } x_i \leqslant \theta_i; \qquad \tilde{x}_i = 1, \quad \text{if } x_i > \theta_i, \qquad (2)$$

where θ_i is a real positive number called the threshold. The dynamics are given by the piecewise linear equations

$$\frac{dx_i}{dt} = \lambda_i B_i[\tilde{x}_1, \tilde{x}_2 \cdots \tilde{x}_{i-1} \tilde{x}_{i+1} \cdots \tilde{x}_N] - x_i, \quad i = 1, N \quad (3),$$

where λ_i is a real positive number chosen so that $\lambda_i > \theta_i$ and B_i is a Boolean function of its (N-1) input variables.

Since each B can be chosen in $2^{2^{N-1}}$ different ways there are $2^{N \times 2^{N-1}}$ different ways to choose the B_i for the entire network.

As in the 2-dimensional example, illustrated in Fig. 1, the phase space for Eq. (3) can be partitioned into 2^N orthants defined by the threshold hyperplanes. Each orthant is labelled by a Boolean N-tuple $\tilde{x}_1 \tilde{x}_2 \ldots \tilde{x}_N$. Furthermore the flows between neighboring orthants are of unique orientation. This results from the condition of no self-input in Eq. (3). Consequently, the state transition diagram for the flows between neighboring orthants can be represented by a directed graph on a Boolean N-cube in which each edge of the Boolean N-cube is assigned a unique orientation (10). The different labellings of directions of the edges correspond to the different ways of choosing the Boolean functions in Eq. (3). The state transition diagram can also be constructed directly from the truth table of the underlying switching network. The state transition diagram gives all allowed transitions in an asynchronous switching network of the same logical structure. It is not necessary to consider all possible state transition diagrams but only those that are distinct under the symmetries of the Boolean N-cube. In 2 and 3 dimensions the number of distinct state transitions diagrams are 4 and 112, respectively (10).

In some cases, the state transition diagram can be used to predict the qualitative dynamics of the associated piecewise linear equation, Eq. (3). First, note that if a vertex on the N-cube has only edges directed towards it (called a stable vertex) there will be an associated stable steady state in Eq. (3). In the limit $t \to \infty$, any initial point in the associated orthant in phase space will approach the steady state in which $x_i = \lambda_i$ (or 0) if $\tilde{x}_i = 1$ (or 0) at the stable vertex.

A more interesting situation arises if there is a configuration of edges on the state transition diagram which is analogous to a limit cycle in an ordinary differential equation. A cyclic attractor is a cycle on the directed N-cube for which there are (N-2) vertices adjacent to each vertex of the cycle and the edge(s) from each adjacent vertex to the cycle is (are) directed toward the cycle. An N-dimensional cycle attractor is a cyclic attractor on an N-cube which is not contained on any lower dimensional sub-cube. The number of distinct, under the symmetries of the N-cube, cyclic attractors in dimensions 2, 3, 4, 5 are 1, 1, 3, 18 respectively (11).

Theorem: Given Eq. (3) in which the state transition diagram has an N-dimensional cyclic attractor, then one of the following two situations holds:

i) There is a stable limit cycle in phase space which passes through the orthants in the same sequence and order as the cyclic attractor in the state transition diagram. The trajectories through the points of orthants represented by vertices of the cyclic attractor and the points of boundaries represented by edges of the cyclic attractor asymptotically approach the limit cycle as t→∞.

ii) The trajectories through the points of orthants and boundaries represented by the cyclic attractor asymptotically approach the intersection of the threshold hyperplanes as t→∞.

A proof of this theorem in a somewhat more general situation is given in (12). I will only briefly discuss here the strategy of the proof and the possibility of extending it. The proof relies on explicit algebraic computation of the Poincaré map. Let S be the open boundary between two neighboring orthants corresponding to two neighboring vertices of the cyclic attractor. Then if $x_i \epsilon S$

$$x_{i+1} = \frac{A x_i}{1 + <\phi, x_i>}, \tag{4}$$

where A is a $(N-1) \times (N-1)$ positive matrix and ϕ is a non-negative non-zero vector. Eq. (4) is called a linear fractional map. The composition of two linear fractional maps is a linear fractional map. Call ρ the leading eigenvalue of A and v the associated eigenvector, and using the Perron-Frobenius theorem it is possible to show that under iteration

$$\lim_{n\to\infty} x_{i+n} = \alpha v, \tag{5a}$$

where

$$\text{i)} \quad \alpha = 0 \quad \text{for } \rho \leqslant 1, \tag{5b}$$

$$\text{ii)} \quad \alpha = \frac{\rho-1}{<\phi,V>}, \quad \text{for } \rho > 1. \tag{5c}$$

For the case in which $\rho > 1$ the fixed point is one point on the attracting limit cycle oscillator. The remainder of the limit cycle is found by integrating Eq. (3) through the orthants associated with the cyclic attractor until S is once again reached.

The algebraic demonstration of the limiting behavior of Eq. (4) is particularly simple for the one-dimensional case. Letting r and b be positive scalars and $x_i \epsilon R^+$, we have

$$x_{i+1} = \frac{rx_i}{1+bx_i}, \qquad (6a)$$

and

$$x_{i+n} = \frac{r^n x_i}{1+bx_i(1+r+\ldots+r^{n-1})}, \qquad (6b)$$

From Eq. (6b), by summing the geometric series, we find

$$\lim_{n \to \infty} x_{i+n} = 0, \quad \text{for} \quad 0 < r < 1, \qquad (7a)$$

$$\lim_{n \to \infty} x_{i+n} = \frac{r-1}{b}, \quad \text{for} \quad r > 1. \qquad (7b)$$

The demonstration of the limiting behavior of Eq. (4) depended on our ability to explicitly compute successive iterates. One problem which is still not solved is to find an appropriate topological version. For the one-dimensional case the following result has been obtained.

Let $x_{i+1} = f(x_i)$ where $x_i \epsilon R^+$ and f has the following properties: i) $f(0) = 0$; ii) there is a c_0 such that for $c > c_0$, $f(c) < c$; iii) $f'(x) > 0$; iv) $f''(x) < 0$.

Theorem: For a function satisfying the four properties listed above, given $x_i \neq 0$

  i) $\lim_{n \to \infty} x_{i+n} = 0$, for $f'(0) < 1$,

  ii) $\lim_{n \to \infty} x_{i+n} = \gamma$, for $f'(0) > 1$,

where γ is a positive number.

Proof: A steady state arises if $x_{i+1} = x_i$. Therefore a steady state is a solution of $g(x) = f(x) - x = 0$. One root occurs at $g(0) = 0$. By differentiating $g''(x) = f''(x) < 0$ and therefore the only extremal points of g are maxima. We know that $g(c) < 0$ for $c > c_0$. In the case $f'(0) > 1$, $g(x) > 0$ for x sufficiently small and there will be exactly one additional root greater than 0, say at $x = \gamma$. The root is unique by Rolle's theorem. By continuity of f, any point,

x_∞, for which $x_\infty = \lim\limits_{n \to \infty} f^n(x)$ is a fixed point. For
$0 < x_i < \gamma$, $\gamma > x_{i+1} > x_i$ and there is monotonic
convergence to the steady state at $x = \gamma$. For $x_i > \gamma$, $x_i >$
$x_{i+1} > \gamma$ and there is again convergence to the steady state
at $x = \gamma$. If $f'(0) < 1$, $g(x)$ is negative for $x > 0$ so there
are no solutions other than at $x = 0$ and $x_{i+1} < x_i$. Thus,
in this case, under iteration the series x_i, x_{i+1}, $x_{i+2} \cdots$
converges to zero.

I have not yet found an extension to higher dimensions,
but the presence of monotonicity and convexity in higher
dimensions should likewise guarantee an orderly dynamics.

3. PHASE LOCKING OF BIOLOGICAL OSCILLATIONS

In the preceding example the dynamics were insensitive to
parameters once the "logical structure" of the system was
set. The next class of examples shows great sensitivity to
parameters. The examples arise from two studies on the
effects of periodic stimulation of biological rhythms. In
one, the respiratory rhythm of paralyzed, anesthetized cats
is entrained to a mechanical ventilator (13,14). In the
second, spontaneously beating cells from embryonic chick
heart are perturbed by a periodic chain of brief electrical
current pulses (15). In both experiments, the goal is to
understand the observed dynamics as the frequency and
amplitude of the periodic stimuli are varied. Both systems
can be approximated by one-dimensional maps of the unit
circle into itself called "circle maps".
First consider the entrainment of the respiratory rhythm
to a mechanical ventilator. "Integrate and fire " models
have been proposed to represent such systems (7,13,16).
One assumes that the respiratory phase is separated into two
phases, inspiration and expiration. During inspiration an
activity rises until it reaches a threshold and during
expiration, activity falls until it reaches a second lower
threshold. Lung inflation acts by either continuously
changing the thresholds or the activities. Since there are
two phases with separate discrete rules for each, such
models have been called "Boolean" models (7). In these
models the time of the start of inspiratory phase, t_{i+1}, can
be given as a function of the time of the start of the
preceding inspiratory phase $t_{i+1} = f(t_i)$. Since the time t_i
is normally considered modulo the period of the mechanical
ventilator, f is a circle map. The function f which can be
called the "time advance map" plays the same role as the
Poincaré map in continuous systems.

Circle maps also arise in the analysis of the effects of
periodic stimulation of strongly attracting limit cycles
by brief perturbations provided: i) the time for relaxation
to the limit cycle is rapid compared to the intrinsic cycle
length and the time between periodic stimuli; ii) the
perturbation does not change the underlying structure of the
oscillator (15). In such situations, calling ϕ_i the phase
before the ith stimulus, one obtains $\phi_{i+1} = f(\phi_i)$, where f
is once again a circle map.

An example of a circle map is the "sine map" (17,18)

$$x_{i+1} = f(x_i,\tau,b) = x_i + \tau + b \sin 2\pi x_i \pmod 1. \quad (8)$$

Under iteration, if $x_{i+n} = x_i$ and $x_{i+j} \neq x_i$ for $1 \leqslant j \quad n$,
there is a cycle of period n. If an extremal point of f is
a point of the cycle than the cycle is called superstable.
The rotation number is defined

$$R = \lim_{n\to\infty} \sup \frac{1}{n} \sum_{i=1}^{\infty} (\tau + b \sin 2\pi x_i) \quad (9)$$

It is well known that if f is monotonic ($0 < b < 1/2\pi$), that
R is independent of initial condition and is a nondecreasing
function of τ which is piecewise constant on rational
values. The situation for $b > 1/2\pi$ is much more complex.
The following result has been obtained.

Theorem: For Eq. (8) with b fixed at a value greater than
$1/2\pi$ there are at least two superstable cycles for each
rational rotation number.

The proof of this theorem in a more general context is given
in (18). It is based on continuity and the topological
properties of Eq. (8). Based on this theorem and numerical
studies a conjecture has been developed for the structure of
zones associated with different rotation number in (b,τ)
parameter space. For
$0 < b < 1/2 \pi$ it is known that there are distinct
non-overlapping regions called Arnold tongues (18). For
$b > 1/2\pi$, it was conjectured that the tongues split apart
and overlap (Fig. 2). In Fig. 2 the zone p:q has a cycle
of period p with $R = q/p$. Further, it was proposed that in
the V-shaped region of each Arnold tongue one observes the
same period-doubling and other bifurcations as described in
detail elsewhere (18). I believe these bifurcations are
"universal" and will be observed in other two parameter
circle maps of topological degree 1 provided certain

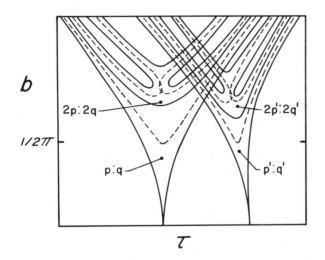

Fig. 2. Conjectured structure for phase locking in Eq. (8). Dashed lines are the locus of superstable cycles (18).

technical conditions (still to be specified) are satisfied. If such were the case, then such bifurcations might be observed in many systems during periodic forcing as the frequency and amplitude of the periodic stimulus are changed.

4. ACKNOWLEDGEMENTS

I thank my collaborators S.A. Kauffman, J. Pasternack, and R. Perez. J. Belair made useful comments on the presentation of the proof. The work is partially supported by a grant from NSERC.

5. REFERENCES

1. Sugita, M. (1963). Functional analyses of chemical systems in vivo using a logical circuit equivalent, J. Theor. Biol. **4**, 179-192.
2. Kauffman, S.A. (1969). Metabolic stability and epigenesis in randomly constructed genetic nets, J. Theor. Biol. **22**, 437-467.
3. Glass, L. and Kauffman, S.A. (1973). The logical analysis of continuous, non-linear biochemical control networks, J. Theor. Biol. **39**, 103-129.

4. Thomas, R. (1973). Boolean formalization of genetic
    control circuits, J. Theor. Biol. 42, 563-585.
5. Kling, V. and Szekeley, G. (1968). Simulation of
    rhythmic nervous activities, Kybernetik 5, 89-103.
6. Glass, L. and Young, R.E. (1979). Structure and
    dynamics of neural network oscillators, Brain Res. 179,
    207-218.
7. Baconnier, P., Benchetrit, G., Demongeot, J. and Pham,
    T. (1983). Simulation of the entrainment of the
    respiratory rhythm by two conceptually different
    models. In "Rhythms in Biology and Other Fields of
    Application" (Eds. M. Cosnard, J. Demongeot and A. Le
    Breton) pp. 2-16. Springer-Verlag, Berlin.
8. Kaufman, M. (1983). This volume.
9. Glass, L. and Kauffman, S.A. (1972). Co-operative
    components, spatial localization and oscillatory
    cellular dynamics, J. Theor. Biol. 34, 219-237.
10. Glass, L. (1975). Combinatorial and topological
    methods in nonlinear chemical kinetics, J. Chem. Phys.
    63, 1325-1335.
11. Glass, L. (1977). Combinatorial aspects of dynamics in
    biological systems. In "Statistical Mechanics and
    Statistical Methods in Theory and Application" (Ed. U.
    Landman) pp. 585-611. Plenum, New York.
12. Glass, L. and Pasternack, J. (1978). Stable
    oscillations in mathematical models of biological
    control systems, J. Math. Biol. 6, 207-223.
13. Petrillo, G.A. (1981). Phase locking - a dynamic
    approach to the study of respiration. Ph.D. Thesis,
    McGill University.
14. Petrillo, G.A., Glass, L. and Trippenbach, T. Phase
    locking of the respiratory rhythm in cats to a
    mechanical ventilator, Can. J. Physiol. Pharmacol.
    61, 599-607.
15. Guevara, M.R., Glass, L. and Shrier, A. (1981). Phase
    locking, period doubling bifurcations and irregular
    dynamics in periodically stimulated cardiac cells.
    Science 214, 1350-1353.
16. Petrillo, G.A. and Glass, L. (1983). A theory for
    phase locking of respiration in cats to a mechanical
    ventilator, Am. J. Physiol., In Press.
17. Perez, R. and Glass, L. (1982). Bistability, period
    doubling bifurcation and chaos in a periodically forced
    oscillator, Phys. Lett. 90A, 441-443.
18. Glass, L. and Perez, R. (1982). Fine structure of
    phase locking, Phys. Rev. Lett. 48, 1772-1775.

LOGICAL ANALYSIS OF LYMPHOCYTES INTERACTIONS DURING AN IMMUNE RESPONSE

M. KAUFMAN

Service de chimie-physique II, Université Libre de Bruxelles, C.P. 231, 1050 – Bruxelles, Belgique.

1. INTRODUCTION

The interest in the modelling of immune reactions has considerably grown. This is related on the one hand, to the accumulation of experimental data which sometimes seem contradictory and, on the other hand, to the great complexity of the regulation mechanisms involved in an immune response (1). During the last years several mathematical models were put forward (see (2) to (5)) based on the concept of **idiotypic networks** (6). Our work uses a simple logical language allowing a dissection of some complexities of the immune response by antibody production. We do not oppose antigen regulation to idiotypic regulation but rather incorporate them both into a more general concept. The model rests on well established interactions and some reasonable assumptions. We want to account for essential aspects of the immune response as the establishment and persistence of an immune memory state, the phenomena of high and low dose paralysis, non responsiveness. However, our purpose is not to explain every detail of these situations and our approach may be schematized as follows (10) :

1. We have considered that a satisfactory model should lead to the occurrence of multiple steady states with the required characteristics ;
2. one possible attitude is to introduce at the outset interactions between immunocompetent cells which give such a multistationarity. Instead of this ad hoc approach we choose as starting point a descriptive and minimal structure involving :
   - a small number of essential components (antigen, B,

T helper (T_h) and suppressor (T_s) lymphocytes of the corresponding specificity, antibody) and,
- a small number of well-documented interactions between these components ;

3. We used two complementary methods of analysis : the logical method described in the paper by R. Thomas, which gives rapidly an idea of the possible behaviours without having to explicit the details of the cellular interactions or to introduce a great number of parameters ; the more classical continuous analysis in terms of differential equations which allows a more quantitative and careful study.

2. THE MODEL

The logical method was essential to conceive the model whose salient features are :
- the well-established feedback suppression between T_h and T_s cells
- the occurrence of T - T interactions both at the level of helper and suppressor populations. These T - T interactions are represented in the model by autocatalytic loops on T_h and T_s.
- the fact that immature B cells are highly sensitive to negative signalling as was initially suggested by Lederberg (7). Antigen inactivates immature B cells preventing further differentiation, while it promotes the differentiation of more mature B cells in conjunction with the activation of T helper cells.

The totality of the interactions which were considered is presented on figure 1.

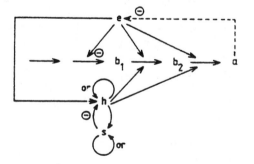

Fig. 1. The symbols are those of table 1. Unless otherwise specified the interactions are positive and the connections are of the AND type.

3. LOGICAL ANALYSIS

3.1. Logical equations

The basic scheme shown on figure 1 indicates which variables are interacting but when two or more variables influence a same element it does not specify how the interactions are connected to each other. For instance, the development of the T_s population is known to be induced from a precursor pool by T_h cells and to involve an autocatalytic component. One could write (see table 1 for the meaning of the symbols) :

$$S = h.s \qquad \text{or} \qquad S = h + s$$

The first possibility implies that the development of T_s lymphocytes requires the simultaneous presence of T_h and preexisting T_s cells. By analogy with other biological situations we choose the second possibility in which the requirement is the presence of T_h OR (inclusive) preexisting T_s cells. This implies two distinct mechanisms : an establisment and an autocatalytic maintenance mechanism. Furthermore we consider that antigen is not involved in the development of T_s cells.

For the development of the T_h population we reason in the same way in terms of an establishment mechanism induced by antigen and inhibited by T_s cells and, a maintenance mechanism based on T_h cells already present :

$$H = e.\bar{s} + h$$

The maturation of fully responsive B cells comprises several steps of which we sorted out **virgin** (b_1) and **more mature** (b_2) B cells. Following the idea of Lederberg, we write that the introduction of antigen blocks the emergence of new virgin B cells of the relevant specificity, while it promotes the differentiation of more mature B cells in the presence of T_h cells :

$$B_1 = \bar{e} \qquad , \qquad B_2 = b_1.e.h$$

Function A, the production of antibody, requires the final maturation of B cells under the influence of antigen and T_h lymphocytes. We simply write :

$$A = e.b_2.h$$

In addition, on the basis of experimental results, we distinguish three levels of antigen described by the binary variable e_1, e_2 such that $e_1 < e_2$. The three levels are : $\bar{e}_1$ (below the first threshold) ; $e_1 \cdot \bar{e}_2$ (between the thresholds) ; e_2 (above the second threshold). The system of logical equations describing the model shown on figure 1 thus becomes :

$$B_1 = \bar{e}_1$$
$$B_2 = b_1 \cdot e_2 \cdot h$$
$$H = e_1 \cdot \bar{s} + h$$
$$S = h + s$$
$$A = b_2 \cdot e_1 \cdot h$$

TABLE 1

Components of the system	logical variable	logical function
virgin B lymphocytes	b_1	B_1
more mature B lymphocytes	b_2	B_2
T_h lymphocytes	h	H
T_s lymphocytes	s	S
Antibody	a	A
Antigen (epitope)	e	no function (input variable)

It should be noted that we did not include here an explicit description of the evolution of the antigen concentration. For sake of simplicity we simply reason that antigen is injected at a given time and persist in the system until there is enough antibody to neutralize it. Thus, we treat antigen as an **input variable** which has the value o initially ($\bar{e}_1$: antigen absent) ; it then takes the value $e_1 = 1$ or $e_2 = 1$ depending on the

dose injected, and returns to the value o after enough
antibody has been produced.

3.2. States tables

From these logical equations one can construct, for the
three antigen levels, the compact states tables given in
table 2, using the method described by R. Thomas.

The model predicts that in the **absence of antigen** ,
the system can persist in either of three stable states,
depending on its past history :
- a **virgin** state (10000) in which only virgin B cells
  are present. This state corresponds to a system
  which has never seen the antigen.
- a **memory** state (10110) in which there is in addition a population of T_h and T_s cells of the proper
  specificity
- a **non-responsive** state (10010) in which besides
  virgin B cells there is also a significant level of
  T_s cells.

For a persisting level of antigen the model predicts
the existence of two stable states :
- a **paralyzed** state (00110) in which, in spite of
  the presence of antigen there are no antibodies of
  the corresponding specificity nor B cells suscepti-
  ble to produce them.
- a **suppressed** state (00010) related to the non
  responsive state of the left column as will be seen
  below.

The positive feedback loops of the T_h and T_s cell populations on themselves play an essential role for the
occurrence of this multistationarity. This is an illustration of the conjecture (8) that the presence of a positive loop in the logical structure is a necessary but
not sufficient condition for multiple states. Let us
briefly recall that what makes a loop positive or negative is simply the parity of the number of negative (inhibitory) interactions in the loop. Negative loops permit homeostatic control while positive loops permit multistationarity (9). We would like to emphasize that
what is crucial here is the positive character of the T_h
and T_s loops, not the detail of their structure. For
instance, they might perfectly well comprise several
steps, including an even number of negative ones,
without a major change of the global behaviour.

TABLE 2

$\bar{e}_1$	$e_1 \cdot \bar{e}_2$	e_2
$b_1\ b_2\ h\ s\ a$	$b_1\ b_2\ h\ s\ a$	$b_1\ b_2\ h\ s\ a$
$\bar{0}0000$	$00\bar{0}00$	$00\bar{0}00$
$\bar{0}000\bar{1}$	$00\bar{0}0\bar{1}$	$00\bar{0}0\bar{1}$
$\bar{0}001\bar{1}$	$0001\bar{1}$	$0001\bar{1}$
$\bar{0}0010$	(00010)	(00010)
$\bar{0}0110$	(00110)	(00110)
$\bar{0}011\bar{1}$	$0011\bar{1}$	$0011\bar{1}$
$\bar{0}01\bar{0}\bar{1}$	$001\bar{0}\bar{1}$	$001\bar{0}\bar{1}$
$\bar{0}01\bar{0}0$	$001\bar{0}0$	$001\bar{0}0$
$\bar{0}\bar{1}1\bar{0}0$	$0\bar{1}10\bar{0}$	$0\bar{1}10\bar{0}$
$\bar{0}\bar{1}1\bar{0}\bar{1}$	$0\bar{1}1\bar{0}1$	$0\bar{1}1\bar{0}1$
$\bar{0}\bar{1}11\bar{1}$	$0\bar{1}111$	$0\bar{1}111$
$\bar{0}\bar{1}110$	$0\bar{1}11\bar{0}$	$0\bar{1}11\bar{0}$
$\bar{0}\bar{1}010$	$0\bar{1}010$	$0\bar{1}010$
$\bar{0}\bar{1}01\bar{1}$	$0\bar{1}01\bar{1}$	$0\bar{1}01\bar{1}$
$\bar{0}\bar{1}00\bar{1}$	$0\bar{1}00\bar{1}$	$0\bar{1}00\bar{1}$
$\bar{0}\bar{1}000$	$0\bar{1}000$	$0\bar{1}000$
$1\bar{1}000$	$\bar{1}\bar{1}000$	$\bar{1}\bar{1}000$
$1\bar{1}00\bar{1}$	$\bar{1}\bar{1}00\bar{1}$	$\bar{1}\bar{1}00\bar{1}$
$1\bar{1}01\bar{1}$	$\bar{1}\bar{1}01\bar{1}$	$\bar{1}\bar{1}01\bar{1}$
$1\bar{1}010$	$\bar{1}\bar{1}010$	$\bar{1}\bar{1}010$
$1\bar{1}110$	$\bar{1}\bar{1}11\bar{0}$	$\bar{1}111\bar{0}$
$1\bar{1}11\bar{1}$	$\bar{1}\bar{1}111$	$\bar{1}1111$
$1\bar{1}1\bar{0}\bar{1}$	$\bar{1}\bar{1}1\bar{0}1$	$\bar{1}11\bar{0}1$
$1\bar{1}1\bar{0}0$	$\bar{1}\bar{1}1\bar{0}0$	$\bar{1}11\bar{0}0$
$101\bar{0}0$	$\bar{1}01\bar{0}0$	$\bar{1}\bar{0}1\bar{0}0$
$101\bar{0}\bar{1}$	$\bar{1}01\bar{0}\bar{1}$	$\bar{1}\bar{0}1\bar{0}\bar{1}$
$1011\bar{1}$	$\bar{1}011\bar{1}$	$\bar{1}\bar{0}11\bar{1}$
(10110)	$\bar{1}0110$	$\bar{1}\bar{0}110$
(10010)	$\bar{1}0010$	$\bar{1}0010$
$1001\bar{1}$	$\bar{1}001\bar{1}$	$\bar{1}001\bar{1}$
$1000\bar{1}$	$\bar{1}0\bar{0}0\bar{1}$	$\bar{1}0\bar{0}0\bar{1}$
(10000)	$\bar{1}0\bar{0}00$	$\bar{1}0\bar{0}00$

3.3. Sequences of states

Starting from any initial state one can develop a complete graph of sequences of states as described in the paper by R. Thomas. Here we shall illustrate some of the main dynamical behaviours when antigen is injected in the system at the level e_2 (this amounts formally to a shift from the left to the right column). A detailed description of all possible pathways can be found elsewhere (10, 11).

3.3.1. **Primary and secondary responses.** Let us start from the virgin state. A typical pathway in which T_h cells, B cells, T_s cells and antibody appear in this order, is shown below :

$$\boxed{10000}$$

$$\underset{\text{e}_2}{\bigcirc} \quad \downarrow \quad \overset{h}{} \quad \overset{b_2}{} \quad \overset{s}{} \quad \overset{a}{}$$
$$1\bar{0}000 \longrightarrow 1\bar{0}1\bar{0}0 \longrightarrow 1\bar{1}1\bar{0}0 \longrightarrow 1\bar{1}1\bar{1}0 \longrightarrow 1\bar{1}111$$

We then reason that once enough antibody has been produced to neutralize the antigen there is a shift back to the left column and the evolution is :

$$\bar{1}1111$$
$$\underset{\bar{\text{e}}_2}{\bigcirc} \quad \downarrow \quad \overset{\bar{a}}{} \quad \overset{\bar{b}_2}{}$$
$$1\bar{1}11\bar{1} \longrightarrow 1\bar{1}110 \longrightarrow \boxed{10110}$$

The system does not return to its initial state but to a new state which has kept memory of the immunisation. If one adds antigen again the pathway is now :

$$\boxed{10110}$$
$$\underset{\text{e}_2}{\bigcirc} \quad \downarrow \quad \overset{b_2}{} \quad \overset{a}{}$$
$$1\bar{0}110 \longrightarrow 1\bar{1}11\bar{0} \longrightarrow 1\bar{1}111$$

Again the system produces antibody and this secondary response follows a more direct and faster pathway. Indeed the time to produce antibody in the primary response above is $t = t_h + t_{b_2} + t_a$, while here it is only $t = t_{b_2} + t_a$.

3.3.2. **High dose paralysis.** Let us now investigate the predictions of the model when, starting from the virgin state, one maintains a constant, high antigen level (this amounts to move to and to stay in the right column) :

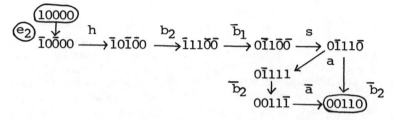

This time the system reaches and remains blocked in the
stable state (00110) which does not produce antibody
anymore despite the presence of antigen. This behaviour
accounts for the **high dose paralysis** observed experimentally. When the antigen disappears, by spontaneous
degradation, the situation is reversed and leads to the
immune memory state from which one obtains an efficient
secondary response :

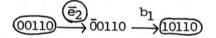

3.3.3. Non responsiveness.

If one adds antigen to a
population in state (10010), the pathway is :

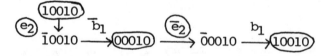

Antigen blocks the synthesis of new virgin B cells; but
here one goes necessarily to the stable state (00010)
and there is no immune response. If antigen eventually
disappears the system returns to its initial state.
However, the origin of this non responsive state must be
looked for outside the scope of the model. There is no
way, except for a major fluctuation, to reach state
(10010) starting from the virgin state.

These few examples show how the model accounts for a
number of features of the immune response.

4. CONTINUOUS DESCRIPTION

The main predictions of our model being known we **translated** the logical equations into corresponding continuous differential equations. Clearly the continuous
analysis could not have taken place without the prior

logical work. However the continuous analysis has been essential for at least two reasons :

a) the evolution of the antigen is treated in a rather primitive way in the logical description : for sake of simplicity antigen was treated as an input variable. In contrast, the continuous description contains an additional equation which describes in a self-consistent way the kinetics of disappearance of antigen taking into account both the formation of an inactive complex with the antibody and the slower spontaneous decay. An obvious advantage is that one can analyse in detail how the predictions depend on the kinetics of the antigen-antibody binding reaction and on the amounts of antigen injected.

b) the relative simplicity of the model allows an analytical treatment of the continuous equations, at least in the absence of antigen. This situation is crucial since, depending on its past history, the system will remain in a different steady state. It is thus important to determine more precisely the range of existence and stability of the various steady states in function of typical parameters.

An excellent agreement between the discrete and continuous description was found when the cellular interactions were represented by highly non linear functions as heaviside functions or steep sigmoids. However we choose to study in more details a simplified system of equations in which a number of non-linear control functions are replaced by linear ones, while the Hill coefficients of the remaining sigmoids are reduced to the low value of 2. The rationale behind this is that very little is known about the actual non linear character, if at all, of the various cellular interactions which occur during immune response. We thus investigated how far one can relax the non-linearities and yet keep the essential qualitative behaviours predicted by the discrete description. This led to the moderately non linear system (4.1) shown below :

$$\frac{dx_1}{dt} = k_1 . \overline{F_1}(x_6) - d_1 . x_1 \qquad\qquad (4.1.1)$$

$$\frac{dx_2}{dt} = k_2 . \overset{+}{F_2}(x_6) . x_1 . x_3 - d_2 . x_2 \qquad\qquad (4.1.2)$$

$$\frac{dx_3}{dt} = k_3 \cdot F_3^+(x_6) \cdot F_3^-(x_4) + m_3 \cdot F_3^+(x_3) - d_3 \cdot x_3 \qquad (4.1.3)$$

$$\frac{dx_4}{dt} = k_4 \cdot x_3 + m_4 \cdot F_4^+(x_4) - d_4 \cdot x_4 \qquad (4.1.4)$$

$$\frac{dx_5}{dt} = k_5 \cdot x_2 \cdot x_6 \cdot x_3 - q \cdot k_6 \cdot x_5^q \cdot x_6^p - d_5 \cdot x_5 \qquad (4.1.5)$$

$$\frac{dx_6}{dt} = - p \cdot k_6 \cdot x_5^q \cdot x_6^p - d_6 \cdot x_6 \qquad (4.1.6)$$

where the x_i's, $i = 1, 6$ are respectively the continuous analogs of the logical variables b_1, b_2, h, s, a and e. The $F_i(x_j)$'s are Hill functions defined by :

$$F_i^+(x_j) = \frac{x_j^n}{\theta_{ij}^n + x_j^n} \qquad \text{for activation,}$$

$$F_i^-(x_j) = \frac{\theta_{ij}^n}{\theta_{ij}^n + x_j^n} \qquad \text{for inhibition.}$$

p and q are the stoechiometric coefficents in the anti-gen–antibody reaction. The existence of a constant level of precursors cells of each required type is taken into account in the rate constants. Equation (4.1.6) describes the kinetics of disappearance of the antigen. There is also an additional term of disappearance of an-tibody through antigen binding in equation (4.1.5) which was not explicit in the logical equations. As x_1 trans-forms in x_2 rather than catalyzing its development, application of the mass action law would lead to an additional negative term in (4.1.1) describing the con-version of x_1 into x_2. We have reasoned however, that virgin B cells produce more mature B cells by a process of unequal division without being depleted.

The sigmoids which were maintained are :
- those corresponding to inhibitory interactions ;

- those describing the autocatalytic maintenance of the T_h and T_s populations which are essential for the existence of stable multiple steady states ;
- the ascending sigmoids reflecting the need for an activation threshold for the antigen.

The detailed analysis of equations (4.1) can be found elsewhere (10,11). Let us only mention that for a wide range of parameter values, they exhibit multiple steady states with the required properties in agreement with our logical study. Examples of dynamical behaviours following antigenic stimulation are shown on figures 2 to 5 for a set of parameters estimated from actual data :

$$k_1=.4 \quad k_2=.4 \quad k_3=.4 \quad k_4=.2 \quad k_5=.08 \quad k_6=.2$$

$$d_1=.2 \quad d_2=.05 \quad d_3=.2 \quad d_4=.2 \quad d_5=.2 \quad d_6=.05$$

$$m_3= m_4=.5$$
$$\theta_{16}= \theta_{36}=.5 \quad \theta_{26}=5. \quad \theta_{33}= \theta_{44}= \theta_{34}=1.$$

$$p=q=1.$$

$n = 2$ for each F_i (x_i) and, whenever possible, we used variables reduced to their threshold value. In the case of the antigen concentration (for which we used two distinct threshold), we introduced the dimensionless variable $X_6 = x_6/\theta_{26}$.

5. CONCLUSION.

The logical analysis and the numerical simulations of the continuous differential equations show that our model explicitly accounts for :

- the occurrence of multiple steady states (a virgin state, a memory state and a non responsive state) in the absence of antigen ;

- the kinetics of primary and seconday responses ;

- the phenomena of high dose and low dose paralysis induced by persisting antigen.

Its fit with real situations is surprisingly good for a model of this simplicity. Nevertheless we do not present it as **the model** but as a starting point of mini-

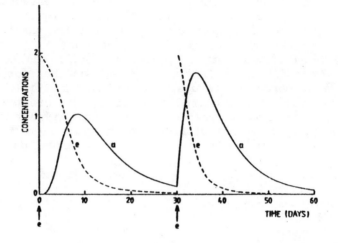

Fig. 2 : Primary and secondary responses for high antigen doses.   Time evolution of antibody (a) and antigen
(e) concentrations.

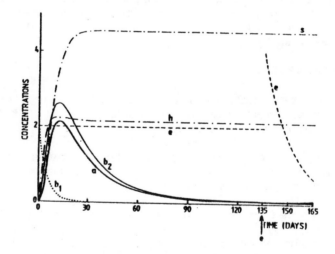

Fig. 3 : High dose paralysis induced by a constant level of antigen.   After a strong response, the system
reaches the paralyzed state characterized by the absence
of antibody despite the presence of antigen.

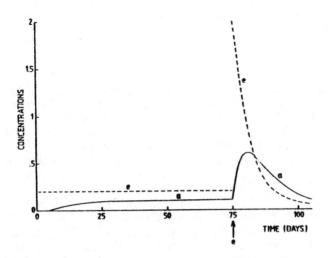

Fig. 4 : Low dose paralysis induced by a constant but low level of antigen, followed afterwards by a high antigen dose. This behaviour has to be compared to the primary and secondary responses obtained for the same antigen amounts without controlling the antigen level in the system as shown in figure 5.

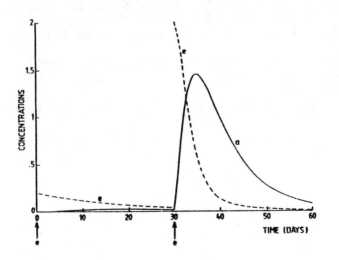

Fig. 5 : Induction of memory with a low antigen dose.

mum complexity to explain major facts of the immune response and as an example of what can now be done in the
field.

References

1. Urbain, J., Wuilmart, C. and Cazenave, P.A. (1981),
Idiotypic regulation in immune networks, **Contemporary
Topics in Molecular Immunol.** , 8, 113–148.
2. Richter, P.H. (1975), A network theory of the immune
response, **Eur. J. Immunol.** , 5, 350–354.
3. Hoffmann, G.N. (1975), A theory of regulation and
self–nonself discrimination in an immune network, **Eur.
J. Immunol.** , 5, 638–647.
4. Hiernaux, J. (1977), Some remarks on the stability of
the idiotypic network, **Immunochemistry** , 14, 733–739.
5. Grossman Z. (1982), Recognition of self, balance of
growth and competition : horizontal networks regulate
immune responsiveness, **Eur. J. Immunol.** , 12, 747–756.
6. Jerne, N.K. (1974), Towards a network theory of the
immune system, **Ann. Immunol.** (Institut Pasteur), 125C,
373–389.
7. Lederberg, J. (1959), Genes and antibodies,
 Science , 129, 1649–1653.
8. Thomas, R. (1979), Kinetic logic : a boolean analysis
of the dynamic behaviour of control circuits., **Lect.
Notes in Biomath.** , 29, 107–142.
9. Thomas, R. (1983), Logical description, analysis and
synthesis of biological and other networks comprising
feedback loops., **Adv. in Chemical Physics** , 55, 247–
282.
10. Kaufman M., Urbain J. and Thomas R. (1984), Towards
a logical analysis of the immune response, submitted for
publication.
11. Kaufman, M. and Thomas, R. (1984), in preparation.

SELECTIVE ADAPTATION AND ITS LIMITS
IN AUTOMATA AND EVOLUTION

Stuart A. Kauffman

Department of Biochemistry and Biophysics
University of Pennsylvania School of Medicine
Philadelphia, Pennsylvania 19104

INTRODUCTION

I wish to sketch a new pattern of inference, both with respect to biological evolution, and with respect to the theory
of adaptive automata: 1) Specific classes, or ensembles of
automata (considered either as models of genetic regulatory
systems (1,2) or as actual electronic circuitry) exhibit ensemble generic structural and dynamical self-organized properties. 2) These generic properties differ systematically in
different ensembles. 3) Biological evolution of genetic regulatory systems is, in fact, exploring some such ensemble via
mutations in the regulatory machinery and architecture. 4)
Biological adaptation occurs via differential proliferation
of more and less "fit" genomic regulatory systems. 5) Critically, the properties of the adapted system will typically
reflect a <u>compromise</u> between the generic properties of the
ensemble, and the properties selection is attempting to max-
imize. More precisely, selection in populations is analogous
to Maxwell's Demon, attempting to enrich the fraction of
fast molecules in the right box of a pair of boxes by operating a window between the boxes. As the faster molecules
accumulate in the right box, pressure builds opposing the
Demon's efforts. If he is finite and rather weak, the balance
struck is a steady state rather close to thermodynamic equilibrium. Similarly, the generic properties of the ensemble
being explored in biological evolution, or by mutation and
population selection in adaptive automata, are the "equilibrium" features expected in the absence of selection. Mutation supplies the variance upon which selection acts, but
simultaneously drives partially selected systems back toward
the generic properties of the ensemble being explored. Selection will attempt to enrich for certain properties, but

DYNAMICAL SYSTEMS
AND CELLULAR AUTOMATA

221

unless it is sufficiently powerful to attain and maintain the
population at any arbitrary member system of the ensemble,
selection, like the weak Demon, will typically result in a
compromise reflecting a balance between the generic properties of the ensemble and the properties being maximized.
From this, follow a number of epistemological and practical
points. First, if the balance struck is not too far from the
generic properties, then those properties are largely predic-
tive of the features found in the selected population. In
consequence, an enormous variety of features in organisms
spread through many phyla may occur, and persist, not due to
selection, but due to common membership in an evolutionary
ensemble. These properties would constitute true biological
universals. Second, since those universals are the generic
properties of the ensemble actually explored by evolution,
they constitute the underlying ubiquitously available features for the adaptive process to build upon. Both in considering the evolution of genomic regulatory systems and in
building a practical theory of adaptive automata based on
mutation and population selection, the generic properties of
different ensembles therefore emerge as the essential beginning data required.

Basic Properties of Finite Automata

In its simplest form, a binary automaton is a deterministic
dynamical system consisting of N elements, or nodes, each
capable of two values (activities, spins) 1 or 0. Nodes receive inputs from nodes in ways which may be geometrically
simple, eg. one, two or higher dimensional lattices with
nearest neighbor interactions, and with all couplings symmetrical; or complex, eg. the nodes may be connected at random with one another by arrows denoting oriented asymmetric
coupling such as model neural or genetic regulatory interactions. Since the elements are binary, the deterministic
dynamical behavior of each element must be given by one of
the possible 2^{2^k} Boolean functions of its K inputs, specify-
ing its next value, 1 or 0, for each of the 2^k combinations
of values of its K inputs. Again, in the simplest case, time
is assumed to occur in discrete clocked moments, at each of
which, each element assesses the values of its K inputs,
consults its Boolean rule, and assumes the proper next value.
   The fundamental dynamical properties of any such automaton with fixed external inputs are simple. An automaton with
N elements has 2^N states, each an N vector listing a possible combination of values, 1 or 0, of the N elements. The
system is state determined, hence at each moment, each state
transforms into a successor state. However, two states may

converge on a single successor state, thus in general the
dynamical behavior is not uniquely time reversible. Since at
each clocked moment the automaton undergoes a transition from
a state to a state, and there are a finite number of states,
it follows trivially that eventually the automaton must re-
enter a state previously encountered on a trajectory. Since
the system is state determined, it will thereafter cycle repeatedly about this reentrant sequence of states, termed a
state cycle. Thus, each state flows into some specific state
cycle. The set of states which lie on trajectories flowing
into, or on, one state cycle constitute the basin of attraction of that state cycle, while the collection of distinct
state cycles constitute the repertoire of asymptotic, recurrent dynamical attractors of the automaton.

Among the natural questions which arise first in considering specific ensembles of automata are these: typically how
many states lie on a given state cycle? How similar are the
states lying on any state cycle? How many state cycles are
in the dynamic repertoire of the automaton? How similar are
states on these different cycles? Is there a stable core of
elements whose values are fixed on or off on all state
cycles? If perturbed in all possible ways by reversing the
value of any single element for each state of a state cycle,
how often does the system return to the perturbed cycle, how
often does it undergo transition to another state cycle, how
many state cycles can be reached by such minimal perturbations from each cycle? If the automaton is mutated, eg. by
removing a binary element, or changing its Boolean rule, how
similar is the resulting dynamical repertoire to the unperturbed one?

These questions about generic properties can be posed for
any class, or ensemble of automata; alternative answers pro-
scribe alternative "self-organized" generic behaviors in diverse ensembles.

Elsewhere (3) I have recently discussed the alternative
generic dynamical properties now known for three alternative
ensembles:

1) Completely connected automata with N elements each receiving inputs from all N elements, and each assigned at
random one of the possible 2^{2^N} Boolean functions on its inputs. Results on this ensemble are a fundamental background
to more specialized ensembles, since all others are included
in it. The basic results are that cycle lengths typically
have a U shaped distribution, and a mean cycle length
$C1=0.5 \times 2^{N/2}$ (1,4,5). Thus cycle lengths increase exponentially as N increases. The expected number of state cycles
is known to increase linearly with system size, i.e. N/e (1).

Because this class of automata is an unbiased random mapping
of the 2^N states into itself, adjacent states along trajectories are unrelated, state cycles show stability properties
to perturbation proportional to the sizes of basins in
which they lie, and proper perturbations can drive a system
from each cycle to any other cycle. Results known for this
ensemble can be generalized to specific classes of biased
mappings of the 2^N states into itself, where the probability
P_i that any element i is assigned a 1 (or 0) value for each
of the 2^N input states is greater than 1/2. On average, the
most probable state of the N elements occurs as a successor
state $2^N P^N = (2P)^N$ times, thus $(2P)^N$ states converge on one
state. For $P > 1/2$ this biased mapping engenders enormous
convergence in state space, yet cycle lengths still increase
exponentially as N increases; $Cl = \frac{1}{2}(1/P)^{N/2}$ (3).

2) A second critically interesting ensemble consists in
automata in which each element realizes a threshold function
of its inputs. Typically these derive from model neuronal
systems, in which each element receives excitatory and inhibitory inputs and "fires" if a weighted sum of inputs exceeds a threshold. Symmetric models based on this design
have been used by Hopfield recently to encode alternative
memories as alternative cycles in basins of attraction (6).
Little is known of the generic properties of asymmetric
threshold automata. My own recent numerical results suggest
that the number of alternative state cycles in such nets is
low, but that the length of state cycles increases faster
than the number of elements in the system. Specifically, in
nets with 100 or fewer elements, the number of cycles is in
the range 1-10, but mean cycle length appears to increase
proportionally to N^2 (3). Further, the stable core of elements fixed active or inactive on all state cycles averages
42% and is inversely correlated with state cycle length.
Careful studies of the similarity of states on each state
cycle, the stabilities of state cycles, the perturbation induced flow among state cycles, and the typical alterations
caused by mutations have not been carried out to my knowledge.

3) The canalizing ensemble arises naturally in automata
models of genetic regulatory circuitry (1,2,7,8). A Boolean
function is canalizing if at least one input has one value
which guarantees one value of the regulated element, regardless of values on other inputs to the element. For example,
OR is a canalizing Boolean function, Exclusive OR is not.
Among the 2^{2^k} functions of K inputs which are canalizing,
is a maximum for K=2, 12 of 16, and declines rapidly as K increases. A maximum bound on the fraction is (17):

$$\frac{4K}{2^{K^{K-1}}} \to 0 \quad \text{as} \quad K \to \infty$$

Abundant work on canalizing automata, eg. the ensemble of K=2 automata, demonstrate their enormously ordered dynamical behavior (1,2,7-17). Typically, state cycles are a mere $N^{1/2}$ in length, thus an automaton with 10,000 elements cycles among a mere 100 states. The number of state cycles is also about $N^{1/2}$, a stable core of about 60%-80% of the elements are fixed active or inactive on all state cycles, most cycles are stable to almost all minimal perturbations, and if unstable, can undergo transitions to only a few neighboring cycles. Mutations of elements cause modest alterations in behavior. With the interpretation of a state cycle attractor as a cell type in the genomic repertoire, and transitions among attractors as a model of differentiation, I have felt for some time that these behaviors are deeply related to the evolutionary emergence of dynamical order in metazoan development (1,2,7-9).

Highly localized dynamical behavior in canalizing automata appears to be based on the "crystallization" of forcing structures (2,7,8) described below and by Fogelman elsewhere in this volume. While space limitations preclude a detailed description of forcing structures and their powerful properties, an intuitive indication can be given by considering a K=2 net in which each element realizes the OR function on its two inputs. Then a (1) value introduced at any element is guaranteed to induce the (1) value of each immediate descendent element, regardless of the values of other inputs to those elements. This guaranteeing relation between nodes is formally characterized as a forcing connection. Forcing connections are transitive relations evident in this (OR) net, where all connections are forcing, and the (1) value, once introduced, propagates throughout the net. If a closed (forcing) feedback loop falls to the state with all elements active (1), it is stably fixed in that forced state, regardless of other inputs to the forcing loop, and radiates fixed values along descendent forcing connections. The expected size and connectivity properties of forcing structures in canalizing automata is a percolation problem given by the number of forcing connections, the total number of elements, and the lattice or random properties of the connectivity graph among the elements of the automaton. Thus, in a random automaton, when the number of forcing connections exceeds the number of elements, large forcing structures "crystallize", and underlie the stable core of elements fixed in active or inactive states on all state cycles (3,8,18,19). Further, this forcing structure core

leaves behind functionally isolated clusters of elements
which are not members of the forcing structure, and cannot
communicate with other functionally isolated members through
this stable core. The independent subsystems generally have
alternative state cycles, and since the subsystems are independent, the entire behavior of the system is given by the
stable core of elements in fixed active or inactive states,
and the combinatorial choices of the alternative modes of
behavior of each of the set of functionally isolated subsystems.

Selective Adaptation in Automata

This brief description of these three alternative ensembles
suffices to show that different types of dynamical behaviors
are generic to the three distinct ensembles (3). Patterns,
capacities, and styles of adaptive modification based on a
population selective paradigm will be expected to differ
among the three. Thus, the existence of functionally isolated subsystems in the canalizing ensemble implies that one
subsystem can be modified without altering others, allowing
piecemeal improvement of behavior. By contrast, some small
modifications anywhere in a fully connected unbiased automaton, such as removal of any single element, will cause
drastic modifications in the entire dynamical flow of the
system, precluding gradual improvement by that mechanism
(although not others) in this ensemble. The fundamental issue to be raised is how to construct a theory of selective
improvement, and its limits, in an ensemble of automata,
where mutations induce exploration of the ensemble and supply the variance upon which selection acts, yet simultaneously mutations (component failures) destroy partially selected systems and drive toward the generic properties of
the underlying ensemble. The following question is basic:
Under what conditions of population size, mutation rate,
fitness rules, and automaton complexity, can selection achieve and maintain the population at the global maximum, or
at least local maxima of the fitness function? I suspect it
shall prove to be general that, for fixed selection rules
and mutation rate, as the complexity of the automaton to be
selected is increased, the capacity of selection to achieve
the maxima or submaxima will fail. If selection is not powerful enough to maintain the population near any arbitrary
member of the ensemble, then necessarily, the properties
seen in the maximally adapted system attainable are a balance between selection and the generic properties of the
ensemble, with implications about universal biological properties mentioned above, and the design of adaptive auto-

mata that can suffice despite component failure.

Population Selection for an Arbitrary "Good" Wiring Diagram

As a simplest beginning to this question, consider a wiring diagram with N elements or "genes", and T "regulatory" connections among them. Let an arbitrary wiring diagram, assigning the T connections among the N be "perfect", let the relative fitness of any wiring diagram be a monotonic decreasing function of its "distance" from perfect:

$$W_i = (1-b)\left(\frac{G_i}{T}\right)^\alpha + b$$

where G_i is the number of proper connections in network i, b ($0 \le b \le 1$) is a basal fitness for $G_i = 0$, and α is a selective exponent, measuring cooperativity among good connections in abetting fitness. Let mutations alter wiring connections at a rate of μ per connection per "genome" per generation. Simulation of this simplest model reveals a number of fundamental properties: 1) For appropriate parameter values, selection is able to attain and maintain the global maximum fitness, holding the population in a narrow dispersion around the "perfect" wiring diagram. 2) For parameters $\alpha=1$, or b=0, the system has a single globally stable stationary state. 3) As the complexity of the wiring diagram specified, T, increases, and for other parameters fixed, selection no longer can attain the global maximally fit wiring diagram. The globally stable stationary state balance between selection and the mutational drive back toward the generic properties of the unselected ensemble, shifts smoothly toward those generic properties. 4) The fitness of wiring diagrams is a monotonic function with a single global maximum, the "perfect" wiring diagram. Nevertheless, for appropriate parameters, $\alpha>1$, b>0, the selection system exhibits two stable stationary states, one close to the global maximum, one near "generic". 5) As the complexity, T, increases, the stability of the nearly perfect stationary state decreases and vanishes, hence increasing complexity leads to a complexity catastrophe, beyond which the population falls to the near "generic" stationary state.

Analytic Approaches

Insight into this simplest system, where only automaton wiring diagrams are considered, must be fundamental to any future development of an adequate theory of adaptation of automaton behavior via a population selection paradigm.

Even this simplest case appears to be difficult to characterize fully. Nevertheless, some limited analytic insight has been achieved. Because I have restricted attention to a haploid model, the entire variance in fitness in the population is heritable and Fisher's fundamental theorem applies in simplified form.(20). The rate of change of relative fitness in the population equals the variance of fitness in the population divided by the mean population fitness.

1)
$$\frac{d\overline{W}}{dt} = \frac{\sigma^2\overline{W}}{\overline{W}}$$

Let a total of T regulatory arrows connect the N nodes or "genes". If the probability that any randomly assigned connection arrow among the N nodes is a "good" connection is (1-P), and P is the probability a connection is "bad", then in a randomly connected wiring diagram, a mean $T(1-P)$ are "good" and TP are "bad". This average reflects the generic properties of this ensemble. Mutations (at rate μ per arrow per generation), provide a restoring force (Rf) reducing the number of good connections toward this mean. The mutational restoring force is 0 at the mean, and linearly proportional to the deviation in the number of good connections above the mean. In the simplest case, where multiple good connections between the same pair of genes is allowed, the restoring force is:

2)
$$Rf = \mu(TP + \overline{G} - T)$$

The effect of selection is to increase the fitness according to Fisher's theorem above, thus the selective force, Se, depends upon the ratio of the variance of the fitness in the population to the mean fitness. I have been unable to obtain an analytic expression for the variance as a function of T, α, b, P, μ, and population size. However, simulation results show that for fixed values of these parameters, the variance in G, $\sigma^2 G$, is approximately independent of the mean number of good connections, $\overline{G}$, itself. This approximation is used next to characterize, in terms of $\overline{G}$ and $\sigma^2 G$, the balance struck between mutation and selection for "good" wiring diagrams.

Consider the simplest fitness law, $W_i = (G_i/T)^\alpha$. One first can show that

3)
$$\sigma^2 W \simeq \left(\frac{dW}{dG}\right)^2 \cdot \sigma^2 G$$

where the derivative is evaluated at $\overline{G}$. The approximation is good if $\frac{\sigma G}{T}$ is reasonably smaller than 1, as observed.

The temporal derivative of $\overline{G}$ due to selection, restated in terms of $\overline{G}$ and $\sigma^2 G$ is approximated by:

4)
$$\frac{d\overline{G}}{dt} \simeq \frac{dG}{dW}\frac{d\overline{W}}{dt} \simeq \left[\frac{dG}{dW}\left(\frac{dW}{dG}\right)^2 \cdot \sigma^2 G\right] \Big/ \left(\frac{\overline{G}}{T}\right)^\alpha$$

where dG/dW and dW/dG are both evaluated at $G=\overline{G}$.

For the simplest law, $W_i=(G_i/T)^\alpha$, after substitution, the temporal derivative to increase $\overline{G}$ due to selection (i.e. effective selection, Se) is:

5)
$$Se = \frac{d\overline{G}}{dt} = \frac{\alpha\sigma^2 G}{\overline{G}}$$

The full differential equation for the rate of change of the population mean number of good connections per network, reflecting selection and mutation, becomes:

6)
$$\frac{d\overline{G}}{dt} = \frac{\alpha\sigma^2 G}{\overline{G}} - \mu(TP + \overline{G} - T)$$

For the more complex fitness law $W_i=(1-b)(G_i/T)^\alpha + b$, where $b>0$ reflects a "basal" fitness, the effective selection force Se increasing $\overline{G}$ per unit time becomes:

7)
$$Se = \frac{d\overline{G}}{dt} = \frac{\alpha(1-b)\overline{G}^{\,\alpha-1}\sigma^2 G}{(1-b)\overline{G}^{\,\alpha} + bT^\alpha}$$

The full differential equation becomes:

8)
$$\frac{d\overline{G}}{dt} = \frac{\alpha(1-b)\overline{G}^{\,\alpha-1}\sigma^2 G}{(1-b)\overline{G}^{\,\alpha} + bT^\alpha} - \mu(TP + \overline{G} - T)$$

For $\alpha=1$, fitness is linearly proportional to the fraction of good connections, and 8) simplifies to:

9)
$$\frac{d\overline{G}}{dt} = \frac{(1-b)\sigma^2 G}{(1-b)\overline{G} + bT} - \mu(TP + \overline{G} - T)$$

The qualitative behavior of the selection system under conditions of no basal fitness, b=0, equation 6), and in the presence of basal fitness, b>0, but for fitness linearly proportional to the fraction of good connections, $\alpha=1$, equation 9), straightforwardly parallels the Maxwell's Demon analogy. The mutational back pressure toward the generic properties of the ensemble, Rf, is a monotonic increasing

function of the mean number of good connections per organism,
$\overline{G}$. In contrast, the selective force, Se, increasing the mean
number of good connections, $\overline{G}$, is a monotonic decreasing
function of $\overline{G}$. For parameter values where the selective
force is greater than the restoring force for all values of
$\overline{G}$ less than T,   Rf and Se do not intersect in the interval $\overline{G}$=0, $\overline{G}$=T, and selection is sufficiently powerful to
achieve and maintain any unique, arbitrary "perfect" wiring
diagram in almost all members of the populations. For parameter values where the Rf and Se curves intersect in the
interval $\overline{G}$=0, $\overline{G}$=T, the intersection corresponds to a single
globally stable stationary population distribution of fit-
ness about $\overline{G}$, and reflects the limited fitness achievable by
selection due to the balancing back pressure of mutation.

These qualitative features imply that there are critical
values of the parameters T, α, b, P, and μ, thus a surface
in that parameter space, on one side of which selection can
attain any arbitrary wiring diagram, while on the other
side, the balance between selection and mutation is less
than $\overline{G}$=T. Since a single steady state exists, as parameters
change beyond the critical surface, the steady state will
shift smoothly away from $\overline{G}$=T.

The position of the globally stable steady state depends,
in particular, on the complexity of the system, T, and the
mutation rate, μ. The true dependence of $\sigma^2 G$ on T and μ is
unknown, but simulation suggests that $\sigma^2 G$ increases linearly with T, but less than linearly and at approximately the
square root of μ. Substituting the approximation
$\sigma^2 G = KT\mu^{\frac{1}{2}}$ into equation 6) and solving for the steady state
G_{ss} yields:

10) $$\overline{G}_{ss} = \frac{T(1-P)}{2} + \frac{1}{2}\sqrt{(T(1-P))^2 + \frac{4\alpha KT}{\mu^{\frac{1}{2}}}}$$

As either T, or μ increases, the position of $\overline{G}_{ss}$ in the interval between $\overline{G}$=0 and $\overline{G}$=T, shifts away from "perfect",
$\overline{G}$=T, toward the unselected generic properties of the ensemble, $\overline{G}$ = (1-P)T.

Multiple Steady States and A Complexity Catastrophe

Under conditions where fitness drops off rapidly as G falls
below T, that is α>1, and where basal fitness b>0, finding
analytic steady state solutions to the full differential
equation 8) becomes impractical. However, graphical solutions are straightforward. The following qualitative features are important. For sufficiently large values of α,
eg. 10, and b, eg. B=.50, the effective selective curve, Se
does not decrease as $\overline{G}$ increases but increases at more than

a linear rate, hence can cross the restoring force curve Rf
at two points. The point closest to "generic", $\overline{G}=T(1-P)$, is
a stable stationary state. The point closer to "perfect",
$\overline{G}=T$, is unstable and repels the population in either direction. The "perfect" state is a reflecting boundary, and
and therefore if the population is above the unstable steady
state, it becomes trapped at a nearly perfect state between
the unstable steady state and perfect. Thus, the population
shows two stable stationary states, with fluctuation driven
transitions between them.

The system has two stationary states, thus, tuning the
parameters increasing mutation rate, μ, or the total number
of connections to be specified, T, or decreasing the cooperativity, α, may lead to bifurcations in which the nearly
perfect stationary state vanishes, catastrophically changing
the maximally attainable population fitness. Explicitly, the
existence of two stationary states requires that, for $\overline{G}$
close to the perfect wiring diagram, $\overline{G}=T$, the selective
force be greater than the restoring force. Substituting $\overline{G}=T$
into 8) and using the approximation $\sigma^2 G=KT\mu^{\frac{1}{2}}$ yields:

11)
$$\frac{d\overline{G}}{dt} = \frac{\alpha(1-b) \cdot KT\mu^{\frac{1}{2}}}{T} - \mu TP$$

Existence of the nearly perfect stationary state requires
Se>Rf, or, simplifying:

12)
$$\frac{\alpha(1-b)K}{P} \geq \mu^{\frac{1}{2}}T$$

As T increases, the selective force is roughly independent
of T, while the restoring force increases proportionally
to T. Therefore as T increases, selection can at first maintain the population in the nearly perfect stationary state
and gradually increase the complexity of an initial highly
adapted small network, but the nearly perfect stationary
state inevitably becomes progressively unstable to fluctuations then disappears and the population collapses to near
generic.

While still approximations, these analytic approaches
seem adequately to account for the major properties found
for this simplest population selection model in simulations
noted above. The very exciting task lying ahead, both from
the point of view of evolutionary theory, and for the practical design of adaptive automata, is to discover proper
extensions of selective adaptation to the dynamical behavior of distinct ensembles of automata. One can hope that
universal features of the adaptive process will be
uncovered.

Summary

Automata can be conveniently classified into diverse ensembles, each with a spectrum of highly typical, or generic structural and dynamical properties. Natural ways of defining ensembles include not only those which are mathematically convenient, but more interestingly, those which are <u>accessable</u> as a class under the mutational processes (genetic, or hardware) driving the population of organisms or automata. Among the ensembles so far studied, the canalizing ensemble, currently the most plausible candidate model for genomic regulatory systems, exhibits the most localized and orderly dynamical behavior. Threshold automata, typical neural models, exhibit rather different, but well ordered dynamical behaviors. Adaptation based on a population selection paradigm, is one attractive general approach to the study of adaptation in complex systems. Among the most fundamental questions raised by this approach is the extent to which selection can attain and maintain a population near local or global maxima of the fitness function. For sufficiently complex systems, selection may prove to be too weak to maintain such maxima, and populations must lie at stationary points reflecting the balance between selection and the mutational drive toward the generic properties of the ensemble. Biologically, existence near "generic" implies those properties are true biological universals. Practically, understanding how performance is adequate, despite existence near generic properties may lead to genuine advances in the design of adaptive automata able to function despite persistent component failure.

ACKNOWLEDGMENTS

This work was partially supported by grants ACS CD-30, ACS CD-149, NIH GM22341, and NSF PCM-8110601. The author thanks the conference organizers, and host institution, Universite Scientifique et Medicale de Grenoble.

REFERENCES

1. Kauffman, S.A. (1969). Metabolic stability and epigenesis in randomly constructed genetic nets. *J. Theor. Biol.* <u>22</u>, 437-467.
2. Kauffman, S.A. (1974). The large scale structure and dynamics of gene control circuits: an ensemble approach. *J. Theor. Biol.* <u>44</u>, 167-182.

3. Kauffman, S.A. (1983). Emergent properties in random cellular automata. *Physica D*, in press.
4. Rubin, H. and Sitgreave, R. (1954). "Tech. Report No. 19A." Applied Math. and Stats Lab, Stanford University.
5. Wolfram, S. (1983). Statistical mechanics of cellular automata. *Rev. of Mod. Phys.*, in press.
6. Hopfield, J.J. (1982). Neural networks and physical systems with emergent collective computational abilities. *Proc. Natl. Acad. Sci.* 79, 2554-2558.
7. Kauffman, S.A. (1971). Gene regulation networks: a theory for their global structure and behavior. *In* "Current Topics in Developmental Biology 6" (Eds. A. Moscana and A. Monroy) pp. 145-182. Academic Press, New York.
8. Kauffman, S.A. (1972). Cellular genetic control systems. *Lectures on Math. in the Life Sciences* 3, 63-166.
9. Kauffman, S.A. (1983). Developmental constraints: internal factors in evolution. *In* "Development and Evolution" (Eds. B. Goodwin and N. Holder), Cambridge Univ. Press, Cambridge.
10. Aleksander, I. (1973). Random logic nets: stability and adaptation. *Int. J. Man/Machine Studies* 5, 115-131.
11. Babcock, A.K. (1976). Logical probability models and representation theorems on the stable dynamics of the genetic net. Doctoral dissertation, State University of New York, Buffalo.
12. Sherlock, R.A. (1979). Analysis of the behavior of Kauffman binary networks. I. State space description and the distribution of limit cycle lengths. *Bull. Math. Biol.* 41, 687-705.
13. Sherlock, R.A. (1979). Analysis of Kauffman binary networks. II. The state cycle fraction for networks of different connectivities. *Bull. Math. Biol.* 41, 707-724.
14. Atlan, H., Fogelman, F., Salomon, J. and Weisbuch, G. (1981). Random Boolean networks. *Cybernetics and Systems* 12, 103-131.
15. Fogelman Soulie, F., Goles Chacc, E. and Weisbuch, G. (1982). Specific roles of the different Boolean mappings in random networks. *Bull. Math. Biol.* 44, 715-730.
16. Walker, C.C. and Ashby, W.R. (1966). On temporal characteristics of behavior in certain complex systems. *Kybernetik* 3, 100-108.
17. Walker, C.C. and Gelfand, A.E. (1979). A system theoretic approach to organizations: management by exception, priority, and input in a class of fixed-structure models. *Behavioral Science* 24, 112-120.
18. Erdos, P. and Renyi, A. (1959). "On the Random Graphs I" vol. 6. Inst. Math. Univ. de Breceniens.

19. Erdos, P. and Renyi, A. (1960). "On the Evolution of Random Graphs, Publ. 6". Math. Inst. Hungarian Acad. Sci.
20. Ewens, W.J. (1979). "Mathematical Population Genetics" Springer-Verlag, New York.

ALGORITHMS, BRAIN, AND ORGANIZATION
Ch. v.d. Malsburg

Max-Planck-Institut für Biophysikalische Chemie
Postfach 2841, D-3400 Göttingen, W.-Germany

1. DOES THE BRAIN FOLLOW AN ALGORITHM ?

It is the essence of an algorithm that all its qualitative
aspects are premeditated and tested so that during its execution no ideas are necessary, no qualitative questions are
left open, and there are no surprises. Only quantitative
decisions must be met, which can be handled in a mechanical
way (e.g. ' take the next digit and add it to the result').
It was this point that led Turing to formulate algorithms
as machines. There is a very clear division between the qualitative and the quantitative aspects of an algorithm : The
former is invented by the human mind and formulated as rules,
the latter refers to the data handled by the rules. The
rules never act on their own structure.

Restriction of consideration to the explicitely visible
structure of algorithms misses important *Hidden Aspects :*
Semantics, which fills symbols with substance, images and
interpretation and is the basis on which to decide whether
rules and operations make sense ; teleology, i.e. motivations,
aims, values and considerations of esthetics and of applicability ; and heuristics, i.e. methods for the invention of
new structures. All these hidden aspects are suppressed
from explicit communication yet are essential constituents
of the structures of which explicit algorithms only form
the visible surface.

When dealing with algorithms in the context of mathematics
one has the choice either to consider the explicit structure
alone, a view which prevails in mathematics, or to include
the hidden aspects and face the situation comprehensively,
which is compulsory when dealing with the brain and its function. Several problems arise for the currently prevalent
view that the brain can be understood as an algorithmically
controlled machine. A 'brain-algorithm', i.e. a full and detailed description of the brain's function, would be too

DYNAMICAL SYSTEMS
AND CELLULAR AUTOMATA

complicated in principle to be handled by humans or serve as
a medium to communicate knowledge about our mind. Secondly,
where does the algorithm come from ? The obvious answer, that
it was developed by evolution, is excluded because the algorithm would be much too complicated to be transmitted by the
genom from one generation to the next, and because evolution
cannot have developed procedures applicable to such new-
fangled ideas as mathematics or chemistry. A third problem
is that the brain evidently modifies itself when it develops
new ideas. This squarely contradicts the algorithmic scheme,
which requires that all qualitative aspects have been premeditated. A fourth problem is the fact that the algorithmic
scheme implies a clear separation of heuristics and teleology from the explicit algorithm. A brain algorithm, however
would have to include all hidden aspects. In summary, it
would be ridiculous to identify the working of our brain
with the execution of an   algorithm.

2. A DISCUSSION OF ARTIFICIAL INTELLIGENCE

Most discussions of algorithmic structures are focussed on
the explicit and suppress mention of the hidden aspects.
The above arguments are therefore in danger of being brushed
away, and a more practical discussion is in order. Many
adherents of artificial intelligence will point to working
computer programs as demonstrations of the success of artificial intelligence. However, these programs are only simple
caricatures of real processes in the brain which are a
thousand times more sophisticated ; the programs depend entirely on the human programmer, and must be modified when
unforeseen situations arise ; and the various specialized
programs lack integration into a full system. Thus the
relevance of artificial intelligence to the brain rests on
an implied extrapolation over orders of magnitude, a rather
doubtful matter.

Defendants of artificial intelligence will point out that
one could conceive programs which modify themselves, in
order to solve new problems or adapt themselves to new
situations. The hope is that a critical mass can be reached
with the aid of human programmers and the rest is then
programmed by the system itself. Practical experience shows
that this hope is not valid. New programs tend to develop
bugs. It takes human attention to diagnose a bug and to mend
it. This necessitates detailed communication between man
and the digital process in the machine, so that self-
modification of programs can only be allowed up to the
point where the communication between man and machine

breakes down. The attempt to go beyond this point and regulate qualitatively unknown processes in a fail-safe way with
the help of administration-type rules is well-known to lead
to Parkinson's condition : a structure busy with nothing but
self-administration.

In conclusion, artificial intelligence is useful in practical applications and valuable in creating interesting models of intelligent processes, but it cannot be extrapolated
to a model of the mind without a major readaptation : aban-
donment of algorithmic control.

3. THE FUNCTIONAL PRINCIPLE OF THE BRAIN

The conclusion of the last two sections creates a dilemma,
because shurely the brain must follow some functional
principle and shurely this principle can be put into the
form of an 'algorithm' which can be transmitted by the
genom from generation to generation.The solution to this
dilemma lies in the scheme of a *trivial algorithm*. All
concrete, goal-directed, specific ideas : rules, values,
concepts, methods, procedures etc., are treated by the
trivial algorithm as data. The algorithm fixes the general
forme of operations on a fundamental level, and makes shure
that organized states instead of chaos arise. This is to
be seen in analogy to the laws of physics which are of a
very general kind, yet allow the build-up of organized
structures : atoms, molecules, crystals, etc.

The trivial algorithm of our brain has been developed by
evolution with the help of its own methods (reproduction,
mutation, selection) and values (survival), which play the
role of hidden aspects of the algorithmic scheme. It can be
hoped that the trivial algorithm is simple enough for
scientific study.

The concrete functional states of the brain then correspond to ordered arrangements in vast arrays of data. The
rules of the system are data configurations themselves and
form a hierarchical system in which more concrete rules
are modified by more fundamental rules. On a level that
is concrete enough so that it can be subject to explicit
communication, the rules are not fixed and the processes
therefore cannot be said to follow algorithms. The connection between the most fundamental level, that of the trivial
algorithm, and the concrete levels with which we are familiar
is enormously complex. It can only be studied with the help
of idealized model structures, with experimental implementations on machines yet to be built, and with phenomenological
descriptions of a dynamical processes involved. Experience

with hierarchically structured physical systems reaches that simple scientific studies are only possible on two levels : that of the phenomenology of concrete thought processes (which has to renounce at the algorithmic aspect), and on the fundamental level of the trivial algorithm (which has to renounce at all direct relevance to concrete thought processes).

4. ORGANIZATION

What was called trivial algorithm in the last section is nothing but the principle of organization in the brain. The following discussion is inspired by the belief that there is a common conceptual basis to all phenomena of organization and that one day we will have a theory of organization in general. First vague outlines of such a theory are under discussion today. They might be formulated by the following list of principles, which have been taken from biological evolution theory and physics. The principles speak of an ensemble of elements of a system. Attached to each element there is a variable measuring its extent. The extent is often interpreted as a population number (of a biological or chemical species, for instance). The list is neither complete nor minimal.

Reproduction : there are positive feed-back loops by which the extent of an element acts back on itself. In the special case of biological evolution this self-amplification is literally brought about by physical reproduction. In the chemical context the term autocatalysis is used.

Mutation : there is noise by which new elements or configurations are brought into the game.

Cooperation and competition : the extent variables of the system mutually influence each other's growth rates. For simplicity, only pair-wise interactions are considered. Influences may be positive (co-operation) or negative (competition). The various influences on one variable may be consistent with each other (all positive or all negative) or they may be inconsistent (partially cancelling each other), depending on the system's state.

Co-operativity : the system is dominated by globally consistent stable configurations. In these, all (or almost all) interactions are locally consistent with each other (residual inconsistency is often termed frustration).

The term organization implies some inherent direction of development. The above principles formulate competitive growth as the fundamental value of the system. Now, 'survival of the fittest' seems to be a rather dull value, especially if the elements have fixed fitness parameters (rates of reproduction), as is assumed in standard Darvinian theory. However, in a network of elements, such as an eco-system, growth rates depend on configurations of extent-variables. Progress therefore depends on the 'invention' (by mutation) of optimal configurations. These are distinguished by their inner consistency, or by the degree to which the loops of inner interactions are of the positive type and don't contradict each other.

In order to render the complicated processes in an organizing system intelligible it is necessary to make the transition to higher levels of description. This is done by identifying important configurations of a lower level as the elements of a higher level, and by formulating a new system of co-operation and competition between them. In many cases the transition form one level to the next is far from being a well-defined procedure. Often the qualitative character of the descriptions on different levels is completely dissimilar. For instance, a deterministic system can give rise to a probabilistic higher-level description or vice versa, or, which is the point of the above discussion, an algorithmically formulated lower level can give rise to non-algorithmic high-level behaviour.

Certain conditions must be fullfilled in a system in order for it to support the stable states required by co-operativity. The problem is exemplified by temperature control in a thermodynamic system, where organization (e.g. crystallization) can only occur when the temperature sinks below a certain critical value. Above that critical temperature fluctuations are violent enough to disrupt the long-range forces by which the organized states could stabilize themselves. A biological example is the mutation catastrophe which sets in when the mutation rate surpasses a certain threshold (1). Organization requires that the rate of local processes be slow enough so that the long-range interaction loops have time to dominate the system. (This manifests itself in the critical slowing-down of fluctuations near a phase-transition point of a physical system.) The control structure of an organizing system should avoid capture of the state in local optima of stability. This could be achieved by appropriate control of the level of fluctuations, analogous to annealing in metallurgy. It is also a very important matter to exclude the occurrence of

trivial superstable states.Instead., the system should always
stay in a state of metastability, ready to respond to
subtle long-range forces from within or outside.

   Also a digital machine is a network of interaction elements
and one might compare it to an organizing system. In order
to function deterministically and with maximal speed the
elements of a digital machine contain positive feed-back
loops which throw their states quickly into one of a few possible stable states. At each pulse of a clock-oscillator
the elements are made sensitive to the state of a small number of function whether to jump into another state. From
the point of view developed above these switching-events act
as violent local fluctuations which keep the elements from
the digital point of view would be terrible wastes of time.
The fact that a digital network nevertheless performs a
useful function (and thereby conforms to some global order)
is the accomplishment of an organization process in the mind
of the designer. The price for this separation of organization from machine, which puts severe constraints on the
design of machines. One of the consequences is the sequential
control of action, which from the organizational point of
view is a terrible wast of time.

5. BRAIN ORGANIZATION

The general scheme of organization can be applied to the
brain on several levels. Two will be discussed in this
section, one in the next but one. Let us first take the
brain as a network of neurons with fixed excitatory and
inhibitory synaptic connections. Elements and extent-
variables are identified with neurons and their
output signal, respectively. Reproduction is achieved
with the help of positive feed-back loops : excitatory cells
excite their excitors, inhibitory cells inhibit their
inhibitors, and so on to longer loops. Ordered states
are stationary activity distributions, each of them characterized by the set of cells active in it. The ordered states
are stabilized by the exchange of excitation among active
cells and the transmission of inhibition from active to
silent cells. The interaction pattern can be such as to
support a large number of ordered states (which implies
some frustration - excitation and inhibition partly cancelling each other). As is shown in the theory of associative
memory (2), a particular ordered state can be reached from
a large number of initial activity distributions, because
already a fraction of the cells which ought to be active
create with their signals the correct distribution of exci-

tation and inhibition in the total system. The ordered
states are thus able to correct deviations or to complete
information which was not specified by the input to the
system. Some authors consider the associative reconstruction of ordered states as the fundamental dynamic process
in the brain.

A more specific application of the scheme of organization
on the level of neural activity is a theory for a how visual
cortex identifies corresponding points in the stereo-images
received by the two eyes. The theory was first described
with the help of an analogy (3) and was later put into neural form simultaneously by several authors. Neurons responding to a particular point in one eye and to different
points (stereo depths) in the other exchange inhibition,
neurons corresponding to the same stereo-depth in neighbo-.
ring points exchange excitation. In the ordered states of
the system only a single cell is active for each point
and it is selected such that stereo-depth varies smoothly
from point to point. In this theory the ordered states form
a multidimensional continuum. When the eyes or the observed
objects move, the ordered state can be deformed or moved
through an ambiguous phase. This demonstrates the establishment of a qualitatively new hierarchical level, the elements
of which are the ordered states of the previous level.

The quality of the ordered states in a network depends
on the degree to which they can attain mutual consistency
between the signals converging on cells. It depends on
the structure of the interconnection pattern in the network. A second king of organization process is necessary
to shape this structure. It has synaptic connections as its
elements. Extent variables are synaptic weights (the
efficiencies with which a synapses transmit signals).
Weights are modified by *synaptic plasticity* according
to wich a synaps is reinforced by its own success in
influencing its target cell. As this success grows with
the weight of the synapse, the mechanism leads to reproduction. Competition between the synapses contacting one
cell is introduced by a regulation mechanism which keeps
the average activity in the cell constant. Success of
a synapse is strongly enhanced by mutual consistency of
its signal with those of other synapses on the same target
cell. This constitutes co-operative interactions. The
main reason for different signals to be correlated with
each other is that they come over short alternative
pathways from the same original source, inside or outside
the network.

The fundamental application of synaptic plasticity is the
stabilization of certain activity states of a network, so t
that they become ordered states of it. If the stabilized
states originally had been projected into the network from
outside then plasticity can be said to form memory traces.
The scheme is called associative memory (a recent reference is 2) Another important application is the formation of feature detector cells, i.e. of neurons which
respond to the statistically prominent patterns in a
stream of input activity. Additional fixed connections between the developing feature detector cells can introduce
constraints to regulate the distribution of features over
cells. Purely inhibitory connections tend to distribute
features homogeneously over cells. Spatially organized
connections with short-range excitation and long-range
inhibition can lead to a continuous mapping from feature
space into cell position. This idea has been applied to
the generation of topological mappings (4) and of continuously distributed representation of stimulus orientation
in visual cortex (5).

6. SEMANTIC SYMBOLS

So far, the methodology of artificial intelligence introduces new data structures for new algorithms. In contrast,
organization in the brain has to proceed on the
basis of an integrated system of semantic symbols. The
neurosciences already have a fairly clear conception of
this system. On a coarse level it speaks of localization
of different aspects of the brain's function in different anatomical regions. On a finer level, each neuron
in the brain is considered as the carrier of a
semantic atom. (Examples are cells in sensory regions
which respond to and symbolize the local stimuli of
a particular quality in a particular point of the sensory
surface.) According to the prevailing view the semantic
 symbols which form the stream of our thoughts are
additively composed by the semantic atoms of all the currently active neurons. This view is at the basis of the
vast majority of current brain theory.

In an earlier paper (6) it was argued that a symbol
system which is based on additive composition of atoms
lacks an important aspect of all familiar symbol systems,
namely a scheme of bindings by which the active atoms
can be grouped dynamically into hierarchically structured
subsymbols. In written language, atoms are letters or
words (or ideograms), and they are grouped geometrically

by positioning them appropriately on paper. If, as is
accepted here, semantic atoms are localized in neurons,
geometrical grouping is not possible and another scheme
must be found. The one discussed in the next section has
yet to be validated experimentally but has the advantage
that it forms a natural basis of organization.

7. CORRELATION THEORY

Neurons emit nervous impulses. Their instantaneous fre-
quency(determined in intervalsof about .1 sec or more)
is an extent variable which expresses the intensity with
which the corresponding semantic atom is active. In
addition the nervous signal has a temporal fine structure
which can be more or less correlated between cells.
Cells with correlated (synchronous) signals express the
fact that they are currently bound to each other, lack
of correlations habe to be evaluated with a resolution
of not better than 5msec, and over periods of 100msec
or more). Activity variables can influence each other
only between cells which are bound to each other. Correlations are induced in sets of cells which receive correlated signals from outside. Inside the network correlations are produced and propagated by synaptic
connections.

It is important that correlations change dynamically
from second to second in order to express varying symbol
structures. Accordingly also the connection structure
has to change on the fast time-scale of cognitive processes, in tenths of a second. It is therefore proposed
in (6) that synapses be characterized by two extent
variable w determines the strength of the effect
of a presynaptic impulse on the postynaptic cell.
It varies on determined by S, wich changes on the
slow time-scale of plasticity (hours to days). The value
of w is controlled by the two signals visible to the
synapse, the presynaptic and the postsynaptic. If the
two are synchronous (and the synapse is excitatory),
w is increased, if the two are asynchronous, w
is decreased. If there are no signals on both sides,
w decays with a time-constant typical of short-term memory, tens of seconds, towards a resting value somewhere
in the middle of range set by S. A synapse whose w is
often strongly activated increases S, which otherwise
is competitively controlled in the way described in the
paragraph on synaptic plasticity above.

The mode of control of w , synaptic modulation, is similar to that of synaptic plasticity. Differences lie in the time-scale, which is faster, and in the mechanism of competition. The simultaneous growth of all w-variables is prevented by an inhibitory system which allows only a small percentage of all the cells to be simultaneously active, so that in any moment the majority of synapses connect cells with asynchronous activity and correspondingly their w must decrease.

The introduction of synaptic modulation establishes a third level of organization within the nervous system, with w as extent-variables. They reproduce, because a positive w in an excitatory synapse between two cells enhances their synchronization, which in turn acts to increase w. Co-operation exists between short alternative pathways between the same pair of cells. For instance, the direct connection from cell a to cell c co-operates with the indirect link a to b to c . Competition was already touched upon. Co-operativity requires that certain configurations of the w-variables stabilize themselves as the ordered states. Let us refer to them as optimal networks. The elementary rules of cooperation and competitiion cause optimal networks to be sparse, i.e. contain few activated synapses (with positive w) and to have maximal co-operation between them. It is obvious that there is a great number of different optimal networks possible on the same set of cells. For a given set of neurons in the central nervous system connectivity in the resting state may be considered as a superposition of different optimal networks. When sufficient activity circulates within the network, an instability sets in and the system comes to rest only after one of the optimal networks on the cells is fully activated and all others are switched off. This surviving optimal network and the corresponding activity pattern represent a well-formed hierarchically structured semantic symbol.

In the conventional semantic symbol system, in wich there are no binding variables, plasticity connects all those cells which are active in the same composite symbol and these connections influence all later dynamical processes. In the system discussed here, connections are themselves subject to a fast dynamical process, and both their formation and their later influence is restricted to cases in which they fit the over-all binding structure. Interactions are therefore much more specific and many brain theoretical problems thereby become tractable (6). Examples are perceptual segmentation (7),

the specific projection between isomorphic symbols, and
especially the invariance problem.

8. THE TWO TYPES OF MACHINE

In conclusion, there seem to be two natural regimes for
information processing machines, the algorithmically
controlled computer, and the organizing machine. They
differ in the location of the instruments ('hidden aspects')
necessay for the development of specific functions. The
organizing machine incorporates these instruments as
integrale part. This has the advantage of making the machine
independent of detailed supervision and the corresponding
constraints. Organizing machines therefore can be massi-
vely parallel, complex beyond comprehension, flexible,
and do not require programming. On the other hand these
machines are indeterministic for practical purposes, are
opaque, and difficult to influence in their exact behaviour.
Furthermore they not only can be complex, but they have
to be complex, in order to include the hidden aspects.

Computers, on the other hand, exclude the instruments
necessary for the development of specific functions, and
are specialized on the excution of algorithms. They therefore can be extremely simple, deterministic, and are
potentially universal and fully transparent. On the other
hand they have to be simple in order to allow for detailed
communication with the engineer and programmer. This
constrains it to be digital deterministic and essentially
sequential. Human intervention is necessary in order to
adapt the computer to specific tasks. Considerable effort
is spent to give the computer some of the advantages of
the other type of machine, the brain. However, these
efforts are wasted as long as they are based ont the
algorithmic scheme. Instead, a quantum jump is required,
into the regime of organizing machines.

1. Eigen, M. (1971). Selforganization of matter and evolution of biological macromolecules, *Naturwiss.* 58,
   465-522.
2. Hopfield, J.J. (1982). Neural networks and physical systems with emergent collective computational abilities,
   Proceedings of the National Academy of Sciences USA, 79;
   2254-2558.
3. Julesz, B. (1971). *"Foundations of Cyclopean Vision".*
   University of Chicago Press.
4. Willshaw, D.J. and Malsburg, C. von der (1976). How
   patterned neural connections can be set up by self-
   organization, *Proc. R. Soc. Lond. B.* 194, 431-445.

5. Malsburg, von der (1973). Self-organization of orientation sensitive cells in the striate cortex, *Kybernetik* 14, 85-100.
6. Malsburg, C. von der (1981). Te correlation theory of brain function, *Internal Report 81-2*, Dept of Neurobiology, Max-Planck-Institute for Biophysical Chemistry, D-3400 Gottingen.
7. Malsburg, C. von der and Schneider, W. (1984). A theory of sensory segmentation, *in preparation*.

COMPLETE DESCRIPTION OF BOOLEAN ASYNCHRONOUS AUTOMATA USING GENERALIZED VARIABLES, AND ITS COMPARISON WITH DIFFERENTIAL EQUATIONS FORMALIZATION

J. Richelle

Laboratoire de Génétique, Université libre de Bruxelles, rue des Chevaux 67, 1640 Rhode-St-Genèse, Belgium.

1. INTRODUCTION

The convergence between the results of the boolean description and the continuous description using differential equations has been repeatedly pointed out (Glass and Kauffman (1), Richelle (2-4), Thomas *et al* (5)). The continuous approach is generally used in view of its quantitative character. Its generalized use is however hampered by the difficulty of applying it to such complex systems as those encountered in living organisms. The boolean asynchronous approach described by R. Thomas in his communication was introduced and developped (see for example Thomas (6)) precisely because it allows a rapid survey of the implications of a model. The boolean approach can be very helpful to delineate a model that is to be studied in more detail with the continuous approach (see M. Kaufman's communication). This mode of working assumes that the essential qualitative features of the boolean treatment will also be found by the continuous treatment. For really complex systems, comprising a large number of variables, only simplified continuous models can be treated analytically. The boolean approach is then very interesting if one may assume that the results will be equivalent to the unobtainable results of the complete continuous approach. Such an equivalence has generally been found for the systems where the comparison can be made, but there is no proof that this must always be the case.

In this communication a compact description of the boolean asynchronous approach is developped. It is used to present evidence that the convergence between the boolean and the continuous approaches originates from a fundamental parallelism of the dynamics of the approaches. We consider it justified to assume that the two approaches are

DYNAMICAL SYSTEMS
AND CELLULAR AUTOMATA

247

equivalent.

2. THE BOOLEAN STATE IS AN INCOMPLETE DESCRIPTION OF THE STATE OF A SYSTEM

A boolean state is a n-uple of boolean $(0,1)$ numbers defining the values of the n (internal) variables used to describe a system. For each internal variable there exists an associated function which value depends on the boolean state of the system. We say that a transition is called for for a variable when its value differs from the value of its associated function. In the asynchronous boolean description (contrary to the synchronous one) the transition times of the different variables are generally different. As a transition can be called for for up to the n variables, a given boolean state can be followed by up to n different states. In this sense the boolean state does not completely define the state of the system. In such situations, which state will actually be reached depends on the relative values of the time delays of the called for transitions. In brief, one can say that the boolean state does not define whether called for transitions are far or near to their realization; in real automata this aspect is dealt with by the state of the retarding devices.

3. TOWARDS A COMPLETE DESCRIPTION

The retarding devices may be seen as (Milgram (7)), and are sometimes in practice (Van Ham (8)), "counters" associated with the boolean variables. One might first imagine that for each variable a counter would count the time from the moment an on-transition of the variable is called for and an other would count when an off-transition is called for. However, for a given variable, these two transitions are mutually exclusive and a single counter can thus do the job : it will count the time from the moment a (on- or off-) transition is called for. When the corresponding delay is over, the variable takes the value of its associated function and the counter is set back to zero (the word counter will be loosely used to designate the value of the counter). In the basic description of the so called kinetic logic, it is assumed that when the value of a function changes, the value of the corresponding variable will change (after a defined delay) *unless* a counter-order has taken place before; in this case one reasons as if the order was merely cancelled. Due to this convention, a counter is also set back to zero if the transition ceases to

be called for before the end of the delay.   Notice   that   a
counter   will   keep   on   counting   even   if the state of the
system changes provided the   associated   function   does   not
change value.   Figure 1 depicts the behaviour of a boolean
variable and its associated counter under the control of the
corresponding   boolean   function (the value of this function
itself depends on the state of the n   boolean   variables   of
the system).

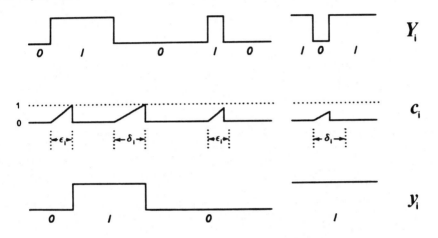

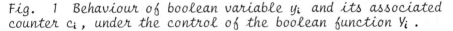

Fig. 1 Behaviour of boolean variable y_i and its associated
counter c_i, under the control of the boolean function Y_i.

Knowing the boolean state and the state of the   counters,
i.e. in short,   the   total   state   of the system, allow the
next state of the system and in particular the next   boolean
state   to   be   calculated   unambiguously.   Notice that this
complete description requires 2n variables for n components.
The   way   the counters function is not restricted to this
mode; figure 2 presents another mode of functioning.
In this case a counter is increased when an on-transition is
called for and it is decreased when   an   off-transition   is
called   for.   In absolute value, the rate of variation of a
counter is equal to the inverse of the corresponding   delay.
When   the   on-transition time   has   elapsed i.e. when   the
counter   reaches   1,   the   transition   of   the   variable   is
executed   and   the   counter   is   set to 2.   When   an
off-transition is due to occur, the counter   decreases   from
2,   reaches   1   after   the   specific   delay   and   then   the
transition of the variable is executed   and the   counter   is
set to zero.   If a transition ceases to be called for before
the corresponding delay has elapsed, i.e. before 1 has   been
reached,   the counter is set back to 0 or 2 according to the

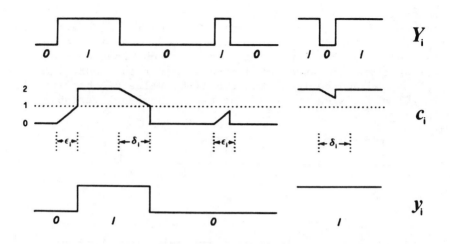

Fig. 2 An other mode of functioning of counter c_i associated to variable y_i under the control of function Y_i .

value (0 or 1) of the associated variable and function.

4. GENERALIZED VARIABLES, A COMPLETE COMPACT DESCRIPTION OF THE STATE OF THE SYSTEM

Here it is important to notice that in our second description (cf figure 2), whenever the counter is below 1 the boolean variable value is 0; conversely whenever the counter is above 1 the boolean variable value is 1. The new idea that is now introduced consists of fusing the boolean variable and the associated counter into a real valued variable endowed with a dual role : such a generalized variable will describe the boolean state of the internal variable and the state of the associated counter at the same time. The way these generalized variables behave is presented in figure 3.
The generalized variables take real values bounded in $[0,2]$; below 1 they are considered to have a boolean value of 0 and above 1, a boolean value of 1. When an on- (off-) transition is called for, the generalized variable is increased (decreased) at a constant rate equal to the inverse of the corresponding delay. When the transition ceases to be called for the generalized variable is set back either to 0 or to 2 according to the value 0 or 1 of the associated function; a transition ceases to be called for not only when the function return to its previous value before the transition occurs but also when the transition is executed. A boolean transition takes place when the value

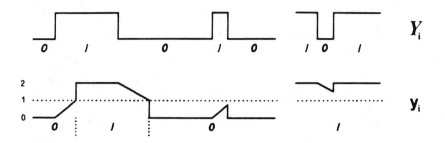

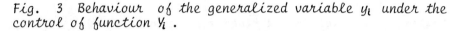

Fig. 3 Behaviour of the generalized variable y_i under the control of function Y_i.

of a generalized variable crosses the threshold value 1 (from a higher or a lower value).

This description is complete for the same reasons as in the preceding cases, but it requires only n variables for an n-component system. The state of the generalized variables is defined in the euclidean space R_n. The boolean state can be considered as the result of a mapping from R_n in B_n, the boolean space of n dimensions. Different boolean pathways can have one (or more) state in common : they represent the mapping of different trajectories in R_n that have no point in common.

5. AN OTHER FUNCTIONING OF THE RETARDING DEVICES

So far we have given a "generalized" description of the kinetic logic. It is possible to imagine boolean asynchronous automata based on retarding devices that have different dynamics. Van Ham (8) has examined a few of them from the automata conceptor point of view. The functioning depicted in figure 4 has been proposed (Richelle (9-10)) in view of biochemical considerations : if one considers the evolution of the generalized variable as an image of the evolution of the component concentration, jumps of value as those the preceding model admits when 1 is reached or after a counter-order (figure 3) are quite unrealistic. In the present model there is no more discontinuity in the value of the generalized variables and so it is a better description. As long as the associated function value is 1 the generalized variable increases or is maintained at 2 (if reached); as long as the associated function value is 0 the generalized variable decreases or is maintained at 0 (if reached). This mode of functioning will not be discussed

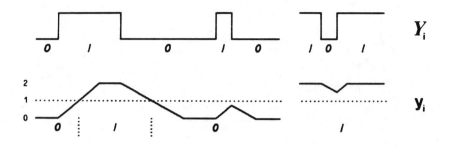

Fig. 4 Behaviour of the generalized variable y_i with more realistic dynamics from a biochemical point of view.

here; notice however that one can expect changes in the behaviour of the automaton.

6. TOWARDS DIFFERENTIAL EQUATIONS

In the preceding models the rate of variation of the generalized variables (and the counters) is discontinuous but constant. A further improvement consists of considering that it is no longer constant; let us now assume that it is proportional to the value of the generalized variable. The generalized variable will have (increasing or decreasing) negative exponential dynamics. One can easily observe that this is exactly the way the "piecewise linear" differential equations introduced by Glass (11) function :

$$y_i = F_i (\{y_i\}) - k_i y_i \qquad \forall\ i,j$$

where $F_i = 0$ or 1.

7. CONCLUSION

In this communication, we proposed a way to completely describe the dynamic behaviour of boolean asynchronous automata by using one generalized variable for each component. This mode of description has allowed us to show the conceptual continuity existing between these automata and the "piecewise linear" differential equations. What has been demonstrate here in qualitative terms has been established in a rigorous and formal way using an algorithm description language (Richelle (9)). This result shows a fundamental reason for the convergence between the boolean and continuous approaches and confirms their assumed equivalence. It makes it possible to understand the

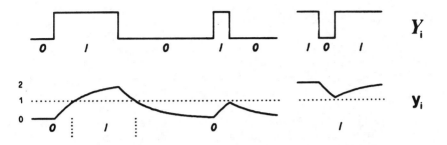

Fig. 5 Behaviour of the generalized variable y_i with negative exponential law; behaviour identical to that of the "piecewise linear" differential equations.

differences which are therefore to be expected between these descriptions, some of which have already been observed.

8. ACKNOWLEDGEMENTS

This work was realized thanks to an "Aspirant" fellowship of the "Fonds National de la Recherche Scientifique" of Belgium.

9. REFERENCES

1. Glass, L. and Kauffman, S. A. (1973) The logical analysis of continuous non-linear biochemical control networks. *J. theor. Biol.*, **39**, 103–129.
2. Richelle, J. (1977) Boucles de rétroaction négatives : un parallèle entre diverses analyses formelles. *Bull. Cl. Sci. Acad. Roy. Belg.*, **63**, 534–546.
3. Richelle, J. (1979) Comparative analysis of negative loops by continuous, boolean and stochastic approaches. *In* "Kinetic Logic", Lecture Notes in Biomathematics, **29**, chap. XIV, pp. 281–325. Springer-Verlag, Berlin, Heidelberg, New York.
4. Richelle, J. (1980) Analyses booléenne et continue de systèmes comportant des boucles de rétroaction I. Analyse continue des boucles de rétroaction simples. *Bull. Cl. Sci. Acad. Roy. Belg.*, **66**, 890–912.
5. Thomas, R., Nicolis, G., Richelle, J., & Van Ham, P. (1979) General discussion on the simplifying assumptions in methods using logical, stochastic or differential equations; the range of applicability and the complementarity of the approaches. *In* "Kinetic Logic",

Lecture Notes in Biomathematics, 29, chap. XVI, pp. 345
-352. Springer-Verlag, Berlin, Heidelberg, New York.
6. Thomas, R. (1973) Boolean formalization of genetic
    control circuits. J. theor. Biol., 42, 563-585.
7. Milgram, M. (1982) Automates probabilistes, Thèse d'Etat
    ès Sciences, Université de Compiègne.
8. Van Ham, P. (1975) Modèles discrets à actions
    différées. Thèse de Doctorat en Sciences, Université
    libre de Bruxelles.
9. Richelle, J. (1984) Analyses booléenne et continue de
    systèmes comportant des boucles de rétroaction   III.
    Description algorithmique des approches booléenne et
    continue. Bull. Cl. Sci. Acad. Roy. Belg., in press.
10. Richelle, J. Unifying description of discrete and
    continuous formalizations. in preparation.
11. Glass, L. (1975) Combinatorial and topological methods
    in nonlinear chemical kinetics. J. Chem. Phys., 63,
    1325-1335.

EVOLUTION OF SELF-REPLICATING MOLECULES -
A COMPARISON OF VARIOUS MODELS FOR SELECTION

P.Schuster

*Institut für Theoretische Chemie und
Strahlenchemie, Universität Wien
Währingerstraße 17, A 1090 Wien
Austria*

1. INTRODUCTION

Darwin's theory of evolution was casted into a quantitative
theoretical frame through the famous works in the "golden
age" of population genetics. The major goals in this field
were to correlate evolutionary optimization to changes in
the distributions of genes in populations, to derive quantitative expressions for the spreading of mutant alleles and
to relate them to abstract phenotypic fitness functions. The
occurrence of mutations and their frequencies were considered
as external empirical factors and not as subjects of the
theory. The discoveries of early molecular biology provided
a new dimension of insight into the evolutionary process: a
molecular picture of the mechanism of genetic replication
emerged and became more and more detailed. Although our
knowledge in molecular genetics is far from being complete it
is nevertheless sufficient to construct a model for polynucleotide replication based on chemical kinetics. The
theory has been formulated first by Eigen (1). Subsequent
studies on the details of primitive replicating systems
were presented later on (2-4). The mathematical details of
this theory of early evolution were developed and published
by several groups. Some key references are found in (5-8).
More recently, the concept was subjected to experimental
test by means of a suitable set-up for studies on RNA replication (9-13).

   In this contribution we present a molecular mechanism
which allows to investigate the prerequisites of selection
and evolutionary optimization by straight forward mathema-

DYNAMICAL SYSTEMS
AND CELLULAR AUTOMATA

255

tical techniques. We shall also discuss the role of replication errors and the problem of mutant propagation within
the frame of this molecular model. Finally, we shall compare several mathematical models of selection and optimization from different fields in biology.

2. Prerequisites of selection

Let us consider an ensemble of replicating macromolecules
$I_1, I_2, \ldots, I_n$ in an open system. The system is driven off
equilibrium by means of an energy and/or material flux. We
denote concentrations of the macromolecules by $[I_1] = c_1$,
$[I_2] = c_2, \ldots, [I_n] = c_n$. The state of the system is characterized by a vector

$$c = (c_1, \ldots, c_n) \tag{1}$$

with non-negative components: $c_i \geqslant 0 \; \forall i=1, \ldots, n$. The total
concentration of macromolecules can be defined by the norm

$$\|c\| = \sum_{i=1}^{n} c_i \tag{2}$$

It is often advantageous to use relative concentration variables

$$x = \frac{c}{\|c\|} \tag{3}.$$

The components of x, thus, span a unit simplex

$$S_n = \{x \in R_n, \; x_i \geqslant 0 \; \forall i=1, \ldots, n; \; \sum_i x_i = 1\} \tag{4}.$$

Selection, now, is defined by the long time behaviour

$$\lim_{t \to \infty} x_m(t)=1 \text{ and } \lim_{t \to \infty} x_i(t)=0 \quad i=1, \ldots, n, \; i \neq m \tag{5}.$$

After long enough time all molecules disappear except the
"master" I_m.

In order to illustrate the prerequisites of selection we
consider first the reaction mechanism

$$A + I_i \underset{f_i'}{\overset{f_i}{\rightleftharpoons}} 2I_i; \quad i=1, \ldots, n \tag{6a}$$

$$I_i \underset{d_i'}{\overset{d_i}{\rightleftharpoons}} B; \quad i=1, \ldots, n \tag{6b}$$

$$B \xrightarrow{g(E)} A \tag{6c}$$

The reactions (6a) are n parallel first order autocatalytic processes – an example of higher order autocatalysis will be discussed later –, n degradation reactions (6b) and a recycling process (6c) which converts the low energy material B back into the high energy compound A. This reaction is driven by an external energy source E. In figure 1 we present a schematic free energy scale for mechanism (6) which will be used to analyse the thermodynamics of selection.

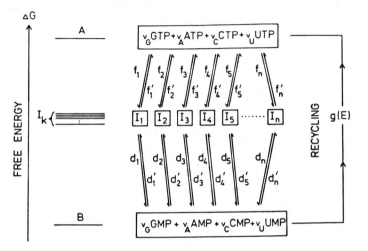

Fig. 1 Relative free energies of a model system of polynucleotide replication and degradation.

In conventional chemical kinetics reaction (6) is described by the differential equation

$$\dot{a} = gb + \sum_{i=1}^{n} (f_i' c_i - f_i a) c_i \qquad (7a)$$

$$\dot{b} = \sum_{i=1}^{n} (d_i c_i - d_i' b) - gb \qquad (7b)$$

$$\dot{c}_i = (f_i a - f_i' c_i - d_i) c_i + d_i' b; \quad i=1,\ldots,n \qquad (7c)$$

As before we denote concentrations by lower case letters, $[A] = a$, $[B] = b$ and $[I_i] = c_i$. A convenient choice of the concentration vector is of the form $c = (a, b, c_1, \ldots, c_n)$ since its norm is constant:

$$\|c\| = a + b + \sum_{i=1}^{n} c_i = c_0 = \text{const.} \qquad (8)$$

Equation (8) evidently allows to eliminate one variable, e.g.

$$b = c_o - (a+ \sum_{i=1}^{n} c_i),$$

and we are dealing with a dynamical system in n+1 degrees of freedom.

Qualitative analysis of equation (7) yields:
(1) In the general case, f_i, f_i', d_i, $d_i' > 0$ ∀ i=1,...,n and g>0 we have two stationary states inside and at the boundary of the simplex S_{n+2} corresponding to c and the conservation relation (8):

$$P_o = (c_o, 0, 0, \ldots, 0) \tag{9}$$

and $\quad P = (a, b, c_1, \ldots, c_n > 0) \tag{10}.$

P_o, called the zero state, is asymptotically stable in the low concentration range

$$0 < c_o < c_{crit}.$$

P is a saddle point in this range. At the critical concentration, $c_o = c_{crit}$, we observe a simple type of bifurcation: the two stationary states exchange their stabilities. P is stable above the critical total concentration, $c_o > c_{crit}$. The system under consideration depends on two external parameters, the total concentration c_o and the rate constant of the recycling reaction g (Thereby we assume that the other general external parameters like temperature and pressure are constant). The ranges of stability of P_o and P in the g, c_o-plane is shown in figure 2. We realize that the stationary state P is stable for all values of g provided c_o exceeds a certain limit. The stationary state P converges to the thermodynamic equilibrium in the limit g→0 and thus represents the thermodynamic branch which becomes unstable at the bifurcation point.

Accordingly, we do not observe selection in this case. At P_o none of the n autocatalysts is present, at P all concentrations are positive.
(2) In the case of irreversible degradation, f_i, f_i', $d_i > 0$, $d_i' = 0$ ∀ i=1,...,n and g>0, we observe n+1 stationary states on the boundary or in the interior of the simplex S_{n+2}. They are uniquely defined and can be enumerated easily if we order the replicating molecules I_1, I_2,...,I_n in a sequence of increasing ratios d/f (we assume absence of kinetic degeneracy):

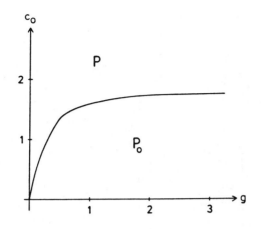

*Fig. 2 Ranges of stability of the two stationary states P
and P_o in the g, c_o-plane*

$$\frac{d_1}{f_1} < \frac{d_2}{f_2} < \cdots < \frac{d_n}{f_n}.$$

We denote a stationary state by P_{ijk} if e.g. only the autocatalysts I_i, I_j and I_k are present at nonzero concentra-
tions ($c_i > 0$, $c_j > 0$, $c_k > 0$ and $c_l = 0$ $l = 1, \ldots, n$ and $l \neq i, j, k$).
Then, the sequence of stationary states at the boundaries
or in the interior of S_{n+2} which are stable for some pairs
of g, c_o values is

$$P_o, \ P_1, \ P_{12}, \ P_{123}, \ldots, P_{123\ldots n}.$$

Again, stability analysis yields a rather simple result: for
a given pair of g, c_o values we have a uniquely defined,
asymptotically stable stationary state – the other n stationary states are unstable. Hence, we can subdivide the positive orthant of the g, c_o plane into regions of asymptotic
stability of one of the different stationary states (figure
3). Now, we observe selection in the region denoted by "P_1".
The replicating molecule with the smallest d/f value is selected in this region. Outside "P_1" there is no selection.

(3) Let us now consider the case of irreversible synthesis
and irreversible degradation, $f_i, d_i > 0$, $f_i' = d_i' = 0 \ \forall \ i = 1, \ldots, n$
and g > 0. Again we assume an ordered sequence of the ratios
d/f. In the limit $f_1' \to 0$ the value of the borderline between
P_1 and P_{12} at $g^{-1} = 0$ (see the g^{-1}, c_o-plot in figure 3)
approaches infinity and hence only two states are asymptotically stable in the positive orthant $g, c_o > 0$: P_o in the range

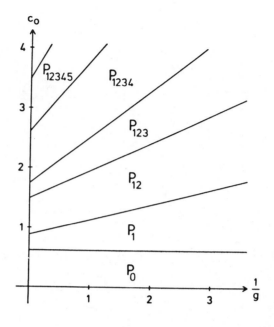

Fig. 3 Ranges of stability of the n stationary states P_0, $P_1,...,P_1$ in the g^{-1}, c_0-plane. The individual regions in this plot are separated by straight lines. Note, that both the values at $g^{-1}=0$ and the slopes increase with increasing number of autocatalysts at the stationary states and, hence, the straight lines do not intersect in the positive orthant.

$0 < c_0 < d_1/f_1$ and P_1 in the range $c_0 > d_1/f_1$. The range of selection, thus, extends towards infinity as f_1' approaches zero.

(4) Finally, we assume the existence of an inexhaustible reservoir of compound A in addition to irreversibility of replication and degradation. In other words, we consider buffered concentrations of A: $a = \bar{a} = a_0 = $ const. Then, the recycling reactions does no longer contribute to the dynamics of replication which is described by the differential equation

$$\dot{x}_i = x_i(f_i a_0 - d_i - \bar{E}); \quad i = 1,...,n \tag{11}$$

Here, we introduce a mean excess production

$$\bar{E} = \sum_i x_i(f_i a_0 - d_i) \tag{12}.$$

Now, we observe selection under all conditions. The criterion of selection is the maximum value of the function w=fa$_o$-d which we call the selective value:

$$w_m = (f_m a_o - d_m) = \max\{(f_i a_o - d_i)_i; \quad i=1,\ldots,n\} \quad \text{and}$$

$$x_m(0) \neq 0 \rightarrow \lim_{t \to \infty} x_m(t) = 1 \quad \text{and} \quad \lim_{t \to \infty} x_j(t) = 0 \quad j \neq m \tag{13}.$$

Selection is accompanied by optimization of the mean excess production (1,2,14):

$$\frac{d\overline{E}}{dt} = \sum_i x_i (w_i - \overline{E})^2 \geqslant 0 \tag{14}.$$

Qualitative analysis has been carried further in order to show that the results presented here are not questioned on a global scale by the occurence of stable closed orbits or strange attractors. This analysis can be found together with the details of the derivations in two recent publications (15,16).

3. Replication with errors

In this section we present a simple model of polynucleotide replication (1,2) which allows to account for replication errors explicitely. Apart from the ordinary autocatalytic process we are now dealing with a whole network of mutations leading, in principle, from every given polynucleotide sequence (I_i) to every other sequence (I_j) in the system:

$$(A) + I_i \quad \xrightarrow{\quad f_i Q_{ii} \quad} \quad 2I_i; \quad i=1,\ldots,n \tag{15a}$$

$$(A) + I_i \quad \xrightarrow{\quad f_i Q_{ji} \quad} \quad I_i + I_j; \quad i,j=1,\ldots,n \tag{15b}$$

Mechanism (15) is confined to the range of most efficient selection as outlined in section 2. Hence, the concentration of A is buffered. For short we incorporate a$_o$ into the rate constant: f=f'.a$_o$. Mutations are introduced by means of the matrix Q. We distinguish the "diagonal terms" Q_{ii}, the quality factors of the replication process and the "off-diagonal terms" Q_{ij}, the mutation frequencies. In particular Q_{ij} denotes the frequency at which the sequence I_i is obtained as an error copy of I_j. All elements of Q are non-negative and confined to the range

$$0 \leqslant Q_{ij} \leqslant 1; \quad i,j=1,\ldots,n.$$

Clearly, we have a conservation law – every copy has to be either correct or erroneous:

$$\sum_{j=1}^{n} Q_{ji} = 1 \tag{16}$$

or alternatively

$$Q_{ii} = 1 - \sum_{j \neq i} Q_{ji} \tag{17}.$$

The degradation process is assumed to be the same as before (see equation 6b). The complete mechanism (15a,b; 6b) is now described by the differential equation:

$$\dot{x}_i = (f_i Q_{ii} - d_i - \overline{E}) x_i + \sum_{j \neq i} f_j Q_{ij} x_j; \quad i,j = 1, \ldots, n \tag{18}$$

The mean excess production $\overline{E}$ is defined as before (12). The selective value has to be modified:

$$w_i = f_i Q_{ii} - d_i \tag{19}$$

Equation (18) has been studied extensively in the literature (1,2,17-19). We present here only the most important results. There are two different scenarios of the replicating ensemble:

(1) Replication is sufficiently accurate to sustain a stable mutant distribution in the long time solution of (18).
(2) The accuracy of replication is below a certain critical limit. The mutant distribution for long times degenerates to equipartition of sequences. Such a distribution as we show later cannot exist in reality.

In order to illustrate the two scenarios we consider a sequence of ν digits as a concrete example. Let the frequency of the correct incorporation of a single base be q. We call this quantity single digit accuracy and have $0 \leqslant q \leqslant 1$. The implicit assumption that the single digit accuracy is independent of the position in the sequence is an approximation the justification of which has been discussed previously (2). Now, we can write the elements of Q in the form

$$Q_{ii} = q^{\nu} \tag{20a}$$

and

$$Q_{ij} = q^{\nu - \Delta_{ij}} (1-q)^{\Delta_{ij}} \tag{20b}$$

where Δ_{ij} is the Hamming distance, i.e. the number of base exchanges between the sequences I_i and I_j. A nonlinear transformation of variables

$$y_i = x_i \cdot \exp\{ \int_o^t \overline{E} \, d\tau \}; \quad i=1,\ldots,n \tag{21},$$

brings equation (18) into the compact form

$$\dot{y} = W \, y; \quad y=(y_1,\ldots,y_n) \tag{22}$$

of a linear differential equation. The matrix W is defined by

$$W = \{ w_{ij} = f_j Q_{ij} - d_i \delta_{ij} \} \tag{23}.$$

All solutions of equation (22) can be expressed in terms of the eigenvalues of matrix W. The stationary mutant distribution is given simply by the eigenvector corresponding to the largest eigenvectors.This distribution has been called "quasispecies" in analogy to the concept of a species in biology. It consists of a master sequence I_m and its most frequent mutants.

In figure 4 we present a numerical example. We realize a remarkably sharp transition between the two scenarios (1) and (2) already at a moderate chain length of $\nu=50$. It becomes still sharper for longer sequences. Perturbation theory allows to derive a simple expression for the critical minimum accuracy

$$Q_{min} = \sigma_m^{-1} = \frac{d_m + \overline{E}_{i \neq m}}{f_m} \tag{24}$$

we call σ_m the superiority of the master sequence; $\overline{E}_{i \neq m}$ is the "mean except the master":

$$\overline{E}_{i \neq m} = \frac{1}{1-x_m} \sum_{i \neq m} x_i E_i \tag{25}.$$

The minimum accuracy of replication corresponds to a maximum chain length ν_{max}:

$$Q_{min} = q^{\nu_{max}} \longrightarrow \nu_{max} = -\frac{\ln\sigma_m}{\ln q} \simeq \frac{\ln\sigma_m}{1-q} \tag{26}.$$

Let us consider the scenario of replication below the minimum accuracy, $Q<Q_{min}$. According to the solutions of the deterministic differential equation (18) all sequences are present in equal concentration. This, however, cannot happen in realistic systems. The number of possible sequences, 4^ν exceeds by far the number of molecules in an experiment or in nature. For a small polynucleotide with a chain length of $\nu=100$ we have already $1 \cdot 6.10^{60}$ different sequences! The

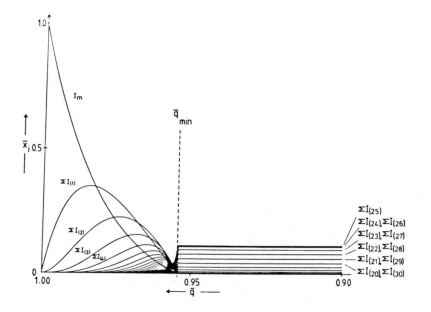

Fig 4 Stationary distribution of mutant classes as a function of the single digit accuracy q for ν=50. ΣI_{(k)} represents the sum of all k error mutants with respect to the master sequence I_o. At q=1 we have a pure population: x_o=1, x_i=0 ∀ i≠0. Mutant frequencies in the quasispecies increase with decreasing q-values. Note the sharpness of the transition from selection to random replication at q=q_{min}. Below this critical value we have equipartition of sequences. Hence, the 25 error mutants (ΣI_{(25)}) dominate.

deterministic approach breaks down since we really cannot have less than a single copy of a given sequence in the system under consideration. We are dealing with a set of sequences which change from generation to generation. New sequences appear due to copying errors and a certain percentage of the old sequences disappears as a consequence of degradation and dilution. The notion of "presence in equal amounts" can be replaced (at best) by "equal probability of realization" in the course of a long term experiment.

Finally, we come back to the problem of optimization in a system with mutations. It is easily verified that the mean excess production $\overline{E}$ is no longer optimized during selection of the quasispecies. Jones (20), however, found a function $\overline{G}$ in the space spanned by the eigenvectors ζ_k of the matrix W which is non decreasing:

$$\overline{G} = \frac{\sum\limits_{k} |z_k| \cdot Re\lambda_k}{\sum\limits_{k} |z_k|} \rightarrow \frac{d\overline{G}}{dt} > 0 \tag{27}$$

Herein, we denote the coefficient for the eigenvector ζ_k by z_k and the corresponding eigenvalue by λ_k (In this notation the quasispecies is characterized by $z_0=1$, $z_i=0$ $i=1,..$.,n-1 and λ_0 is the largest eigenvalue). Thus, optimization is easily understood in the space of eigenvectors of W.

4. Optimization in models of evolution

So far we have discussed selection and evolution as optimization processes of mean net production in the space of concentrations or, if mutations are admitted, in the space of the eigenvectors of the matrix W which describes replication and mutant synthesis. How does this simple model of asexually replicating populations compare to the evolutionary models used in population genetics and theoretical ecology? These models are more elaborate with respect to the mechanism of replication but do not account for mutations explicitely. Accordingly, we encounter constraints on the optimization process. We mention the two most important of them briefly:
(1) A genetic constraint operates in populations of sexually replicating diploid organisms. To give a simple but illustrative example we consider two alleles $(a_1 . a_2)$ on a single locus. We have three genotypes $(a_1 a_1)$, $(a_1 a_2)$ and $(a_2 a_2)$. Assume that the heterozygote $(a_1 a_2)$ carries the fittest phenotype. The population, nevertheless, does not approach a pure state because the two homozygotes, $(a_1 a_1)$ and $(a_2 a_2)$, inevitably appear in the progeny of $(a_1 a_2)$. This genetic constraint leads to a restriction of the optimiuation process onto a subspace of the space of genotypes, the so-called Hardy-Weinberg surface. On this surface we observe optimization of the mean fitness according to Fisher's fundamental theorem of evolutionary optimization.
(2) In the models presented here we applied a constant selective value. It is constant in the sense that it does not depend on the distribution of the replicating elements in the population. In classical population genetics constancy of fitness is commonly assumed in the mathematical models analysed (21). Cases of frequency dependent selective values $w_i = w_i(x_1,...,x_n)$ have been studied in the literature as well. Most examples use linear functions w_i. (For some recent systematic investigations see 22-24). Commonly, one finds

enormously rich dynamics including cases of multiple stable
stationary states and more complex attractors leading to
stable oscillations or chaotic dynamics. Another important
feature of systems with frequency dependent selective values
concerns evolutionary optimization: in general, stable stationary states do not coincide with maxima of the mean excess production. The same is true for more complex attractors. Thus, systems with frequency dependent fitness
functions do not allow optimization in the sense of Darwinian evolution.

Acknowledgements: This work has been supported financially
by the Austrian Fonds zur Förderung der wissenschaftlichen
Forschung, project no.5286.

REFERENCES

1. Eigen, M. (1971). Selforganization of matter and the
   evolution of biological macromolecules, *Naturwissenschaften* 58, 465-526.
2. Eigen, M. and Schuster,P. (1979). "The Hyperycle - A
   Principle of Biological Self-Organization". Springer,
   Berlin.
3. Eigen, M. and Schuster,P.(1982). Stages of emerging life
   - five principles of early organization. *J.Mol.Evol.*
   19, 47-61.
4. Schuster, P. (1981). Prebiotic evolution. *In* "Biochemical Evolution" (Ed. H.Gutfreund). pp.15-87. Cambridge
   Univ.Press., Cambridge, U.K.
5. Schuster, P. (1981). Selection and evolution in molecular systems. *In* "Nonlinear Phenomena in Physics and
   Biology" (Eds. R.H.Enns, B.L.Jones, R.M.Miura and S.S.
   Rangnekar). pp.485-548. Plenum Press, New York.
6 Schuster, P., Sigmund, K. and Wolff, R. (1979). Dynamical systems under constant organization III. Cooperative
   and competitive behaviour of hypercyclces. *J.Diff.Equ.*
   32, 357-368.
7. Schuster, P., Sigmund, K. and Wolff, R. (1980). Mass
   action kinetics of selfreplication in flow reactors.
   J.Math.Anal.Appl. 78, 88-112.
8. Phillipson, P.E. and Schuster, P. (1983). Analytical
   solution of the coupled nonlinear rate equations. II.
   Kinetics of positive catalytic feedback loop. *J.Chem.
   Phys.* 79, 3807-3818.
9. Biebricher, C.K., Eigen, M. and Luce, R. (1981). Product analysis of RNA generated de novo by Qß replicase.
   *J.Mol.Biol.*148, 369-390.
10. Biebricher, C.K., Eigen, M. and Luce, R. (1981). Kinetic
    analysis of template-instructed and de novo RNA synthe-

sis by Qß replicase. *J.Mol.Biol.*148, 391-410.

11. Biebricher, C.K., Diekmann, S. and Luce, R. (1982). Structural analysis of self-replicating RNA synthesized by Qß replicase. *J.Mol.Biol.* 154, 629-648.

12. Biebricher, C.K., Eigen, M. and Gardiner, jr., W.C. (1983). Kinetics of RNA replication. *Biochemistry* 22 2544-2559.

13. Biebricher, C.K. (1983). Darwinian selection of self-replication RNA molecules. *In* "Evolutionary Biology" (Eds. M.K.Hechet, B.Wallace and G.T.Prance), Vol.16, pp.1-52. Plenum, New York.

14. Schuster, P. and Sigmund, K. (1983). From biological macromolecules to protocells - The principle of early evolution. *In* "Biophysics" (Eds. W.Hoppe, W.Lohmann, H.Markl and E.Ziegler) 2nd edn, pp.874-912. Springer, Berlin.

15. Hofbauer, J. and Schuster, P. (1984). Dynamics of linear and nonlinear autocatalysis and competition. *In* "Stochastic Phenomena And Chaotic Behaviour In Complex Systems" (ed.P.Schuster), Vol.21, pp.159-171, Springer, Berlin.

16. Schuster, P., Sigmund, K., Hofbauer, J. and Kemler, F. (1984). Energetic boundary conditions and optimization in evolution, submitted.

17. Thompson, C.J. and McBride, J.L. (1974). On Eigen's theory of the self-organization of matter and the evolution of biological macromolecules. *Math.Bioscience.* 21, 127-142.

18. Jones, B.L., Enns, R.H. and Rangnekar,S.S. (1976). On the theory of selection of coupled macromolecular systems. *Bull.Math.Biol.* 38, 15-28.

19. Swetina, J. and Schuster, P. (1982). Self-replication with errors. A model for polynucleotide replication. *Biophysical Chemistry.* 16, 329-345.

20. Jones, B.L. (1978). Some principles governing selection in self-reproducing macromolecular systems.*J.Math.Biol.* 6, 169-175.

21. Ewens, W.J. (1979). "Mathematical Population Genetics". Springer, Berlin.

22. Hofbauer, J., Schuster, P., Sigmund, K. and Wolff, R. (1980). Dynamical systems under constant organization II. Homogeneous growth functions of degree p=2. *SIAM J.Appl.Math.* C38, 282-304.

23. Zeeman, E.C. (1981). Dynamics of the evolution of animal conflicts. *J.Theor.Biol.* 89, 249-270.

24. Bomze, I.M., Schuster, P. and Sigmund, K.(1983). The role of Mendelian genetics in strategic models on animal behaviour. *J.Theor.Biol.*101, 19-38.

KINETIC LOGIC AS A FORMAL DESCRIPTION OF ASYNCHRONOUS AUTOMATA AND BIOLOGICAL MODELS.

R. Thomas.

Laboratoire de Génétique, Université libre de Bruxelles, Belgique.

1. INTRODUCTION

Essential qualitative aspects of complex systems can be kept in a *logical* description, i.e. a description in which variables can take only a limited number of values, typically two only (0 and 1) ; for biological systems see (1) to (10). Our method, known for some time as "kinetic logic", is described elsewhere (recent review (10)). It will be restated here, mostly to serve as an introduction to the papers by M. Kaufman and by J. Richelle, who use it, and also show its relation to automata.

2. RESTATEMENT OF KINETIC LOGIC, AS AN INTRODUCTION TO THE PAPER BY MARCELINE KAUFMAN

The classical way to introduce time into logical equations consists of relating the values of logical variables at time $t + 1$ to their values at time t :

$$\underline{y}_{t+1} = \Phi (\underline{y}_t) \qquad (1)$$

In this description, when vectors $\underline{y}_{t+1}$ and $\underline{y}_t$ differ by the values of two or more variables, these variables have to commute (i.e. change their values) simultaneously as the system proceeds from t to t+1 . This *synchronous* description is easy to handle. However, the need for simultaneous commutations appears very irrealistic in many systems ; in addition, each state has only one possible next state.

Instead, we associate with each internal component (i) of the system both a logical *variable* (y_i) aimed to describe the *level* (e.g. concentration) of the component and a logical *function* (Y_i) aimed to describe its *rate of production*, and we write :

$$\underline{Y} = \Phi (\underline{y}) \qquad (2)$$

DYNAMICAL SYSTEMS
AND CELLULAR AUTOMATA

269

Time is no more present in an explicit way in (2). However, since Y represents a *rate*, it has the dimensions of a time derivative of concentration ; time is thus present implicitly in equations (2), as it is in the ordinary differential equations used in chemical kinetics.

The precise relation between our logical variables and functions can be best understood from a concrete case. Consider a set of genes, each of which directs the synthesis of a product. We say that gene $\underline{i}$ is "on" (Y_i=1) when it synthesizes its product at a significant rate, "off" otherwise ; and we say that the corresponding gene product is "present" (y_i=1) when its concentration exceeds a characteristic threshold (θ_i) , "absent" (y_i=0) otherwise. These genes interact in the sense that each gene is on or off depending on which products are present or absent. When a gene has been off (Y_i=0) for a sufficient time, its product, which is perishable, is absent (y_i=0). If now the gene is switched on (Y_i=1) the product will appear (y_i=1) but only after a certain delay t_{yi} . As long as the gene remains on , the product will remain present. If the gene is switched off, the product will disappear, but only after a delay $t_{\overline{yi}}$. In other words, when the logical values of a function and its variable are the same they have no reason to change (state of regime). But when they are different the value of the function behaves as an *order* to change the value of the variable and this order is *executed* after a characteristic time delay. The delay is usually different depending on whether the function has been switched on or off.

We would like to stress that in this *kinetic logic* description vector $\underline{\Phi}$ ($\underline{y}$) does no more define the next state of $\underline{y}$. Rather, comparison of $\underline{\phi}$ ($\underline{y}$) with $\underline{y}$ tells which variables have an order to commute ; but usually only one commutation will take place at a time, depending on the values of time delays.

More generally, the relation between our logical variables and functions is the following. On the one hand the value of each function is at any time a function of the values of variables at that very time, as seen in equations (2). On the other hand, the value of each variable is a deferred image of the value of its function . This relation between our functions and variables may seen somewhat involved, as the value of each function depends in an instantaneous way on the values of variables, while in turn the value of each variable can be computed from an earlier value of the corresponding function. In fact, this circular relationship is not very different from that of a time derivative which depends on variables whose value in turn depend on the value of their time derivative during the preceeding period.

Strictly speaking , the variables and functions described
so far should be called *internal* variables and functions ;
there are in addition *input* variables, whose value can in
principle be decided at will from outside the system (change
the temperature, add a product ...).

One important precision : in our description, the order
to change the value of a variable is executed after a characteristic time delay *unless a counter-order*(i.e. a new change
of the value of the function) *has taken place before comple-
tion of the delay.*   More precisely, we reason in this case
as if the order was merely cancelled, as if it had never
taken place   (but in the absence of a counter-order, an
order may persist for a number of logical states if the corresponding delay is long enough).   This simplifying assumption represents a choice (inertial delay) between various possibilities, which have been analyzed by Van Ham in "Modèles
logiques à action différée" (11,12).It results in a *selective
loss of memory,* which may be objectionable in some particular
cases but has a decisive advantage; it greatly simplifies the
iterative process which permits to compute the sequence
of logical states from the logical equations and values of
the delays (this computation is the logical equivalent of the
integration which provides the trajectory in continuous
systems).   A *a posteriori* justification for using this
simplifying assumption is that it provides a simple relation
between the "logical behaviour" and the behaviour predicted
by homologous systems of ordinary differential equations describing the same logical structure.

Another possibility, already considered by Van Ham (11)
and recently developed by Gill* (13) assumes that whenever
an order has been given it will be executed after the proper delay.   This assumption may lead to exceedingly complex
behaviours even for rather simple logical structures, because
at a given time a given variable may be subject to a whole
string of orders to commute.

The system we are dealing with can be described by *"graphs
of interactions"* (or "logical structures") which are oriented,
signed graphs completed by mentions concerning the connections
(and, inclusive or) between the interactions. In what follows, we will use variables a, b, c ... and functions A, B,
C, ... rather than y_i and Y_i , in order to avoid indices.
For instance, the interaction graph of Fig.1 describes a three-
component system whose elements interact as a positive loop
(i.e., a loop with an even number of negative interactions).
Component a is inactivated at high temperature, hence the
introduction of an input variable T.

* with, however, the limitation of equal "on" and "off" delays.

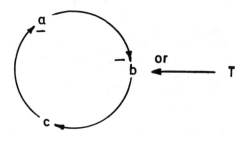

Fig.1

Product b is synthesized iff product *a* is absent *or* temperature is high (thus inactivating product a), etc ...   This system can be described by the *logical equations* :

$$A = \bar{c}$$
$$B = \bar{a} + T \qquad (3)$$
$$C = b$$

In a n elements system, the state of the (internal) variables and the state of the functions are represented each by a string of n binary digits which can be treated as a vector. From the logical equations, one can derive a *states table*, which gives the value of the vector of functions for each of the 2^n possible values of the vector of internal variables. If in addition m input variables are involved, one uses as entries :
- the 2^n_m values of the vector of internal variables (*rows*)
- the 2^m combinations of values of the input variables (*columns*)

Table 1 gives the states table for the system described in Fig.1.

One sees, for instance, that at low temperature (T = 0), when the vector of variables is 000 the vector of functions is 110 .   Thus, variables a and b disagree with their function ;   both products are absent but as the corresponding genes are on there is an order to change the values of a and b.

TABLE 1

a b c	T = 0 A B C	T = 1 A B C
0 0 0	1 1 0	1 1 0
0 0 1	0 1 0	0 1 0
0 1 1	0 1 1	0 1 1
0 1 0	1 1 1	1 1 1
1 1 0	1 0 1	1 1 1
1 1 1	0 0 1	0 1 1
1 0 1	0 0 0	0 1 0
1 0 0	1 0 0	1 1 0

Complete states table for the system described by Fig.1 and equations (3).

Instead of describing this situation by 000/110 (the vector of variables followed by the corresponding vector of functions) one can write in a more compact way 000; the dashes show which variables disagree with their function and have thus an order to change their value. From Table 1 we derived a *compact states table* (Table 2), which provides for each value of the input variable (columns) a list of the combinations of values of the internal variables completed by the appropriated dashes. One sees that most states carry dashes and are thus transient states. Two states in the first column (low temperature) and one state in the second column(high temperature) have no dash , thus no order to change the value of any variable : these are *stable states* (circled in the Table).

Note that within a column the system proceeds *spontaneously* from a transient state to another transient state ,or to a stable state where it is blocked. To leave a stable state, one has to change an input variable ; this is usually a voluntary change, seen as a column shift.

TABLE 2

T = 0	T = 1
a b c	a b c
$\bar{0}$ $\bar{0}$ 0	$\bar{0}$ $\bar{0}$ 0
0 $\bar{0}$ $\bar{1}$	0 $\bar{0}$ $\bar{1}$
(0 1 1)	(0 1 1)
$\bar{0}$ 1 $\bar{0}$	$\bar{0}$ 1 $\bar{0}$
1 $\bar{1}$ $\bar{0}$	1 1 $\bar{0}$
$\bar{1}$ $\bar{1}$ 1	$\bar{1}$ 1 1
$\bar{1}$ 0 $\bar{1}$	$\bar{1}$ 0 $\bar{1}$
(1 0 0)	1 $\bar{0}$ 0

The compact states table for the same system.

For instance, let us start from state 100 and shift to high temperature (shift from the left to the right column of Table 2). The sequence of states is :

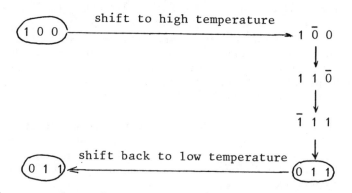

Note the occurence of two stable states at low temperature ; multistationarity , a characteristic potentiality of systems comprising positive feedback loops, will be found in concrete situations in the paper by M. Kaufman.

Note also the hysteresis   if one starts from state $\widehat{100}$,
switches to high temperature, remains for a sufficient time
at high temperature, then switches back to low temperature,
the system is trapped in the other stable state $\widehat{011}$.
    Let us now start from state $0\bar{0}0$ at low temperature.
The next state may be $\widehat{100}$ or $0\bar{1}0$ *:

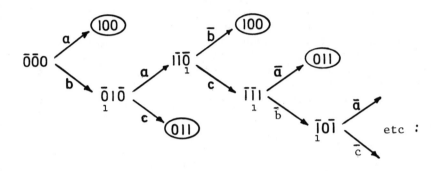

What determines which sequence of states (pathway) will be
followed ? Clearly, from state $\bar{0}00$ the system will proceed to 100 or to $0\bar{1}0$ depending on whether a or b reaches first its threshold value, in other words, on whether
or not the time delay $t_a < t_b$. From state $0\bar{1}0$, the
situation is slightly more complicated, as the order to synthesize c has just been given at the onset of state $0\bar{1}0$
while the order to synthesize a was already present from
the preceeding step** ; thus, the question is whether or not
$t_a < t_b + t_c$ etc ...

_____

* The possibility that variables a and b commute simultaneously ($\bar{0}00 \Longrightarrow 1\bar{1}0$) is not considered explicitly.
Should it take place, it would be taken into account automatically in simulations.

** In such cases, we use subscript digits which tell how
many steps ahead an order has been given : $0\bar{1}0_1$

3. THE RELATION TO AUTOMATA.

Our formalism, first described in Thomas (5) and Thomas
& Van Ham (6) had received a decisive impulse from the ideas
of Huffman(15) and Florine (16). In turn,it has provided a
generalized way to describe asynchronous sequential automata
(see Van Ham (11, 12) ; Milgram (16).

Sequential automata are logical systems whose outputs
depend not only on the present values, but also on earlier
values of the inputs. In order to deal with these situations
Huffman(14) and Florine (15) used, in addition to the
input variables (x) and output functions (Z), *auxiliary* (or
internal) functions (Y_i) and variables (y_i) which serve
for memorization. This allows them to write equations of the
type

$$Y_i = f_i \ (x_1, \ \ldots \ x_m \ ; \ y_1, \ \ldots \ y_n)$$

in which time is no more seen in an explicit way,but Y_i
and y_i are related by a time shift Δt :

$$(y_i)_t = (Y_i)_{t+\Delta t}$$

The great interest of this formalism is that a sequential
system using m input variables is formally reduced to a
combinational system comprising m input and n internal variables.

One often finds situations where the values of two or
more internal functions differ from those of the corresponding variables. In other words, two or more variables
have received an order to change their value . One attitude consists of reasoning as if these orders should be obeyed
in exact simultaneity. However, in materialized automata, unless one uses an *ad hoc* technology, one of the variables will
change its value first, depending on minute, ill-controlled
differences between the electronic organs of the logical machine. In this case, one speaks of a "race" between the variables. Often, the final state reached by the system will depend on the result of such a race; in this case one speaks of
a "critical race".Clearly,in a real sequential machine, the
critical races must be avoided by all means, as otherwise
the system will behave in unpredictable ways. In practice,
one devises the auxiliary variables and functions in such
a way that they will not produce critical races. Alternatively, one can deliberately treat the system as asynchronous.

The formalism just described allows an asynchronous treatment but, initially at least, it was used mostly to avoid critical races ; only more recently was the asynchronous treatment developed per se : Van Ham (11), Milgram (16).

For a biologist, the preoccupations are very different, because instead of trying to conceive (synthesize) a reliable logical machine, we try to analyze existing machines (biological systems). On the one hand, we cannot adopt a synchronous description because even if several biological orders (e.g. switching on or off two or more genes) have been given at the same time, their execution has no reason to take place simultaneously. On the other hand, we need a formalism such that some states may have two or more possible next states (depending on the values of parameters). Thus we adopted the notion of coupled internal functions (Y_i) and variables (y_i). But instead of being abstract auxiliary devices, these internal functions and variables are aimed to represent the situation of the relevant components of the system to be described. Accordingly, critical races are no more technical difficulties which have to be eliminated. Rather, they become essential features of the description, because many biological systems actually have a choice between two or more next states depending on the values of parameters.

The main novelty probably resides in providing the time delays with a parametric significance ; concretely, our delays represent the time between an order (change of the value of a function) and its execution (change of the value of the corresponding variable). This results in a new type of asynchronous automata whose behaviour depends not only on the sequence of input states, but also (and in many cases exclusively) on the values of the delays. Note also that since our delays can take any value, *time is continuous* in this description.

As briefly mentioned, one can use various structures of delays (cf. Van Ham (11)) but so far we use always inertial delays. Apart from that, one can use them in various ways:
- in the simplest version we ascribe to each delay a defined value. As the delays are usually different from each other, our description suits for asynchronous automata*
This description may be compared with the determinist continuous description with well-defined values of the parameters ; the future of the system is entirely determined

---

* In fact, even if the delays were all equal, we might have an asynchronous behaviour, provided one or more delay is already partly elapsed at the initial state.

by the initial state.

– even in a population of supposedly identical cells (e.g. some bacterial populations), it seems unreasonable to ascribe a defined value to each time delay. Rather, one is tempted to ascribe it *an average value and a distribution*. This description provides *probabilistic automata* ; see for instance Van Ham (12,) Thomas (9).

– as mentioned repeatedly, the most interesting attitude would consist of using delays whose value (or, in the probabilistic version, average value) would be a function of the state of the system. As our description of time is continuous, the values of the delays might be continuous functions of the "total" state (see the last § of section 4) of the system.

To close this section, I would like to recall the possibility to use our method in a *synthetic* (or *inductive*) way, i.e. proceed from an observed or desired behaviour towards the simplest logical structures which could produce this behaviour. Another possibility, poorly explored so far, consists of building hierarchized structures in which one would use as input variables of a system internal variables of a "metasystem" (a clear formulation of this idea was greatly helped by reading (19),in a distant field) ; here, the very logical structure of the initial system may change depending on the state of the metasystem (see also (20)).

4. INTRODUCTION TO THE PAPER BY JEAN RICHELLE.

Relation between our logical equations and differential equations.

The differential equations most commonly used to describe regulatory processes are of the type :

$$\dot{y}_i = f_i (y_1 , \ldots y_n) - k_{-i} y_i$$

(see for instance ref (18) and (4) in which f_i describes the rate of synthesis of y_i and $k_{-i} y_i$ is a term of decay. When we describe the same process by the logical equation :

$$Y_i = \Phi_i (y_1 , \ldots y_n)$$

the term Φ_i is the logical description of the conditions of synthesis ; it corresponds to the term f_i of the differential equation. Thus, the logical function Y_i does not correspond to the time derivative $\dot{y}_i$ but rather to $\dot{y}_i + k_{-i} y_i$.

When we translate a logical equation into a "homologous" differential equation, we translate the logical function Φ_i into a continuous function f_i. Insofar as Φ_i is a sum of products of the variables or their complement, we use as f_i a sum of products of increasing or decreasing sigmoid functions of the variables. For instance, the logical expression

$$y_1 \cdot \bar{y}_2 + y_3$$

is translated into the continuous expression

$$F^+ (y_1) \cdot F^- (y_2) + F^+ (y_3)$$

in which F^+ and F^- are, respectively, increasing and decreasing sigmoids. Thus, the logical equation :

$$Y = y_1 \cdot \bar{y}_2 + y_3$$

generates the differential equation

$$\dot{y}_i = F^+ (y_1) \cdot F^- (y_2) + F^+ (y_3) - k_{-i} \, y_i$$

The use of sigmoids to describe regulatory interactions (18,4) is justified by the fact that such biological processes as rates of gene expression usually depend on the concentration of regulators in a non-linear way ; in particular, these rates have an upper bound, if only because the supply of building blocks (nucleotides, aminoacids) is limited. In addition (4), sigmoid functions such as $\dfrac{x^n}{\theta^n + x^n}$ tend to the step functions used in the logical description if one uses high value for the Hill number n .

Delays versus parameters

In the continuous description, evolution is entirely
determined by the *initial state* and the values of the *parameters* present in the differential equations. At first
view, one could tell that the situation is entirely compa-
rable in the logical system, in which any logical state can
be arbitrarily chosen as the *initial state,* and in which
the *time delays* play the role of the parameters*. However,
the comparison is not that simple, because logical states,
and in particular, the initial logical state, are not punc-
tual ; in a system with two continuous variables y_1 and y_2,
the logical state 00 corresponds to a box of the space of
variables limited by the axes and the threshold values
θ_{y_1} and θ_{y_2} .

More generally, the logical state of the system only
tells us in what box of the space of variables the system
is (and towards which box it is proceeding). In many
cases, and in particular when we want to compute the sequence of logical states or to perform a stability analysis,
we have to be more precise. The way to do it consists of
expressing (in addition to the logical states) what fraction
of the running time delays is evolved. This is made concrete in simulations by associating with each relevant delay
a counter**.

It is essential to carefully distinguish the *values* of
the time delays and the *states of the counters.* If one
tries to compare from that viewpoint the continuous and the
logical description, one could say that the *parameters* of
the continuous description are represented in the logical
description by the *time delays* and that the state
in the space of variables (and, in particular, the initial
state chosen) is represented by the *logical state, completed
by the state of the counters ("total" state).* This aspect is
treated in a new perspective by Jean Richelle.

---

* For an estimation of the values of the time days of the
logical description from the values of the parameters of a
continuous description, and *vice versa* see Thomas (17).

** The use (see p.7, second footnote) of subscripts mentio-
ning how many logical states ahead an order has been given,
is useful but less precise.

References

1. Rashevsky, N. (1948). "Mathematical Biophysics". University of Chicago Press.
2. Sugita, M. (1963). Functional analysis of chemical systems *in vivo* using a logical circuit equivalent. II. The idea of a molecular automaton, J. *Theor. Biol.* 4, 179-192.
3. Kaufman, S.A. (1969). Metabolic stability and epigenesis in randomly constructed genetic nets, J. *Theor. Biol.* 22, 437-467.
4. Glass, L. and Kauffman, S.A. (1972). Co-operative components, spatial localization and oscillatory cellular dynamics, J. *Theor. Biol.* 34, 219-237.
5. Thomas, R. (1973). Boolean formalization of genetic control circuits, J. *Theor. Biol.* 42, 565-583.
6. Thomas, R. and Van Ham, p. (1974). Analyse formelle de circuits de régulation génétique : le contrôle de l'immunité chez les bactériophages lambdoïdes, *Biochimie* 56, 1529-1547.
7. Glass, L. (1975). Classification of biological networks by their qualitative dynamics, J. *Theor. Biol.* 54, 85-107.
8. Thomas, R. (1978). Logical analysis of systems comprising feedback loops, J. *Theor. Biol.* 73, 631-656.
9. Thomas, R. Ed. (1979). "Kinetic logic : A boolean approach to the analysis of complex regulatory systems". Lecture Notes in Biomathematics 29, Springer-Verlag, Berlin, 507 pp.
10. Thomas, R. (1983). Logical description, analysis and synthesis of biological and other networks comprising feedback loops, *Adv. Chem. Phys.* 55, 247-282.
11. Van Ham, P. (1975). Modèles discrets à action différée, Thesis, University of Brussels.
12. Van Ham, P. (1979). Delphin : a logical machine with incrementable phase delays. *In* "Kinetic Logic" Lecture Notes in Biomathematics, 29, 143-148.
13. Ghil, M. (1984). Boolean difference equations, I : Formulation and dynamic behavior, *Siam J. Appl. Math.* (*in press*).
14. Huffman, D.A. (1954). The synthesis of sequential switching circuits, J. *of the Franklin-Institute* 257, 161-189.
15. Florine, J. (1964). "La Synthèse des Machines Logiques et son Automatisation". Dunod, Paris.
16. Milgram, M. (1982). Contribution aux réseaux d'automates. Thèse d'Etat, Compiègne, France.
17. Thomas, R. (1984). Logical vs continuous description of systems comprising feedback loops : the relation between time delays and parameters, Proc. Symp. "Chemical

applications of Topology and Graph Theory", Athens, Georgia (1983). Elsevier, in press.

18. Griffith, J.S. (1968). Mathematics of cellular control processes, *J. Theor. Biol.* 20, 202-216.

19. Watslawikc, P., Weakland, J. and Fish, R. (1975). "Changements, Paradoxes et Psychothérapie". Le Seuil, Paris.

20. Van Ham, P. (1984). Modelling by asynchronous sequential variables. This symposium.

MODELLING BY ASYNCHRONOUS SEQUENTIAL VARIABLES

Ph. Van Ham

University of Brussels
Brussels, Belgium

1. Introduction

During the last ten years, it has been shown that asynchronous boolean sequential machine analogy is a fruitful method
to represent complex regulatory systems. (Thomas 73, Thomas
and Van Ham 74, Van Ham 75, Richelle 77, R. Thomas et al. 79,
Leclerq & Thomas 81).

Such a modelization approach uses principaly asynchronously
time-delayed boolean variables and functions to mimic qualitatively the behaviour of physical entities.
However, in some cases, a same physical entity may display
more than two different regulatory effects. Therefore multivalued variables and functions have to be used instead of
the binary ones.
When the multivalued variables can be related to an ordered
set, then transition rules must be defined to avoid jumps
between nonadjacent values. (Van Ham 79).
The aim of this paper is to extend the multivalued modelization to "non-numerical" variables.
In this "sequential variables" approach and on the contrary
of multivalued algebras, I tried to preserve the "user-
friendly" aspects of the classifical binary modelization.
A general expression of asynchronous discrete models may be
given by the following equations :

(Change of variable i) = Function (Present values of va-
     at time t   riables 1, 2, ... n at time t)

(Value of variable i) = Operator Delay (History of the
     at time t   changes of i between $t-dt_i$ and
                            t)

The "delay operator" does not work as a shift in time but
acts as an integrator wich accumulates the changes of a variable during a time interval specific with respect to the

variable and the excited transition.
However, for the sake of simplicity, we shall use the so-called
inertial delays for which a transition occurs only if the
change of the variable remains the same at the input of the
delay operator during a time interval called shortly "delay".
Otherwise, the transition is simply aborted and the delay
operator is reseted. (Van Ham 1975).
In this paper, we make the assumption that the models we want
to study are "input-free" (Van Ham 82), that is, the external
variables are kept constant during the evolution of the system. For each external state (external variables configuration), a separate study should be made.
Figure 1 shows a schema of an asynchronous discrete system.
It is easy to imagine situations where more than one transition are in preparation in the different delays (races). The
values of the delays will then be used to compute the first
change in a variable value. In the resulting next state of
the system some transitions will be confirmed, started or
aborted.
A state for which no transition is requested (old ones aborted and no new ones) will be called "stable state".

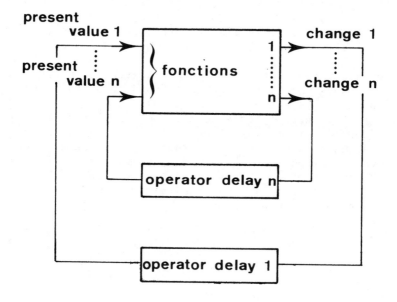

Figure 1. A discrete asynchronous automaton.

2. Sequential variables

For many systems, it should be interesting to be able to
handle completely non-numerical entities (computer protocols,
psychology, etc.). In systems where the variables are related to multi-treshold phenomena, the modeller may assign for
each descriptive variable, a finite set of consecutive integers (Van Ham 79 a, 80). No attempt is made to represent any
quantitative aspect except the fact that no jump may occur in
any evolution between non-adjacent values.
Our aim here is to handle variables which does not correspond
to an ordered set (color, behavioural characteristics, etc.).
Figure 2 shows that the passage from a classical multivalued
numerical variable to a non-numeric one corresponds to a simple generalization of the transition matrix associated with the
variable. One can see that, for the sequential variable
graph, there is no a priori structure.
The similarity between such a graph and the state diagram of
a sequential machine is the reason why I called such variables : "sequential variables".

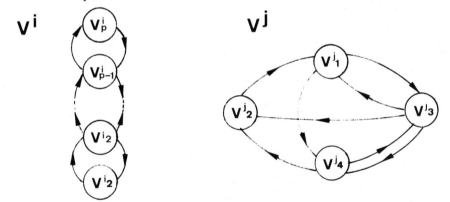

Figure 2. *Transition graphs for numerical (v^i) and non-*
numerical (v^j) variables. Each arrow is associa-
ted with a specific delay.

To build a model with sequential variables, it is necessary
first to define a graph for each variable.
One may see here that such a system is an assembly of n interacting sequential machines performing n parallel evolutions in an asynchronous manner, that is, with races between
several transitions.
Next the modeller has to formalize the interactions between
variables mapping $v^1x \ldots v^n$ on itself. This mapping genera-

tes the "next superstate function" where a superstate is defined as a n-uplet of sequential variable values (states). The "verbal expressions", currently used by the various modellers in order to define the mapping are sentences like : "that variable executes that transition if this one is (or is not) in this state and if ... etc." That is the reason why we prefer to write a particular transition table for each variable.

3. Transition tables

Let us call $v_1^i \ v_2^i \ \ldots \ v_{p_i}^i$ the p_i values of variable v^i.
Let us define the following boolean function :

$$\text{if} \quad (v^i = v_j^i) \qquad ; \quad \text{then} \quad y_j^i = 1$$

$$\qquad\qquad\qquad\qquad ; \quad \text{else} \quad y_j^i = 0$$

It is now easy to write any transition function T_{rs}^i of v^i between v_r^i and v_s^i as a boolean function of the y_p^q 's.
For example, if we say that the transition of v^i from v_r^i to v_s^i is excited if and only if $(v^k = v_1^k)$ AND $(v^n = v_t^n)$ OR $(v^a \neq v_b^a)$, we may simply write $T_{rs}^i = y_1^k \cdot y_t^n + \overline{y}_b^a$.

So the modeller can build transition tables ($p_i \times p_i$ matrices) where :

- if the transition depends on the system superstate, he will write the corresponding function (T_{rs}^i)
- if the transition is unconditional, he will write a "1"
- if the transition do not exist, he will write a "0".

4. Delay tables

We must now, to preserve the asynchronous charachter of our system, associate a "time delay" with each non-zero entry of the transition tables (section 1). If experimental data are sufficient to give a delay for each transition, the system becomes deterministic. However, if a set (eventually all) of delays are only known by their distribution, there are different ways to compute the system behaviour.

First one may assume that each delay is constant during each particular behaviour of the system from a given initial su-perstate. It is then possible to accumulate a lot of such evolutions and to make an histogram of the final observed be-haviour (a "final behaviour" may be a stable superstate or a stable cycle of superstates). An other way to proceede con-sists in the "refreshing" of the randomized values of the de-lays each time the associated transition is excited by a T^i_{rs}. It must be pointed out that such system are not immediately similar to the classical probabilistic automata. In fact the transition probabilities between superstates are not computed and no care is made to know if the sum of those probabilities from each superstate, is equal to 1.

Finaly, it is possible, as in the Boolean Asynchronous models case, to consider that the delays are some function of the superstate itself or other external parameters (Van Ham 75, 79 b).

5. Practical aspects

It is obvious that, for a somewhat complicated model, the number of superstates will prevent the modeller to make any use of a "paper and pencil" method as in the Boolean case. Our formalism makes it easy to define the sequential variables graphs and their instantaneous interactions. To compute the various behaviours of the system, it is necessary to use computerized algorithms similar to those already written for the Boolean models (Van Ham 75, Verhamme and Van Ham 81) :

*   - TABET, V.S.   Computes an equivalent of the state table (the superstate table) and indicates the stable superstates. Only the T^i_{rs} are given.

     - PRF01, V.S.   Computes superstates sequences from a given initial superstate. The T^i_{rs} 's and all the delays are given.

     - PRAN, V.S.   Computes the "final behaviours" and their occurence probability from a given initial superstate. The T^i_{rs} 's and the delays dis-tributions are given.

---

* These programs, written by Didier de Formanoir de la Case-rie, are available at the Logical & Numerical Department, C.P. 165, University of Brussels.

6. A (very) simple example

Let us make a model of the interactions between three objects : one central processor (1) and two peripherals (2 and
3). A very simplified vew about the interactions of these
devices needs only three sequential variables. We shall assume that the communication is not a polling procedure but
completely asynchronous.
Let us assume that the variable v^1 describing the central
processor may take four values :

- v_1^1 (the central processor is in its "stand-by" state) ;
- v_2^1 (the processor requests a link with <u>a</u> peripheral) ;
- v_3^1 (the processor is linked with peripheral v_2) ;
- v_4^1 (the processor is linked with peripheral v_3).

On the other side, the peripherals are described each by a
three-valued variable (v^2 and v^3) :

- v_1^2 (or v_1^3) (the peripheral is in its "stand-by" state) ;
- v_2^2 (or v_2^3) (the peripheral is linked with the processor) ;
- v_3^2 (or v_3^3) (the peripheral is performing a clearing
  procedure).

One may see on figure 3 the associated graphs and the modelled transitions.

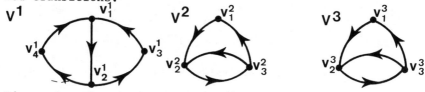

Figure 3. The transitions graphs of the three devices.

Following the definitions mentioned in section 3, we may model the T_{rs}^1 transition tables, that is, the rules of the game.

T_{rs}^1	v_1^1	v_2^1	v_3^1	v_4^1	to s
v_1^1	0	$y_2^{-2}+y_2^{-3}$	0	0	
v_2^1	0	0	y_2^{-2}	y_2^{-3}	
v_3^1	y_3^2	0	0	0	
v_4^1	y_3^3	0	0	0	

from r

T^2_{rs}	v^2_1	v^2_2	v^2_3
v^2_1	0	y^1_3	0
v^2_2	0	0	1
v^2_3	y^1_1	$y^1_2\,\bar{y}^3_2$	0

T^3_{rs}	v^3_1	v^3_2	v^3_3
v^3_1	0	y^1_4	0
v^3_2	0	0	1
v^3_3	y^1_1	$y^1_2\,\bar{y}^2_2$	0

One may see in table of T^1_{rs} that the central processor cannot change its linked peripheral without passing througt the "stand-by" state.

On the other hand, we have consider that the peripherals are busy during a specified time interval. Then, they return unconditionnaly to their "clearing state".

If we don't make any assumption about the values of the delays, one may be interested by the complete set of behaviours logicaly generated from a given initial superstate (for example : $v^1v^2v^3 = v^1_1v^2_1v^3_1$ where the three devices are in standingby).

This set is given by another graph in which each node corresponds to a superstate and is represented by a box containing two informations :

- the values of the sequential variables constituting the superstate ;

- the images of each value ("changes") computed from the transition tables.

The image(s) ("changes") are indicated as suscripts of the present values of the sequential variables.

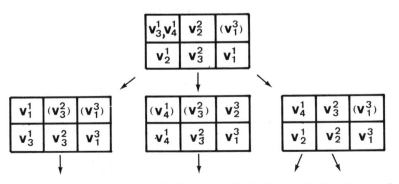

Figure 4. Superstates and "changes" indicated in the nodes of a little part of the "behaviours graph".

Figure 4 shows that races may not only occur between transitions of different variables but also between non-exclusive transitions inside a same sequential variable.

7. Conclusions

The Asynchronous Sequential Boolean Machine (ASBM) used as a formalism to represent complex regulatory systems may be generalized to the case where variables are multi-valued and non-numerical : Asynchronous Sequential Variables Machine (ASVM). The "delay operator" concept associated with the transitions of a variable is conserved.
A A S V M may be seen as an assembly of interacting automata where each of them describes a particular sequential variable.

8. Acknowledgments

Thanks to D. De Formanoir for many fruitfull discussions and to A. Verhamme for his help in writing this paper.

9. References

1   R. THOMAS, 1973, J. Theor. Biol. 42, 563-585.

2   R. THOMAS & Ph. VAN HAM, 1974, Biochimie 56, 1529-1547.

3   R. THOMAS, Ed. 1979, "Kinetic Logic", Lectures notes in Biomathematics, vol. 29.

4   Ph. VAN HAM, 1975, Thèse de Doctorat en Sciences, Université libre de Bruxelles.

5   J. LECLERCQ & R. THOMAS, 1981, Bull. Classe des Sciences, Acad. Roy. Belgique, 5e série, Tome LXVII, 190-225.

6   Ph. VAN HAM, 1982, Bull. Classe des Sciences, Acad. Roy. Belgique, 5e série, Tome LXVIII, 267-294.

7   A. VERHAMME, 1979, Mémoire de Licence, Université libre de Bruxelles.

8   A. DE PALMA, 1979, in 3, chap. XVIII, 402-439.

9   A. VERHAMME & Ph. VAN HAM, 1981, Numerical Methods in
     the Study of Critical Phenomena, chap. 3, 194-210.
     Springer Verlag, Ed. J. Della Dora, J. Demongeot &
     B. Lacolle.

10   J. RICHELLE, 1977, Bull. Classe des Sciences, Acad.
     Roy. Belgique, 5e série, Tome LXIII, 534-546.

11   Ph. VAN HAM, 1979, in Thomas ed. "Kinetic Logic", Lec-
     tures Notes in Biomathematics, vol. 29, ch. XV, 326-
     345.

MODELLING NATURAL SYSTEMS WITH NETWORKS OF AUTOMATA :
THE SEARCH FOR GENERIC BEHAVIOURS

G. Weisbuch

*Groupe de Physique des Solides de l'Ecole Normale Supérieure
24 rue Lhomond, 75231 Paris Cedex 05, France[+]*

This paper has two parts : a review of several attempts to model natural systems using ideas from spin glass theory and my own contribution to the subject, concerning the natural evolution of species.

During the seventies physics of condensed matter has progressed in two directions of interest to us. First in understanding the universal character of scaling laws observed in phase transitions, with the help of renormalization theory (1). Second with the building of simple systems in order to model disordered structures. The spin glass model (2) for instance, exhibits as generic dynamical properties the existence of a large number of different stable configurations. These properties of diversity and stability are typical of natural systems.

ASSOCIATIVE MEMORIES

The Hopfield's model (3) for associative memories is a good illustration of this approach. Hopfield considers a network of threshold automata modelling neurones. The state V_i of automaton i changes in time according to

$$V_i = 1 \quad \text{if} \quad \sum_{ij} W_{ij} \, V_j \geq U_i$$
$$V_i = 0 \quad \text{otherwise} \tag{1}$$

where W_{ij} is the strength of the connection between automata i and j, and U_i is the threshold of automaton i. If the W_{ij} matrix is symetric, starting from any initial configuration of states, several sequential iterations of Eq.(1) drive the network into a stable configuration, or, occasionnally, into

[+]*permanent address : Département de Physique, Case 901,
Faculté des Sciences de Luminy, 13288 Marseille Cedex 9.*

DYNAMICAL SYSTEMS
AND CELLULAR AUTOMATA

293

a limit cycle of periode 2. By a suitable choice for the W_{ij}, the system can be made to reach a selected set of stable configurations. These configurations can be considered as the binary representations of the items to be memorized. When the initial configuration differs from a stable configuration by only a few bits, iterations drive the system into the later. This system can thus be interpreted as an implementation of a content addressable memory, for which any item composed of a series of bits can be retrieved from a partial knowledge of the series.

The rule of construction of the W_{ij} is

$$W_{ij} = \sum_{s} (2V_i^s - 1)(2V_j^s - 1) \qquad (2)$$

where the summation is extended to all the states to be memorized $\{V^s\}$. The success of the iteration procedure is due to the existence of a decreasing energy function

$$E = -\frac{1}{2} \sum_{ij} W_{ij} V_i V_j$$

The performance announced by Hopfield for such a system are the possibility to retrieve up to 5 items with at most 5 errors in the initial state with a network of 30 automata.

PATTERN RECOGNITION

The system proposed by Hinton and Sejnowski (4) for pattern recognition has many similar features to that of Hopfield. They consider a network of threshold automata a proportion of which called retinal cells, receives external information in addition to input from other automata. Furthermore, their approach is probabilistic, with the probability for an automaton to be in the state 1 given by :

$$p(V_i = 1) = \left[1 + \exp - \left(\frac{\sum_{j} W_{ij} V_j + I_i - V_i}{T} \right) \right]^{-1} \qquad (3)$$

where I_i is the external input, which is non zero only for retinall cells, and T the temperature parameter explained further. Pattern recognition in this model is related to the fact that in the presence of a given external input on the retinal cells, the system evolves toward some stable configuration. The internal automata are interpreted as a set of a priori hypotheses that are recognized as true or false according to the external input. Once more the iteration process correspond to the decrease of an energy function

$$E = - \frac{1}{2} \sum_{ij} W_{ij} V_i V_j - \sum_i (I_i - U_i) V_i \qquad (4)$$

But a further refinement is added. In order to reach the absolute minimum for the energy (and not only one of the local minima) an annealing technique is applied, first proposed by Kirkpatrick et al. (5) for operationnal research problems. Parameter T can be varied. T = 0 corresponds for instance to a deterministic threshold automaton. Increasing T corresponds to increase the region where the probability function differ from 0 or 1. This non zero probability for processes increasing the energy function can drive the system out from local minima. The annealing process consists in starting iterations with large value of T and to carry them while the value of T is decreased until the system reaches the most stable configuration.

Although this model is only under preliminary tests, it already shows some interesting features.

- it establishes the equivalence between the Monte Carlo approach to search configurations of minimal energy for Hamiltonian of spin systems and the iteration of probabilistic threshold functions.

- the expression (3) given for the probability function is similar to the Bayesian inference probability

$$p(h/e) = \left[1 + \exp - \left(\log\frac{ph}{p\bar{h}} + \log\frac{p(e/h)}{p(e/\bar{h})} \right) \right]^{-1} \qquad (5)$$

where h is an hypothesis and e the experience. We can thus interprete V_i as an hypothesis the truth value of which depends of the other automata and upon inputs, thus fixing the summation terms in (3), and upon an a priori probability fixed by the threshold.

- a learning mechanism is proposed.

PREBIOTIC EVOLUTION

At a different time scale, the problem of evolution from a primeval soup containing only small organic molecules to the self replicating biopolymers shows similar properties of stability of the selected chemicals species and of diversity of the stable products. A number of mathematical models have been proposed, among which the most well-known is the hypercyle model of Eigen and Schuster (6). Anderson and Stein looking for a mathematical construction that exhibits both stability and *diversity*, have proposed the following algorithm. Their model supposes the existence of only one type of biopolymers, the polynucleotides, built from only two bases Guanine and Cytosine, corresponding to "spin" states $S = +1$ and -1.

- There exists in the primeval soup at time $T = 0$ a finite concentration of some oligonucleotides (small chains containing a few bases).

- A constant supply of mononucleotides is present.

- The reproduction mechanism may use any polynucleotide as a template allowing the concatenation of partially complementary chains such as in the following exemple :

- In the course of reproduction a small mutation rate changing S_i into $- S_i$ is allowed.

- The most important step in the model is the choice of a death probability, function of chemical composition of a chain, which limits the growth of concentration. Since our knowledge about the degradation rate of single stranded polynucleotide is too limited to calculate this function, Anderson proposes to look for a family of death functions that exhibits a dynamical behaviour similar to that of natural systems. More precisely, since we observe today a large variety of stable natural polynucleotides, Anderson proposes a death function which closely resemble the expression of the energy of spin glasses :

$$D_N (S_1..S_i..S_N) = \sum_{i \neq j}^{N} J_{ij}^{N} S_i S_j \qquad (6)$$

Excluding linear death functions, he claims that only such an expression, with long range interaction among spins, is able to generate the properties he is looking for. The first numerical simulations that were made do indeed exhibit non zero limit concentrations for a few chemical species.

EVOLUTION OF SPECIES

Present theories of natural selection do not allow a clear understanding of important phenomena like the apparition of new species and the existence of several possible dynamics, phyllogenic series in some cases, or punctuated equilibria in others.

Most models are based on population dynamics. The problem

is that natural selection can only be related to properties
of the phenotype (physical properties of the living organism),
while only genotypes are transmitted. Since nobody is able to
predict the phenotype from the genotype, most models are
built on the simplifying assumption that each phenotypic
character is determined by one simple gene. Such models can
only yield quasi-continuous changes in physical characters in
the presence of constant mutation rate.

On the opposite, we present here an attempt to take into
account the fact that phenotypic characters are determined as
we know by the interactions of a large number of genes. Our
starting point is the proposal by Kauffman (8) to model the
interactions among the genes by networks of Boolean automata.
Populations dynamics obeys a set of differential equations,
as usual in such models. The coefficients of these equations
are derived from the global properties of a set of several
genes. Numerical simulations exhibit as a generic property
of such systems behaviours that mimics natural evolution of
species.

Population dynamics

One of the most simple differential system that can be
proposed in order to describe the dynamics of populations of
organisms of genome i is the following :

$$\dot{P}_i = a_i\, P_i - b \sum_j P_j\, P_i + \sum_k m(t)\, P_k \quad , \quad \forall_i \qquad (7)$$

Population changes are due to three terms :
- a growth term $a_i\, P_i$, where a_i is the fitness corresponding
to genome i.
- a term which is the differential expression of a
constraint : all the organisms share a common ressource,
which in facts limits the total population.
- a stochastic term corresponding to mutations from
organisms of neighbouring genomes, in the sense that they
differ by only one gene. The mean value of $m(t)$ is m, which
is always taken as much smaller than any a_i.

This is of course a very simplistic model which does not
take sex into account and where interaction between species
is restricted to competition for a common resource.

The dynamics of such a system is quite simple if all
populations are non zero : pretty soon the population of the
organisms with the highest fitness becomes predominant. The
only non zero populations are those of neighbouring genomes
because of the finite mutation rate.

As a matter of fact, the problem of the evolution of

species corresponds to initial conditions such that few
populations are non zero, none of them having maximum fitness.
Because of mutations, new organisms appear, the population
of which depends upon their fitness, available ressources and
history which has allowed the existence of their neighbours.

At this point, we would like to establish some relation
between the fitnesses and the genome. Since we have no model
for the development of the organism, we shall only use the
fact that fitness like any other phenotypic character is a
global property of the genome. More precisely, we conjecture
that the qualitative properties of the dynamics obeyed by
the differential system are the same for a large class of
distribution of the fitnesses, such as the ones that are
obtained by using global properties of a system made of
interacting discrete components, similar to the genes.

Networks of Boolean automata as models for the genome

Using an idea of Kauffman, we model the set of interacting
genes by a network of automata. Every gene is represented by
a Boolean automaton. The interaction among genes due to the
fact that proteins coded by a gene may control by
repression or by cooperation the expression of other genes,
is represented by the interactions among the different
automata belonging to the network representing the genome.
At each discrete time interval, each automaton changes its
state according to the application of its Boolean law to its
inputs. This model described in a number of references (9,
10) is further illustrated by a simple example.

S. Kauffman has investigated the dynamical properties of
these networks, in particular the large time limit sets,
which are stable states or limit cycles. They are
interpreted as different cellular types that can be
expressed from the same genome. Of interest to us is the
possibility of computing global properties characteristic of
the whole network, like the periods of limit cycles and
their number.

The idea of our model is to investigate the dynamical
properties of the differential system when the fitness is
computed from some global property of a network composed of
six Boolean automata.

The simulated system

Let us start with the set of "genomes" composed of six
"genes" modelled by six automata, the interaction rules being
of four possible types : NAND (# 0), OR (# 1), AND (# 2) and

NOR (# 3). The connection structure is the following :

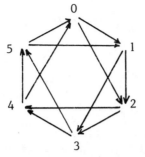

Automaton # 2 for instance receives two input signals from automata # 1 and # 0 at time T. According to its own Boolean law, it computes its output signal which is sent at time T + 1 towards automata # 3 and # 4. The total number of possible organisms is then 4^6 = 4096. Each network can have 2^6 = 64 different states, some of them belonging to different limit cycles. Let us take as the fitness coefficient a term proportional to the period of the longest limit cycle.

The only allowed mutations between genes correspond to an arrow in the following scheme :

$$3 \rightleftarrows 0 \rightleftarrows 1 \rightleftarrows 2 \rightleftarrows 3$$

Approximate analytical solutions

Population dynamics derived from the differential system (7) can be qualitatively described as follow. We are considering a lattice of all possible organisms on a six dimensional hypercube, and a population function on the hypercube. To simplify the choice of initial conditions, let us suppose that at the origin of time only one population is non zero. Because of mutations, the 12 neighbouring nodes are soon populated, and those of higher fitness become predominant. The narrow cloud of populations starts shifting towards the region of maximum fitness that can be reached from the starting genome by climbing the slope of increasing fitness. If the mutation rate is sufficiently high so that new favorable genomes appear before saturation of the population of the ancestor is reached, each time step between the apparition of a descendant in a series is given by :

$$\int m P_i(t)\, dt = 1$$

where i is the ancestor, P_i its population, i + 1 the descendant and Δt the time step. Because of the exponential increase of P_i far from saturation

$$\Delta t = [\log (a_i/m)] / a_i$$

For smaller m, saturation is reached before mutation and :

$$\Delta t = mb/a_i$$

In both cases the time intervall varies as $1/a_i$. Since the fitness increases along the evolutionary path, the time step decreases and the dynamical regime is accelerated evolution. This is reminiscent of the dynamics of phyllogenic series observed in natural evolution.

Around the node of maximum fitness, genetic polymorphism is maintained by mutations, and populations distribution has a minimum width. This narrow cloud is what we call a species. In this quasi-equilibrium state, the population of the different organisms can be obtained by recursion. Starting from the maximum, equating to zero the derivatives of populations, and neglecting in the mutation term at each step of the recursion populations far from the maximum with respect to those which are closer, we obtain :

$$P_m = a_m / b$$

where index m stands for maximum

$$P_1 = m P_m / (a_m - a_1)$$

where index 1 stands for first neighbour

$$P_i = P_m m^d \sum_C \prod_j (a_m - a_j)^{-1}$$

where the sum is extended along all paths C of length d going from the maximum to i. At a distance d from the maximum the population is decreased by a factor of the order of $m^d/\prod_j \Delta a_j$, where Δa_j is the difference in fitness with the maximum.

For small m, the width of the distribution is smaller than the distance between local maxima. The valleys are depleted : the probability of appearance of mutants is small and they rapidly extinct. As a consequence, the population cloud remains fixed on the local peak and one should wait very long before a new mutant of fitness higher than that of the local peak arises. The order of magnitude of this time is $(P_m m^{d+1})^{-1} \prod_j (a_m - a_j)$, where the product is evaluated along the saddle point joining the peak to the fittest mutant. After this event, the population cloud shifts rapidly towards the new maximum. A new species has appeared.

The dynamics we have just described is known in paleontology as punctuated equilibria and has been proposed by Gould and Eldredge (11,12). For long periods the population cloud remains fixed on a local peak of fitness. The set of neighbouring genomes belongs to a natural species. Organisms which genome is too different from that of the maximum appear for short periods but extinguish because of the competition for the common resource. The transition towards a new equilibrium is very fast, missing links appear in small number and even extinguish when the population cloud centers on the peak corresponding to the new dominating species.

This analysis is confirmed by the numerical simulations that we have done.

Numerical simulations

The first step is to compute the largest period of all 4096 genetic networks. When looking at the results, several secondary maxima can be observed which are at mutation distances of the order of 5. Let us simulate the displacement of a population cloud from one peak to the next one. Two peaks which are in the same hyperplane xy3200 can be taken as examples. The notation indicate that the Boolean rule for the first automaton is x, y for the second, 3 for the third... The three tables represent in the hyperplane the periods, the fitnesses, and the populations at the different stages of the simulation.

TABLE 1

Periods

x/y	2	3	0	1	2
2	7	1	4	1	7
3	10	1	3	8	10
0	4	4	11	9	4
1	1	1	1	1	1

Column number correspond to y, line number to x. Some rows are repeted in order to illustrate the vicinity of rules 0 and 3. We can then figure all possible mutation passways from genome 323200 of period 10 to genome 003200 of period 11.

Table 2

Fitness

x/y	2	3	0	1	2
2	0,35	0,05	0,2	0,05	0,35
3	0,5	0,05	0,15	0,4	0,5
0	0,2	0,2	0,55	0,45	0,2
1	0,05	0,05	0,05	0,05	0,05

Fitness coefficients in table 2 are computed by dividing the periods by 20, which fixes time scale.

Table 3

Populations

x/y	3	0	1	2	
3	8	0	11	186947	T=30
0	0	0	0	9	
3	222	0	998	4998446	T=90
0	0	0	0	333	
3	222	0	999	4998445	T=85998
0	0	1	1	333	
3	0	0	275	0	T=96658
0	314	5498312	1100	0	

The table 3 shows the result of a numerical simulation with b=10^{-7} and m=2 10^{-5}. At time T=0 all populations are 0, except that of genome 323200 wich is 1. At time=100, the quasi-equilibrium is reached, with populations close to the analytic expressions we have given. They remain unchanged until time 86988, when the first 003200 mutant appears. From this moment the cloud moves toward this mutant and the new equilibrium is reached within 660 time units.

Such punctuated equilibria are observed for small values of m such that the population of the neighbours of 003200 is less than 0 when the cloud is centered on 323200. During this simulations we have not taken into account mutations toward neighbouring hyperplanes.

This behaviour can be observed in several regions of the space of genomes for small values of m, depending upon the fitness geographical map. It is also observed if fitness is chosen according to an other global property like the number of different limit cycles : the map of fitness is not identical, but the general character of population dynamics are the same.

Scope of the model

As in the case of Anderson model, there is no direct relation between every step of our mathematical construction and definite biological facts. The idea is rather to show that from very general considerations on the fact that genotypic characters depend upon global properties of a set of genes, one readily obtains a simple differential system, with dynamical properties similar to behaviours observed in the natural evolution of species. An interesting result is that different dynamical regimes similar to phyllogenic series with accelerated evolution or punctuated equilibria can be the results of the same elementary evolutionary mechanisms. No catastrophy of any kind is needed in order to observe the fast changes between the long stasis in the case of punctuated equilibria.

CONCLUSIONS

The four models presented here are not achieved constructions. Rather, they are under extensive testing by numerical simulations. Although they differ in their field of application and their formalism :

deterministic	/	probabilistic automata
parallel	/	stochastic iterations
threshold	/	Boolean automata

they present common features such as :
- the search for dynamics allowing the existence of stable but numerous attractors.
- the importance of frustration to obtain such dynamics.
- the equivalence between seemingly different formalisms: Monte Carlo procedure to minimize energy of spin systems and probabilistic threshold automata.
- the use of energy functions.

REFERENCES

1. Toulouse, G. and Pfeuty, P. (1975). "Introduction au Groupe de Renormalisation et à ses applications". Presses Universitaires de Grenoble, France.
2. Balian, R., Maynard, R. and Toulouse, G. Editors, (1979) "Ill Condensed Matter", North Holland.
3. Hopfield, J.J. (1982). *Proc. Natl. Acad. Sci. USA* 79, 2554-2558.
4. Hinton, G.E. and Sejnowski, T.J. (1983). "Proc. of the IEEE Conference on Computer Vision and Pattern Recognition Recognition", 448-453.
5. Kirkpartick, S., Gelatt, C.D. Jr. and Vecchi, M.P. (1983). *"Science"* 220, 671-680.
6. Eigen, M. and Schuster, P. (1977). *Naturwissenschaften* 64, 541-565.
7. Anderson, P.W. (1983). *Proc. Natl. Acad. Sci. USA* 80, 3386-3390.
8. Kauffmann, S. (1969). *J. Theor. Biol.* 22, 437-467.
9. Kauffmann, S. (1970). *In* "Towards a Theoretical Biology" Vol.3, (Ed. C.H. Waddington). pp.18-37. Edinburgh University Press.
10. Fogelman-Soulié, F., Goles-Chacc, E. and Weisbuch, G. (1982). *Bull. Math. Biol.*, 44, 715-730.
11. Eldredge, N. (1971). *Evolution*, 25, 156.
12. Gould, S.J. and Eldredge, N. (1977). *Paleobiology*, 2, 115.

Applications to Computer Science

PARTIAL OBSERVATION ON A PETRI NET

C. ANDRE

Laboratoire de Signaux et Systèmes – ERA 835 du CNRS
Université de Nice, 41 Bd Napoléon III, 06041 Nice cedex
France

I. INTRODUCTION

In 1962 C.A. PETRI suggested a model for communications between automata (1). He introduced transition nets. A more convenient representation was soon proposed (2) and it became
the popular model known as Petri net.

Petri nets (PNs) are a widely used model for concurrent
systems. The structure of the net expresses the constraints.
Starting from a given initial marking an evolution rule
(firing rule) allows to deduce the possible states. For PNs
and derived models, the firing rule is easy to apply and it
is a good simulation of the evolutions of many real world
systems, (synchronisations in operating systems, protocols,
industrial control processes...). Moreover some analysis methods are available (3) or are now in progress.

In this paper we consider a system as a collection of subsystems. Each subsystem is modelled by a PN (not necessary a
sequential process). The coordination between subsystems is
made by "rendez-vous", that is represented by the operation
of FUSION of PNs. Generally the behaviour of the whole system is complex. We have chosen firing sequences to represent
the evolutions. This Petri net language approach is easy to
understand, convenient to deal with, though it can mask inherent concurrency. We are mainly interested in the analysis
of the influence of a local change on the global behaviour.
Our method is based on a partial observation of transition
firings. Only some transitions (distinguished transitions)
are observed, others are hidden. An equivalence relation on
PNs can be defined using this partial observation. The behaviour of the system can be analyzed by multiple substitutions
of subnets by equivalent ones.

In this paper we first introduce a model : the capacity
Petri net (CPN) a variant of PN. In the second part we introduce the fundamentals of our method : definition of a subclass
of PN, for which the partial observation is a significant
abstraction of the possible evolutions, and the fusion

theorem. The third part is oriented to practical uses of this method.

II. CAPACITY PETRI NET (CPN)

A CPN is a couple $R = (N, CM)$ where $N = < P, T, pre, post>$ is the
<u>structure</u> and $CM = (C, M°)$ is the couple capacity-initial
marking. P is a finite set of places, T a finite set of transitions, pre (precondition) and post (postcondition)   P:
$T \rightarrow \mathbb{N}$. Let $\mathbb{N}' = (\mathbb{N} - \{0\}) \cup \{\infty\}$. $C : P \longrightarrow \mathbb{N}'$ and $M° : P \rightarrow \mathbb{N}$.
A valid CM is such that $M° \leqslant C$.

The dynamics of the system is represented by the "token
game". The marking may change by transition firing. A transition t is firable at a marking M if (firing rule):
$\forall p \in P \quad M(p) \geqslant pre(p, t)$ "preconditions"
$\quad M(p) - pre(p, t) + post(p, t) \leqslant C(p)$ "post conditions"
The firing of t at M leads to a new marking M' (written
$M(t>M')$, such that $\forall p \in P \quad M'(p) = M(p) - pre(p, t) + post(p, t)$. Let
$L(R)$ be the set of firing sequences and $\mathbb{M}°$ be the set of
reachable markings from $M°$.

To introduce partial observation we assign a label to each
transition. Let $\mathcal{A}$ be a set of label. A labelled CPN (LCPN) is
a couple (R, E) where R is a CPN and E a labelling function :
$E : T \longrightarrow \mathcal{A} \cup \{\lambda\}$ (λ = empty word of $\mathcal{A}^*$). E is extended to
sequences. Let Label $(R) = \{a \in \mathcal{A} | \exists t \in T, E(t) = a\}$ and $T' = \{t \in T |
E(t) \neq \lambda\}$ (subset of distinguished transitions). For the sake
of simplicity, in this paper, we only consider the case of
E restricted to T' being injective, the general case is anal-
yzed in (4). Let $\mathcal{R}_{\mathcal{A}}$ be the class of LCPN on $\mathcal{A}$.

III. PARTIAL OBSERVATION

Behaviour
Suppose we observe only a subset of transitions T' for a given
CPN. The language of (R, E) is the set of the "observable sequences". This set is $E(L(R))$. Generally this set is not a
good abstraction of the actual firing sequences. We showed in
a previous paper (5) that if we consider only CPN satisfying
the "behaviour condition" then this set is significative of
the constraints existing on the distinguished transitions.
The behaviour condition (BC) is :
$\forall X, Y \in L(R) \quad \forall t \in T' \quad ((E(X) = E(Y) \Longrightarrow (X(.t> \Longleftrightarrow Y(.t>))$
where $X(.t > \Longleftrightarrow \exists S \in (T - T')^* \ E(S) = \lambda \ XSt \in L(R)$, therefore the
dot stands for a non observable firing sequence. When (BC)
is fulfilled by (R, E) the behaviour of (R, E) denoted by $B(R, E)$
is $E(L(R))$. Let $\mathcal{B}_{\mathcal{A}}$ the subclass of $\mathcal{R}_{\mathcal{A}}$ fulfilling (BC).

Behaviour equivalence

An equivalence relation is defined on $\mathcal{B}_{\mathcal{A}}$:

$(R_1,E_1),(R_2,E_2) \in \mathcal{B}_{\mathcal{A}}$ are B-equivalent iff

Label(R_1) = Label (R_2) and $B(R_1,E_1) = B(R_2,E_2)$

A consequence of this definition is the strong preservation of liveness and synchronic relation properties for the distinguished transitions (5).

Fusion of LCPN :

Let (R_1,E_1), $(R_2,E_2) \in \mathcal{R}_{\mathcal{A}}$. Suppose $t_1 \in T_1$ and $t_2 \in T_2$ bear the same label, say "a"$\in \mathcal{A}$. In the LCPN $(R,E) \in \mathcal{R}_{\mathcal{A}}$ modelling the interconnection of the two subsystems we have a transition $\widehat{t_1 t_2}$ labelled "a" such that its preconditions in R are both the preconditions of t_1 in R_1 and the preconditions of t_2 in R_2, its postconditions in R are both the postconditions of t_1 in R_1 and of t_2 in R_2. (R,E) is obtained from (R_1,E_1) and (R_2,E_2) by FUSION. This operation (for LCPN with no repeated labels) consists in merging the transitions bearing the same label.

Let $T''_i = \{t \in T'_i \mid \exists t' \in T'_j \quad E_i(t)=E_j(t')\}$ $(i,j)\in\{(1,2),(2,1)\}$ T''_i is the set of transitions of T_i to be merged with transitions of T_j.
$\widehat{T} = \{\widehat{tt'}\mid \exists t \in T_1 \; \exists t' \in T_2 \quad E_1(t)=E_2(t')\}$ is the set of merged transitions. The transitions of $(T_i - T''_i)$ are unchanged (connections and label) by the fusion. Therefore we have
Label (R) = Label (R_1) $\cup$ Label (R_2) and $T'=(T'_1-T''_1)\cup(T'_2-T''_2)\cup\widehat{T}$.
We denote the fusion by X .

Properties of the fusion

By induction on the sequence length we can prove (4) :

- $\mathcal{B}_{\mathcal{A}}$ is closed for X .

- the B-equivalence is a congruence on $(\mathcal{B}_{\mathcal{A}} , X)$

i.e : $\forall (R_0,E_0),(R_1,E_1),(R_2,E_2)\in\mathcal{B}_{\mathcal{A}}$ $(B(R_1,E_1)=B(R_2,E_2) \Longleftrightarrow$
$B((R_1,E_1)\in (R_0,E_0)) = B((R_2,E_2)\in (R_0,E_0))$
Moreover let $\mathbb{M}_i$ be the set of reachable marking of $(R_i,E_i)\in$ (R_0,E_0). We have {restriction M to P_0 $|M\in \mathbb{M}_1\}$ = {restrictions M to P_0 $|M \in \mathbb{M}_2 \}$. An important consequence of this theorem is the *substitution theorem*. Let (R_1,E_1) be a LCPN with no repeated label (λ excepted). Consider R_1 a closed subnet of R_1, $\widehat{R}$ the complementary subnet such that $(R_1,E_1)=(\overline{R}_1,\overline{E}_1)$ X $(\widehat{R},\widehat{E})$ if we substitute $(\overline{R}_1,\overline{E}_1)$ by $(\overline{R}_2,\overline{E}_2)$ we obtain $(R_2,E_2)=$ $(\overline{R}_2,\overline{E}_2) \in (R,E)$. This operation is called the substitution of the closed subnet $\overline{R}_1$ by $\overline{R}_2$ in R_1. Suppose that

$B(\bar{R}_1,\bar{E}_1)=B(\bar{R}_2,\bar{E}_2)$ and $\forall t \in \hat{T}$ $\hat{E}(t) = E_1(t) \neq \lambda$. $B(\hat{R},\hat{E})$ is trivially defined because $\hat{T}'=\hat{T}$, hence applying the fusion theorem we get : $B(R_1,E_1) = B(R_2,E_2)$ and the set of reachable markings restricted to $\hat{P}$ is the same. Therefore the substitution of a closed subnet by a B-equivalent one doesn't change properties of transitions and markings on the unmodified part of the net.

IV. STRUCTURAL EQUIVALENCE

The substitution theorem leads to an analysis method consisting in replacing a closed subnet by an equivalent but simpler one. The problem is that, generally, the B-equivalence is undecidable. For practical uses we propose transformations giving B-equivalent nets. The conditions required to apply such transformations are dependent on the structure of the net and occasionally upon the initial marking or the capacity of the net. Analyses on the set of reachable markings or on the set of firing sequences are avoided. Thus those transformations are *quasi-structural transformations*. Such transformations use the B-equivalence of structures.

Let $CM_i \subseteq \mathbb{IN}'^{P_i} \times \mathbb{IN}^{P_i}$ a set of capacity-marking pairs. N_1 and N_2 two structures the transitions of which are labelled by E_1 and E_2 respectively, such that $\{E_1(t) | t \in T_1\} = \{E_2(t) | t \in T_2\}$ are B-equivalent for CM_1 and CM_2 iff $\forall i,j \in \{1,2\}$.

$\forall (C_i,M^\circ_i) \in CM_i$ $\exists (C_j,M^\circ_j) \in CM_j$ such that $B(R_1,E_1)=B(R_2,E_2)$.

A transformation is defined by a condition to be met, a transformation of structure, a transformation of capacity and a transformation of marking. In (6) we proposed a sufficient condition for a transformation to give B-equivalent structures. A dozen of such transformations are now available in a software package on Petri nets, they are elementary transformations with a wide application domain.

VI CONCLUSION

In this paper a system is considered as a collection of communicating subsystems. Each subsystem is modelled by a capacity Petri net (CPN). Most of the events have local influence only, but others, associated with communications for example, have an effect upon the whole system. A partial observation of the significant events (distinguished events) is suggested and labelled CPN (LCPN) are used as model. The communications between subsystems are expressed by the fusion (internal operation on LCPNs). The unobserved transitions are labelled with the empty word.

Of course, only certain labelling functions lead to a correct abstraction of the behaviour of the system. A solution, using a class of LCPN is proposed. A congruence (behaviour-equivalence) is introduced for this class (characterized by the behaviour condition) with the fusion. A substitution theorem can be deduced. Moreover these exist transformations giving behaviour-equivalent nets. Therefore this class of LCPN is of practical interest in the analysis of systems described as intercommunicating subsystems.

REFERENCES

1. Petri C.A. (1962). Fundamentals of a theory of asynchronous information flow. Proc IFIP Congress 62. North Holland Pub Co (63), 386-390.
2. Holt A.W, Commoner F(1970). "Events and conditions". Applied Data Research Inc.
3. Brams G.W (1983). "Réseaux de Petri : théorie et pratique". Masson
4. André C (1983). Comportement d'un réseau de Petri étiqueté. RR 83-6. LASSY, NICE.
5. André C (1980). Behaviour of a place-transition net on a subset of transitions. Springer-Verlag IF 52, 1982, 131-135.
6. André C (1982). Structural transformations giving B-equivalent P/T nets - Springer -Verlag IF 66, 1983, 14-28.

FOLDING OF THE PLANE AND THE DESIGN OF SYSTOLIC ARRAYS

C. CHOFFRUT, Université Paris 7, U.E.R. de Mathématiques
           Tour 55-56, 1er étage, 2 Place Jussieu
           75251 Paris Cedex 05 France

K. CULIK II, Department of Computer Science
            University of Waterloo,
            Waterloo, Ontario
            N2L 3G1 Canada

I. INTRODUCTION

Systolic arrays were first introduced by Kung (see e.g. [2] and [3] as devices composed of processors of a few different types, which are regularly and locally connected. These processors are activated in a synchronous way by a unique clock which is the only global communication between them.

This paper is a continuation of the work presented in [1] where "folding" has been proposed as a technique for the design of systolic arrays. Here we first study the power of folding as a general geometric transformation. Then as an application we show that two congruent sequences on a "regular" grid can be identified by a limited number a foldings (cf. Theorem 2). This result can be used in the design of systolic arrays, in particular when iteration of some computation is involved. Indeed, in such cases it is necessary at each step of the iteration to feed the inputs with the outputs of the previous computation. Drawing wires between inputs and outputs would violate the regularity of the lay out. Instead, we can fold the array, as we would do with a sheet of paper, in such a way that each output matches with the corresponding input.

As far as practical implementations of foldings are concerned, a few solutions may be imagined : implementation in two layers, multiplexing in time, etc...

II. PRELIMINARIES

Throughout this paper, P is a fixed (counter-clockwise) oriented plane.

DYNAMICAL SYSTEMS
AND CELLULAR AUTOMATA

313

For i=1,2 let L_i be an oriented line, a vector or a ray. The angle $0 \leq \theta < 2\Pi$ they define is denoted by $\measuredangle (L_1,L_2)$.

By a displacement (resp. antidisplacement) of the plane we mean a 1-1 transformation of the plane which preserves the distances and the orientations (resp. preserves the distances and changes the orientations).

We recall that a displacement of the plane is either a translation defined by one vector, or a rotation defined by one invariant point and an angle. A rotation is characterized by the fact that it leaves one point invariant. In particular any composition of a translation and a rotation by a non zero mod 2Π angle is again a rotation.

Let I be an ordered set of indices. We say that the two sequences $(a_i)_{i \in I}$ and $(b_i)_{i \in I}$ of points of P are congruent (resp. anti congruent) if there exists a displacement (resp. antidisplacement) of the plane which takes every a_i into b_i.

By a square-grid we mean a (infinite) planar graph yielding a tesselation of the whole plane by squares of a fixed size. More formally it is defined as a set of two orthogonal infinite families of parallel lines, each family intersecting any line of the other family at points defining a two way infinite sequence, two consecutive points being at a fixed distance $d > 0$. The nodes of the square grid are the different intersecting points of the lines in the two families. In the same way we would define the triangular (resp. hexagonal) grid where the tesselation of the plane is defined by equilateral triangles (resp. hexagons). By a regular grid we mean either a square, a triangular or a hexagonal grid.

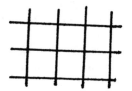

Fig 1. Square grid

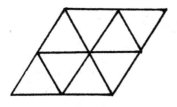

Fig 2. Triangular grid

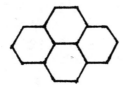

Fig 3. Hexagonal grid

Given any oriented line L we denote by P_{L_+} (or simply
P_+ when L is understood) the positive half plane pointed
at by any vector with origin on L and making with L an
angle $0 \le \theta \le \Pi$). We denote by P_{L_-} (or simply P_-) the
subset $P - P_+$.

We define the <u>folding</u> of P along L as the mapping
of P into P_- which leaves P_- point by point invariant
and which takes each point of P_+ into the point of P_-
which is its reflection in L.

Thus the image of the plane after one folding is a
half plane, after two foldings along two orthogonal lines it
is a quadrant, etc...

We define a <u>wrapping</u> of the plane as either the identity transformation or as a composition of an arbitrary finite
number of foldings.

III. MAIN RESULTS

Let us say that a wrapping w of the plane <u>realizes</u>
a length-preserving transformation τ if it identifies
any point M and its image τ(M) :

$$w(M) = w(\tau(M)).$$

We first determine which length-preserving transformations of the plane can be realized by some wrapping. It must be stressed that we are interested, for the moment, in global transformations of the plane and not in transformations of figures. In particular corollary 1 says that no non-trivial translation can be realized by any wrapping : however, it does not say that any sequence of points and its image by a translation can never be identified by some wrapping.

We consider first the rotations.

Theorem 1

Let $0 < k < n$ be two integers which are relatively prime and ρ a rotation of the plane by an angle $2\frac{k}{n}\Pi$. Then ρ can be realized by the composition of $\lceil \log n \rceil + 1$ foldings (where $\lceil \log n \rceil$ stands for the ceiling of $\log n$).

Proof

Consider a fixed oriented line L_o containing the invariant point O of the rotation. We divide the positive half-plane into $\lfloor\frac{n-1}{2}\rfloor + 1$ (where $\lfloor\frac{n-1}{2}\rfloor$ stands for the floor of $\frac{n-1}{2}$ equal (except the last one when n is odd) regions, by setting for $0 \leq i \leq \lfloor\frac{n-1}{2}\rfloor$:

$$R_i = \{M \in P \mid \frac{2i\Pi}{n} \leq \measuredangle(L_o, OM) < \min(\Pi, \frac{2(i+1)}{n}\Pi)\}$$

For $0 \leq i \leq \lfloor\frac{n-1}{2}\rfloor$ we denote by L_i the line containing O and satisfying : $\measuredangle(L_o, L_i) = \frac{2i\Pi}{n}$. Let $m = \lceil \log \frac{n+1}{2}\rceil = \lceil \log n \rceil - 1$. We are now ready to define a wrapping u as the composition of the following foldings: we first fold along L_o oriented in the opposite direction, obtaining thus the $\lfloor\frac{n+1}{2}\rfloor$ regions $R_o,\ldots,R_{\lfloor\frac{n-1}{2}\rfloor}$ defined above. Then we fold succesively along the lines $L_2m-1, L_2m-2,\ldots,L_1$ which results in piling the regions on top of one other (some of them in reverse orientation).

Given an arbitrary point M we set $M' = u(M)$, $\theta = \measuredangle(L_o, \overrightarrow{OM})$ and $\theta' = \measuredangle(L_o, \overrightarrow{OM'})$. It is straightforward to verify that :

(1) $OM = OM'$, $0 \le \theta' \le \frac{2\Pi}{n}$ and $\theta \equiv \pm \theta'$ mod $\frac{2\Pi}{n}$

Now denote by v the folding of the plane along the oriented line D defined by $\sphericalangle (L_o, D) = \frac{\Pi}{n}$ and let w be the composition of u and v. We claim that the wrapping w realizes the rotation ρ, i.e. $w(M) = w(\rho(M))$ for each point M.

Indeed, let us set $N = \rho(M)$, $N' = u(N)$, $\alpha = \sphericalangle (L_o, \overrightarrow{ON})$ and $\alpha' = \sphericalangle (L_o, \overrightarrow{ON'})$. Since $\theta - \alpha = \frac{2k\Pi}{n}$ holds by hypothesis, we have $\theta = \alpha$ mod $\frac{2\Pi}{n}$ and by(1) $OM' = ON'$ and either $\alpha' = \theta'$ or $\alpha' + \theta' = \frac{2\Pi}{n}$. In the first case we obtain : $u(M) = M' = N' = u(N) = u(\rho(M))$ in the second case the point $M' = u(M)$ is the reflection of the point $N' = u(\rho(M))$ in the line D, i.e. $vu(M) = vu(\rho(M))$.

Finally, we note that the wrapping w is the composition of $m+1 = \log\left\lceil\frac{n+1}{2}\right\rceil + 1 = \lceil\log n\rceil$ foldings. $\square$

Lemma

Let R and R' be two parallel rays starting at 0 and 0'. We assume $0 \ne 0'$ and $\sphericalangle (\overrightarrow{00'}, R) = \sphericalangle (\overrightarrow{00'}, R') \ne \frac{\Pi}{2}$ mod Π. For each point M of R we denote by $\varepsilon(M)$ the point M' of R' such that : $\overrightarrow{OM} = \overrightarrow{O'M'}$.

Then there exists no wrapping of the plane identifying the points M and $\varepsilon(M)$ for each M in R.

Proof

We prove by induction on the number of foldings defining the wrapping w that there exist two points $O_1 \in R$ and $O_1' \in R'$ with $\varepsilon(O_1) = O_1'$ such that, denoting by R_1 and R_1' respectively the two rays collinear to R and R' starting at O_1 and O_1', the two images $w(R_1)$ and $w(R_1')$ are two parallels rays with the same direction, such that $w(O_1) = M_1 \ne M_1' = w(O_1')$ and $\sphericalangle (M_1 M_1', w(R_1)) \ne \frac{\Pi}{2}$ mod Π.

When the wrapping is the identity, there is nothing to prove.

Assume now that w is a wrapping and f a folding of the plane along an axis A. If A is parallel to $w(R_1)$ and $w(R_1')$, then the induction hypothesis is still verified with the same O_1 and O_1'. Otherwise, we consider two points $O_2 \in R_1$ and $O_2' \in R_1'$ with $O_2' = \varepsilon(O_2)$, such that $w(O_2)$ and $w(O_2')$ lie in

the half plane defined by A and containing unbounded subsets
of R_1 and R_1'. Setting $M_2 = fw(O_2)$ and $M_2' = fw(O_2')$ we
get :

$M_2 \neq M_2'$ and $\measuredangle(\overrightarrow{M_2M_2'}, fw(R_1)) = 2\,\Pi - \measuredangle(\overrightarrow{M_1M_1'}, w(R_1)) \neq \frac{\Pi}{2}$ mod Π.$\square$

Corollary 1

Let τ be a translation of the plane different from
the identity mapping. Then it can not be realized by any
wrapping of the plane.

Proof

Let O and O' be two (distinct) points of the plane
such that $\overrightarrow{OO'}$ is the vector defining the translation τ.
Denote by R and R' respectively the rays starting at O
and O' and with direction $\overrightarrow{OO'}$. It then suffices to apply
the previous lemma. $\square$

Corollary 2

Let τ be an antidisplacement of the plane which is
not reduced to a reflection in a line. Then it can not be
realized by any wrapping of the plane.

Proof

The transformation τ is obtained as the product of a
rotation by an angle θ and a reflection in a line L which
we orient arbitrarily. Given any ray R, its image through τ
will define an angle :

$\measuredangle(L, \tau(R)) = -\,(\theta + \measuredangle(L,R))$ mod $2\,\Pi$

In particular, if $\measuredangle(L,R) = -\frac{\theta}{2}$ then R and $\tau(R)$ are
parallel. However $\tau(R)$ may not be the reflection of R in
a line since otherwise τ would itself be reduced to a
reflection. The result follows from the Lemma. $\square$

We now turn to transformations of figures on a regular
grid. We will say that a wrapping w preserves a regular
grid G (as a union of line segments in the plane) if
$w(G) \subseteq G$.

Theorem 2

Consider two congruent sequence of points $(a_i)_{i \in I}$
and $(b_i)_{i \in I}$ on a regular grid. Assume they can be matched
under a rotation of the plane around a non-0 mod Π angle
which preserves the grid. Denote by w the wrapping of the
plane identifying the two sequences : $w(a_i) = w(b_i)$ for all
$i \in I$.

Furthermore, in the case of a square grid, w is the
product of at most 3 foldings. In the case of a triangular
or an hexagonal grid it is the product of at most 4 foldings.

Proof It is always possible, if necessary by adding new points, to assume that the sequence $(a_i)_{i \in I}$ contains at least two different points so that the rotation taking $(a_i)_{i \in I}$ into $(b_i)_{i \in I}$ is uniquely determined.

We shall only consider the hexagonal grid, the other two cases being treated in the same way. Since the points a_i and b_i belong the grid, ρ preserves in fact the whole grid. Thus the fixed point 0 of ρ is either a node in which case the angle of the rotation is necessarily $2\frac{\Pi}{3}$ or $4\frac{\Pi}{3}$, or it is the center of an hexagon, in which case the angle is necessarily equal to $\frac{\Pi}{3}$, $2\frac{\Pi}{3}$, Π, $\frac{4}{3}$ or $\frac{5}{3}\Pi$.

In the first case, as in Theorem 1, we define an oriented line L_0 containing 0 and collinear to any one of the three arcs incident to 0. Now all foldings involved in Theorem 1 are along oriented lines containing 0 and forming an angle with L_0 which is a multiple of $\frac{\Pi}{3}$. The wrapping thus defined is the product of $\lceil \log 3 \rceil + 1 = 3$ foldings preserving the grid.

In the second case, we define an oriented line containing 0 and any one of the six vertices of the given hexagon. Once again, all foldings involved as in Theorem 1 are along oriented lines containing 0 and forming an angle with L_0 which is a multiple of $\frac{\Pi}{6}$. The wrapping thus defined is the product of at most $\lceil \log 6 \rceil + 1 = 4$ foldings preserving the grid. □

REFERENCES

[1] CULIK K. II and J. PACHL, "Folding and Unrolling Systolic arrays", ACM SIGACT-SIGOPS Symposium on Principles of Distributed Computing, Ottawa, August 1982.

[2] KUNG H.T., Why Systolic Architectures ?, Computer Magazine, Vol. 15, n° 1, 1982, pp. 37-46.

[3] MEAD C. and L. CONWAY, "Introduction to VLSI Systems", Addison Wesley, 1980.

TWO-LEVEL PIPELINED SYSTOLIC ARRAYS FOR MATRIX MULTIPLICATION, POLYNOMIAL EVALUATION AND DISCRETE FOURIER TRANSFORM

H. T. Kung

Department of Computer Science
Carnegie-Mellon University
Pittsburg, Pennsylvania 15213, USA

1. INTRODUCTION

In recent years many systolic algorithms have been designed and several prototypes of systolic array processors have been constructed. Major efforts have now started in attempting to use systolic array processors in large, real-life applications. Practical issues on the implementation of systolic array processors have begun to receive substantial attention (1).

One of the important implementation issues relates to the efficient use of pipelined functional units in the implementation of systolic cells. For example, high throughput floating-point multiplier and adder circuits typically employ three or more pipeline stages (5). Systolic cells implemented using these units form a *second level of pipelining* in the pipelined organization of systolic arrays. This additional level of pipelining can greatly increase the system throughput.

Two-level pipelined systolic arrays have been proposed previously for performing one and two dimensional convolution (2). Recently a general theory was established to systematically transform certain existing systolic array designs based on single-stage cells to ones with pipelined cells (3). This paper describes several two-level pipelined systolic array designs for matrix multiplication, polynomial evaluation and discrete Fourier transform, using *both* real and complex arithmetic. These designs illustrate some general techniques on the use of pipeline arithmetic units to implement complex arithmetic which is essential to signal processing applications.

DYNAMICAL SYSTEMS
AND CELLULAR AUTOMATA

The two-level pipelined designs of this paper will be implemented on a high-performance floating-point systolic array processor currently being built at CMU, using off-the shelf components.

2. POLYNOMIAL EVALUATION

Given an $(n-1)$-st degree polynomial,

$$x_0 w^{n-1} + x_1 w^{n-2} + \ldots + x_{n-1},$$

the polynomial evaluation problem is to compute the values of the polynomial at $w = w_0, w_1, \ldots$. Figure 1 depicts a systolic array for the problem with $n = 8$. The $(i+1)$-st cell form the left accumulates the value y_i of the polynomial at $w = w_i$, using Horner's rule :

$$y_i = (\ldots((x_0 w_i + x_1)w_i + x_2 + \ldots + x_6)w_i + x_7.$$

When the final value of y_i is obtained, it is sent to the latch in the bottom of the cell, represented by a small rectangle, to be output from the systolic array.

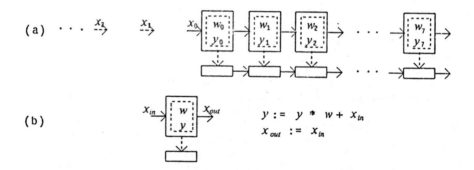

Fig.1 (a) Systolic array for polynomial evaluation, and
(b) cell specification.

Figure 2 shows the internal structure of two consecutive cells for computing y_i and y_{i+1} . Each cell performs a multiply-add operation every cycle and the result is kept in the y register. After n multiply-add operations, the contents of the y register is transferred to the corresponding latch on the y-data stream. These computed results on the y-data stream shift to the right "systolically" at the rate of one cell every cycle. When they reach the

end cell of the systolic array, they are output. Note
that on the x-data stream two rather than one latch is
provided for each cell. This ensures that computation for
y_i be completed two cycles earlier than that for y_{i+1}
and therefore the computed y_i and y_{i+1} will not collide
with each other on the y-data stream.

Consider now the situation that we use pipelined multi-
pliers and adders. Throughout the paper, we assume that
the pipelined multiplier and adder each have five pipeline
stages, as in the case of some recent 32-bit floating-point
arithmetic chips (5,6). It is straightforward to generalize
our results to any numbers of pipeline stages for the multiplier and adder. Figure 3 shows that after x_i enters the
adder, it will take ten cycles (five cycles for the addition and five cycles for the multiplication) to compute
a result that can be added to x_{i+1} . In order to make full
use of the adder and multiplier, we interleave computations
for ten independent polynomial evaluations on the systolic
array. This allows a new task to enter the adder every
cycle. However, with this interleaving, cells will output
results in bursts. That is, at every n-th cycle a cell
will start out-puting results for ten consecutive cycles.
To avoid collisions on the y-data stream, among outputs
from different cells, we use 11 (=10+1) latches on the
x- data stream for each cell, as depicted by Figure 3.

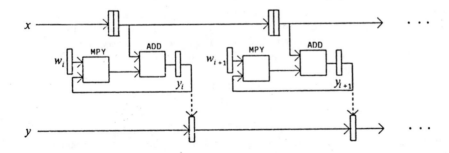

*Fig. 2 Cell structure for the polynomial evaluation array
of Figure 1.*

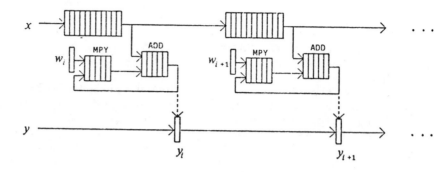

Fig.3 Two-level pipelined systolic array for polynomial evaluation.

3. DISCRETE FOURIER TRANSFORMS

The problem of computing an n-point discrete Fourier transform (DFT) is as follows :

given $x_0, x_1, \ldots, x_{n-1}$,

compute $y_0, y_1, \ldots, y_{n-1}$ defined by
$$y_i = x_0 w^{i(n-1)} + x_1 w^{i(n-2)} + \ldots + x_{n-1}$$

where w is a primitive n-th root of unity.

The n-point DFT problem is the polynomial evaluation problem with evaluation points $w_i = w^i$. Since w is a complex number, in order to use a systolic array for polynomial evaluation to compute DFTs the array must handle complex arithmetic.

A complex multiply-add operation .

$$(y_r + j y_i)(w_r + j w_i) + (x_r + j x_i),$$

$$(= (y_r w_r - y_i \cdot w_i) + x_r + j (y_r \cdot w_i + y_i \cdot w_r) + x_i .)$$

involves four real multiplications and four real additions. They will thus be done in four cycles using the multiplier and adder of each cell. The real and imaginary parts of a complex number are processed in separate cycles ; in particular, the real part of a result is computed two cycles earlier than its imaginary part. In the scheme of Figure 4, the four multiplications, $y_r \cdot w_r$, $y_i \cdot w_i$, $y_r \cdot w_i$ and $y_i \cdot w_r$, occupy four consecutive stages of the multiplier. The resulting products enter directly into the adder to form $y_r \cdot w_r - y_i \cdot w_i$ and $y_r \cdot w_i + y_i \cdot w_r$ and these two additions occupy two stages of the adder. Interleaved with

these two stages are stages for performing the other additions involving x_n and x_i .

After the real (or imaginary) part of x_i enters the adder
it will take five cycles to do the complex addition and
11 cycles to do the complex multiplication (five in the
multiplier, one in the latch following the multiplier,
and five in the adder), before the result can be added
to the real (or imaginary, respectively) part of x_{i+1}.
In these total of 16 cycles, up to four independent complex multiply-add operations can be interleaved on the
systolic array and be carried out simultaneously. Again,
to avoid collisions on the y-data stream, we use 17
(=16+1) latches on the x-data stream for each cell,
as depicted by Figure 4.

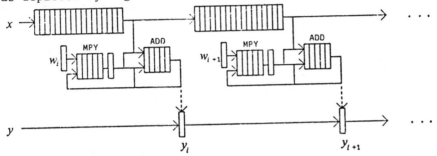

*Fig.4 Two-level pipelined systolic array for DFT or complex
polynomial evaluation.*

The systolic array of Figure 4 can be multiplexed to
compute DFTs of arbitrarily large sizes. Suppose that
the array has k cells, and that we want ot use it to
compute an n-point DFT, with $n > k$. it is well known
that based on the FFT graph, the n-point DFT problem can
be done with $\log n/\log k$ stages of n/k k-point DFTs.
Figure T depicts an n-point FFT graph an a decomposition
scheme for $n=16$ and $k=4$. The n/k k-point DFTs to be performed at each stage of the decomposition are evaluations
of n/k-polynomials at the same set of k points. Therefore
the systolic array of k cells can conveniently carry out
all these evaluations, by taking serially all the
$n=k.n/k$ inputs to the array.

A k-point DFT can also be viewed as a vector-matrix
multiplication, where the vector is composed of the
input elements $(x_0, x_1 \ldots, x_{n-1})$,   and the matrix has
$w^{(i-1)(j-1)}$ as its (i,j) entry. The n/k k-point
DFTs discussed above can be considered as a matrix multiplication, presented in the following section, represents
an alternative method for computing DFTs.

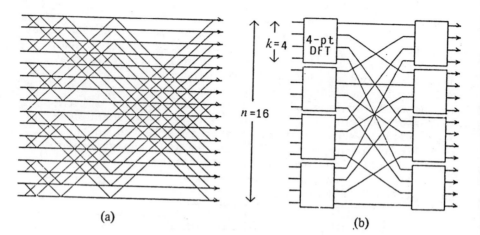

Fig.5 (a) 16-point FFT graph, and (b) decomposing the FFT with n=16 and k =4.

4. MATRIX MULTIPLICATION

Given $n \times n$ matrices $X = (x_{i,j})$ and $W = (w_{i,j})$ we want to compute their product $Y = (y_{i,j})$. Figure 6 depicts a simple linear systolic array for the matrix multiplication problem (4), with $n = 8$. Each cell of the array performs a multiply-add operation every cycle. The j-th cell from the left computes the inner product y_j of vectors $(x_{i,1}, x_{i,2}, \ldots, x_{i,n})$ and $(w_{1,j}, w_{2,j}, \ldots, w_{n,j})$ for each i. By pumping the entries of X into array serially in the row-major ordering and by recirculating $(w_{1,j}, w_{2,j}, \ldots, w_{n,j})$ at cell j for each j , entries of the product $Y=XW$ will be computed and output in the row-major ordering.

Figure 7 shows the internal structure of the cells. Note that the accumulated result is fed directly to the adder rather than the multiplier after each cycle. In order to facilitate the recirculation of $(w_{1,j}, w_{2,j}, \ldots, w_{n,j})$ we store at cell j these values in a shifter implemented by a RAM. The particular w to be used in any given cycle is picked up by an address *(addr)* , that is input to the cell every cycle. Since the address patterns for all the cells are the same, they are passed systolically form cell to cell.

Assume now that the matrices have complex entries and that the multiplier and adder each have five pipeline stages. By techniques similar to those used in the designe of Figure 4, we can derive a two-level pipelined

systoclic array of Figure 8 for complex matrix multiplication. At each cell, two independent (complex) accumulations,
that use a total of four real numbers, are maintained inside the eight-stage pipeline made of the five-stage
adder and the following three-stage FIFO.

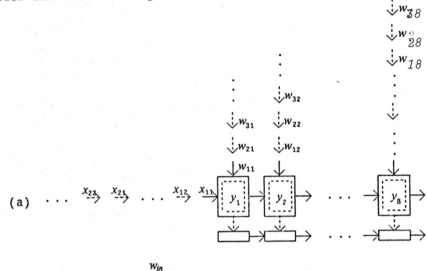

(a)

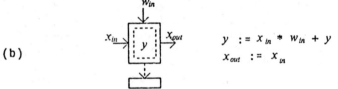

(b)

$$y := x_{in} * w_{in} + y$$
$$x_{out} := x_{in}$$

Fig.6 (a) Systolic array for matrix multiplication
    (b) Cell specification

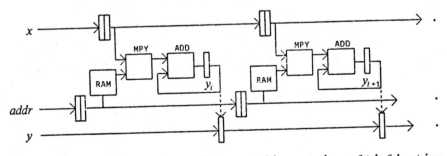

Fig.7 Cell structure for the systolic matrix multiplication
    array of Figure 6.

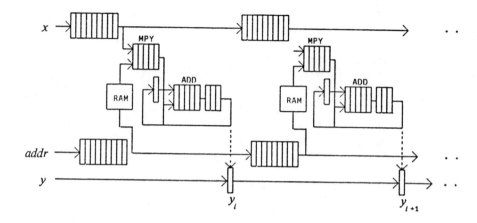

Fig.8 Two-level pipelined systolic array for complex matrix multiplication.

5. ALTERNATIVE DESIGNS FOR POLYNOMIAL EVALUATION

For the systolic designs of Section 2 the w_i , where the given polynomial is evaluated, are stored inside the systolic array. Thus these schemes are efficient when the w_i do not have to be changed for a relatively long time. This is the case when a large degree polynomial or a large set of polynomials are to be evaluated at a relatively small number of points.

We consider here another scenario where a polynomial is to be evaluated at a relatively large or unbounded number of points. In this case it is advantageous to store the coefficients x_i of the polynomial inside the array and let the evaluation points w_i flow conti--nously along the array. Figure 9 depicts such a design.

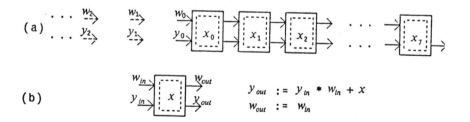

Fig.9 (a) Systolic array for polynomial evaluation, and (b) cell specification.

In contrast with all previous designs of this paper, in
the cell structure of Figure 10 we see that the input
of a multiplier (or an adder) does not depend on any
of its earlier outputs. This acyclic property makes the
two-level pipelined implementation relatively simple and
straighforward, as shown in Figure 11. In fact, one can
check that this scheme works for both real and complex
polynomial evaluation.

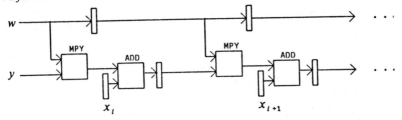

Fig.10 Cell structure for the polynomial evaluation of Fig.9

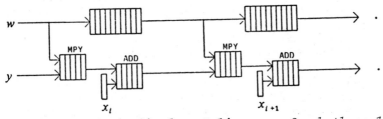

*Fig.11 Two-level pipelined systolic array for both real
and complex polynomial evaluation.*

6. CONCLUDING REMARKS

Systolic array designs in the literature are typically
given in the abstract level of Figures 1,6 and 9, which
assumes that all the cell operations can be performed in
a single cycle. However, for efficient implementation
two-level pipelined designs are often necessary. An
important future research is to automate the process of
transforming designs in the abstract level to ones in
the level of Figures 4,8 and 11. Design examples of this
paper should provide some useful examples for the eventual
construction of such tranformation tools.

ACKNOWLEDGMENTS

The research was supported in part by the Office of Naval
Research under Contracts N00014-76-C-0370, NR 044-422
and N00014-80-C-0236, NR 048-659, in part by the
Defense Advanced Research Porjects Agency (DoD), ARPA
Order No, 3597, monitored by the Air Force Avionics

Laboratory under Contract F33615-81-k-1539, and in part
by a Guggenheim Fellowship.

REFERENCES

(1)   Kung, H.T.
         On the Implementation and Use of Systolic Array
         Processors.
         in *Proceedings of International Conference on
         Computer Design : VLSI in Computers*, pages 370–
         373. IEEE, November, 1983.

(2)   Kung, H.T., Ruane, L.M., and Yen, D.W.L.
         Two- Level Pipelined Systolic Array for Multi-
         dimensional Convolution.
         Image and Vision Computing 1 (1) : 30–36,
         February, 1983.
         An improved version appears as a CMU Computer
         Science Department technical report, November 82.

(3)   Kung, H.T., and Lam, M;
         Fault-Tolerance and Two–Level Pipelining in
         VLSI Systolic Arrays.
         In *Proceedings of Conference and Advanced Research
         in VLSI*. Massachusetts Institute of Technology,
         Artech House, Inc., Cambridge, Massachusetts,
         January, 1984.

(4)   Kung, H.T. and Yu, S.Q.
         Integrating High–Performance Special–Purpose
         Devices into a System. In Randel, B. and
         Treleaven, P.C. (editors), *VLSI Architecture*,
         pages 205–211. Prentice/Hall International 1983.
         An earlier version also appears in *Proceedings
         of SPIE Symposium, Vol.341, Real–Time Signal
         Processing V*, May 1982, pp 17–22.

(5)   Woo, et al.
         ALU, Multiplier Chips Zip through IEEE Floating-
         Point Operations.
         Electronics 56 (10) : 121–126, May 19, 1983.

SYSTOLIC ARRAYS FOR CONNECTIVITY AND TRIANGULARIZATION PROBLEMS

Lamine Melkemi and Maurice Tchuente

Université de Grenoble Laboratoire I.M.A.G.

B.P. 68 38402 Saint Martin d'Hères cédex

F R A N C E

1. INTRODUCTION

We are first interested in the design of a systolic array for
the computation of the connected components of an undirected
graph $G = (V,E)$, where $V = \{1,2,...,n\}$ is the set of vertices
and $E \subset VxV$ is the set of edges. Savage(4) has proposed a
one-dimensional systolic network of size n, where signal
flows in two directions and which computes a spanning forest
(resp. the connected components) of an undirected graph in
time $\nu(2m+n)$ (resp. $2\nu(m+n)$) where n is the number of vertices, m is the number of edges, and ν is the cycle-time of the
basic cell. Here, we present an alternate pipeline solution
which is based on the same idea, and whose running-time is
$\tau(m+n)$ (resp. $\tau(m+2n)$), where τ is the cycle-time of our basic
cell. The second part of this paper, deals with the triangularization of band matrices, i-e square matrices of order n
such that $a_{ij} = 0$ for $i-j \notin [-q,p]$ where p and q are fixed
integers, $1 \leq p,q \leq n$; such matrices arise for instance in
the numerical solution of partial differential equations.
Kung and Leiserson(3) have proposed an hexagonal array of size $(p+1)(q+1)$ which performs L.U decomposition in time
$T = 3n+1+minimum\{p,q\}$, but such a network is applicable only
when Gaussian elimination can be done without pivoting. Subsequently, direct VLSI implementation of numerically stable
schemes, i-e triangularization with neighbor pivoting or Gi-
vens rotations have been studied by Gentleman and Kung(1),
Heller and Ipsen (2). The network introduced here requires
less area than all the previously known solutions.

DYNAMICAL SYSTEMS
AND CELLULAR AUTOMATA

331

2. CONNECTIVITY PROBLEM

Any undirected graph of order n is supposed to be coded as a couple G = (V,E) where V = {1,2,...,n} is the set of vertices and E ⊂ VxV is the set of edges. As explained in (4), a basic idea for constructing a spanning forest H of G is to maintain as the edges of G are examined sequentially, a collection of disjoint sets of edges corresponding to the connected components of H; more precisely, when an edge e = (u,v) is processed, there are four possible situations:

- if it is incident to two preceding edges (u,u'), (v,v') belonging to the same connected component of H, then it is not inserted into H.

- if it is incident to two preceding edges (u,u'), (v,v') belonging to two disjoint components of H numbered h,k, h < k, then it added to H and these components are combined into a single one numbered h.

- if it is incident to exactly one preceding edge of H, then it is added to the component containing it .

- if it is not incident to any preceding edge, then a new component {e} is created .

It is easily verified that, starting from an empty forest, such an algorithm works correctly .

The network presented here consists of a linear array of identical cells (see the figures below).

(u,v,cu,cv,cmax)	
r	s
Nmin	Mmax

Fig. 1 The basic cell

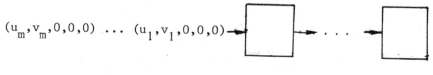

$(u_m,v_m,0,0,0)$ ... $(u_1,v_1,0,0,0)$

cell 1   cell n

Fig. 2 The general structure of the array

Any cell contains two registers r,s which will store an edge
of the spanning forest H, and two registers Nmin, Nmax which
contain the numbers of the component containing that edge;
Nmin is different from Nmax when the insertion of e = (r,s)
into H has merged two components previously numbered Nmin and
Nmax; we shall maintain the condition Nmin ≤ Nmax. As an edge
e = (u,v) travels right, it determines the numbers of the
components containing u,v and carries them along as cu,cv;
cmax denotes the maximum component-number encountered by e;
nitially, cu = cv = cmax = r = s = Nmin = Nmax = 0. When the
5-tuple (u,v,cu,cv,cmax) goes through a cell, it is processed
as follows:

case 1. r = s = 0; e has been compared with all the edges of
the current spanning forest H.

    subcase 1.a. cmax ≥ 0; e does not belong to H;

        subcase 1.a.1. cu = cv = 0; e is not incident to any
preceding edge, and, as a consequence, a new component {e} is created as follows :
r := u; s := v; cu := cv := Nmin := Nmax := cmax+1;
cmax := -1;
the fact that cmax is turned to -1 prevents from any
further insertion and indicates, as e is output at
the rightmost cell, that it belongs to H.

        subcase 1.a.2. (cu = 0 and cv ≠ 0) or (cu ≠ 0 and cv ≠
0); e is added to the component of H containing one
of its end-points as follows :
r := u; s := v; Nmin := Nmax := maximum{cu,cv};
cmax := -1;

        subcase 1.a.3. 0 < cu < cv or 0 < cv < cu; the inser-
tion of e into H merges the components of H previously numbered cu, cv as follows:
r := u; s := v; Nmin := minimum{cu,cv};
Nmax := maximum{cu,cv}; cmax := -1;

case 2. (r,s) ≠ (0,0); the cell has already stored an edge
e = (r,s) of H; therefore cmax must be updated as follows:
cmax := maximum{cmax,Nmax};

    subcase 2.a. u = r or u = s or cu = Nmax; u belongs to the
component containing (r,s) and cu must be updated in
order to take into account tha fact that the components
numbered Nmin and Nmax have been merged into a single
component numbered Nmin :
cu := Nimn;

    subcase 2.b. v = r or v = s or cv = Nmax; (similar to
subcase 2.a.); cv is updated as follows:
cv := Nmin;

In all other cases, the variables cu,cv,cmax,r,s,Nmin,Nmax
are not changed .

It is easily shown by induction that this algorithm proces-
es any edge e = (u,v) of G correctly. Let τ denote the cycle-
time of the basic cell, and let us consider an undirected
graph G = (V,E) with n vertices and m edges, such that
E $\cap$ {(i,i), 1 $\leq$ i $\leq$ n} = $\emptyset$.
- if the edges of G are input sequentially into the leftmost
cell of an array of size n-1, then the edges of a spanning
forest are output at the rightmost cell with cmax = -1, and
the running-time of the algorithm is $\tau(m+n)$.
- if the edges of G, followed by (1,1),(2,2),...,(n,n) are
input sequentially into the leftmost cell of an array of size n, then any couple (u,u) is output at the rightmost cell
with cu indicating the number of the connected component of
G containing u, and the running-time is $\tau(m+2n)$; here we
need n cells in order to handle the case where G contains n
connected components, i-e all its edges are loops.

3. TRIANGULARIZATION OF BAND MATRICES

We are now interested in the triangularization of band matrices, i-e square matrices A = (a_{ij}) such that a_{ij} = 0 for
i-j $\notin$ [-q,p] where p,q are two fixed integers less than n. We
are looking for a network whose size depends on p and q. A
triangularization process is an algorithm which, starting
from a system of linear equations Ax = b, transforms it into
a new equivalent one TAx = Tb where the new matrix TA is triangular; this is usually done by successively combining couples of rows of the original matrix A. Triangularization is
a very important problem because it is the first step of the
most commonly used direct algorithms for the solution of linear systems of equations.

When two rows $(a_{kk},...,a_{kn})$, $(a_{ik},...,a_{in})$ are combined in
a triangularization scheme in order to create a zero element
in position (i,k), the coefficients of the 2x2 matrix which
performs the transformation

$$M_{k,i} := \begin{pmatrix} a_{kj} \\ a_{ij} \end{pmatrix} \mapsto \begin{pmatrix} a & b \\ c & d \end{pmatrix} \begin{pmatrix} a_{kj} \\ a_{ij} \end{pmatrix} = \begin{pmatrix} a'_{kj} \\ a'_{ij} \end{pmatrix} \quad k \leq j \leq n$$

must verify $a'_{ik} = c.a_{kk} + d.a_{ik} = 0$; hence, in a systolic solution, this matrix must be generated by the couple (a_{kk},a_{ik})
and must also be propagated to the couples (a_{kj},a_{ij}), for

$k < j < n$. As a consequence, the corresponding array must consist of generation cells which will generate the matrices $M_{k,i}$, $k < i < n$ and apply them to (a_{kk}, a_{ik}), combination cells which receive the $M_{k,i}$, and apply them to the couples (a_{kj}, a_{ij}) $k < j, i \leq n$, and eventually delay cells. the characteristics of the previous solutions the systolic triangularization of band matrices are summarized in the following table (only the cells corresponding to the transformation $A \longmapsto TA$ are counted) .

TABLE 1

Performances of Prevoiusly Known Solutions

reference	generation. cells	combination cells	delay cells	time performance
(3)	p	$p(p+q)$	$3p(p+q)-2q$	$3n+6p+q-2$
(2)	p	$p(p+q)$	0	$2(n+q-1)$

We are now going to present an array with p generation cells and $pq+p(p+1)/2$ combination cells, whose running time is $T = 3n+p+q$.

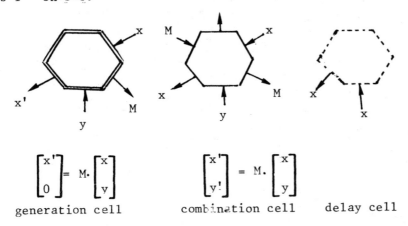

$$\begin{bmatrix} x' \\ 0 \end{bmatrix} = M \cdot \begin{bmatrix} x \\ y \end{bmatrix} \qquad\qquad \begin{bmatrix} x' \\ y' \end{bmatrix} = M \cdot \begin{bmatrix} x \\ y \end{bmatrix}$$

generation cell   combination cell   delay cell

Fig. 3 The structure of the three types of elementary cells.

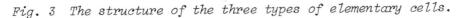

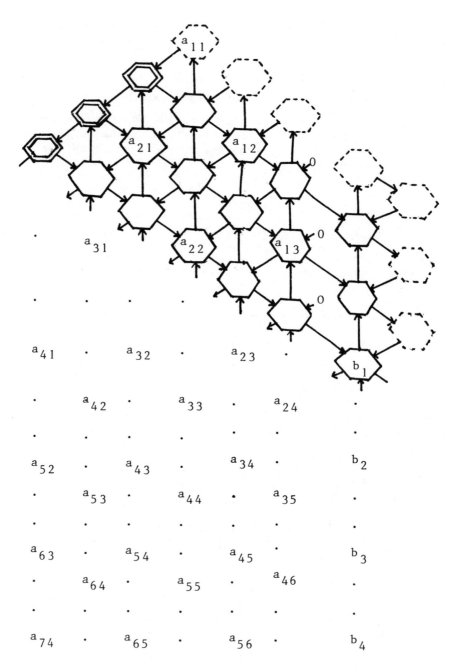

Fig. 4   The general structure of the array

Any row numbered i is input into the array on the low boundary; as it travels upwards, it is combined with the rows numbered i-k, $1 \leq k \leq p$; when it reaches the delay cells at the
upper boundary, it is sent back into the network in order to
be combined with the rows numbered i+k, $1 \leq k \leq p$; as shown
in figure 4, additional cells must be added to the array in
order to perform the transformation $b \longmapsto Tb$.

4. CONCLUSION

In a context where the algorithms are programmed on an array
of microprocessors, the solution proposed by Savage(4) for
the computation of the connected components of an undirected
graph, may run faster than the algorithm presented here, because of the more complex structure of our basic cell. However, in a VLSI implementation, the design of this paper is
probably better because of the following interesting properties of one-dimensional networks where signal flows in only
one direction: first of all, the basic cell can be constructed in a pipeline way, and, as a consequence the time-perfor-
mance for computing a spanning forest and the connected components of an undirected graph with n vertices and m edges
will be $\tau n + \tau' m$ and $\tau n + \tau'(m+n)$, where τ is the cycle-time of our
basic cell and τ' is the delay necessary to go through one
level of the pipeline structure of a cell; secondly, as shown
by Kung (see the paper in these proceedings), the entire network is easily reconfigurable whan an elementary cell fails.

REFERENCES

1. Gentleman, W.M. and Kung, H.T. (1981) Matrix triangularization by systolic arrays, In "Proceedings of SPIE Symposium on Real Time Processing IV " Vol. 298 The Society of
   Photo-optical Instrumentation Engineers.
2. Heller, D.E. and Ipsen,C.F. (1983) Systolic networks for
   orthogonal decomposition, SIAM J. Sci. Stat. Comput.
   Vol. 4 No. 2, 261-269
3. Kung, H.T. and Leiserson, C.E. (1980) Systolic arrays
   for VLSI In "Introduction to VLSI" (Eds. C.A. Mead and
   L.A. Conway), Section 8.3
4. Savage,C. (1981) A systolic data structure for connectivity problems, In "Proceedings of CMU Conference on VLSI
   Systems and Computations", (Eds. H.T.Kung et al.) pp 296-
   300, Computer Science Press.

SYSTOLIC ARRAY FOR INTEGER AND LINEAR PROGRAMMING

NAKECHBANDI[*] M. and TREHEL[*] M.

[*] *Laboratoire d'analyse numérique et informatique*
Université de Franche-Comté - Besançon - FRANCE

1. INTRODUCTION

This paper is concerned with applications of the systolic arrays to Combinatorial Optimization. After decomposition of multidimensional problem in a collection of bidimensional subproblems, we must create a simple systolic array to resolve these subproblems. The Optimator SYBCO (2), studied at Ecole des Mines de St Etienne for solving schedule problem, is a typical exemple.

Our paper presents a network to solve bidimensional integer linear programming.

2. DETAILED PRESENTATION

a) *Outline and work of the network* : Consider the following problem :

$$\text{Max } f = a\,x + b\,y \tag{1}$$

$$\alpha_i\,x + \beta_i\,y \leqslant \gamma_i \;\; ; \; 1 \leqslant i \leqslant m \tag{2}$$

$$x_o \leqslant x \leqslant u_1 \;,\; y_o \leqslant y \leqslant u_2 \;,\;\; x,y : \text{integer} \tag{3}$$

We propose a cellular array to compute x,y verifying (2), (3), and maximizing the function **f**. This array is composed of 5 blocks A_1, A_2, B, C, D, each containing a set of identic cells and 3 registers R_1, R_2, R_3 for communication (see figure 2).

The bloc A_1 is interfacing the transmission of x_o, y_o and Δ (the order of discretisation). The blocks A_2, B are interfacing the transmission of a,b and $\alpha_i, \beta_i, \gamma_i$. The block C is a block of calculus. Each cell of C can:

- compute $f = a x + b y$

- test the inequalities $\alpha_i x + \beta_i y \leqslant \gamma_i$

where x,y are coordinates of the cell. The values α_i, β_i, a, b are propagated in the array. There are two steps :

The first step consists in desactivating all cells which don't verify (2), see figure 1.

Fig. 1

● Cellule non active
○ Cellule active

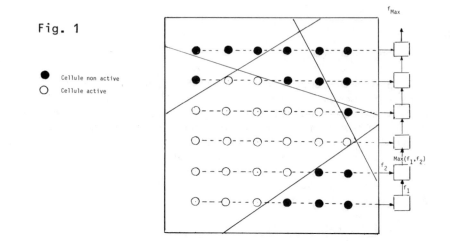

. At the second step, the cell computes $f = a\,x + b\,y$.

Each active cell compares the value of f which has been calculated with the value coming from the next left cell, and transmits the best value to the next right cell. The optima are transmitted from cell to cell to the right column (bloc D) where they are compared between them. The best is the global optimum.

In the figure 2 we find the initial state of the circuit. After m + 2 tops, all cells of block C under the m th diagonal are explored. The cells which don't verify a constraint are noted. When the $\alpha_m, \beta_m, \gamma_m$ of the last constraint are given to the register R_3, the value of a,b are sent from the register R_2 to the block A_2.

Fig.2

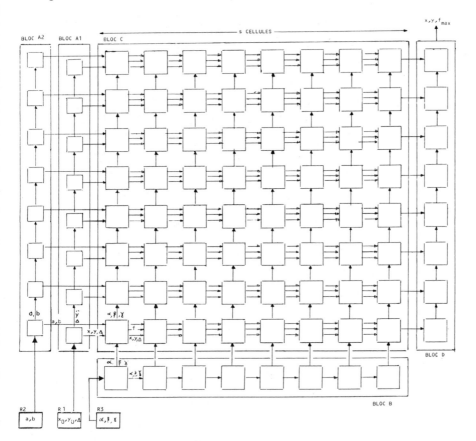

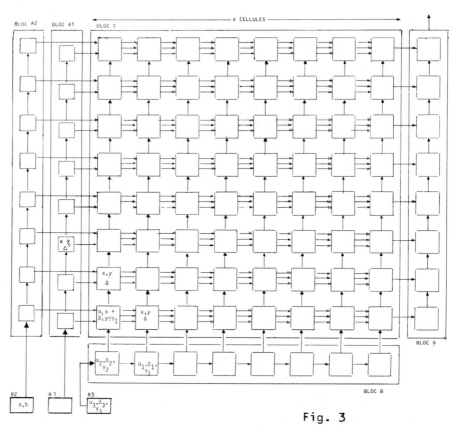

Fig. 3

x,y sont les coordonnées de la cellule.

TOP D'HORLOGE NUMERO **4**

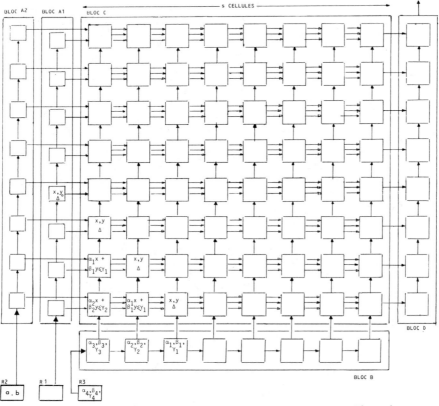

x,y sont les coordonnées de la cellule.

Fig. 4

The figures 5,6 detail some pulsations. The results $f_{max}, x_{max}, y_{max}$ are found after m + 2 + 2s tops.

b) *Complexity* : The time necessary to acquire the m constraints is O(m). The array is s x s cells for C and 4s for A_1, A_2, B, D. The time necessary to obtain $f_{max}, x_{max}, y_{max}$ is O(m + 2s).

For comparison, the algorithm by enumeration of the s^2 nodes would be $O(m s^2)$. Our array is better.

The period (time between the resolution of two problems, Vuillemin (3)) is O(m).

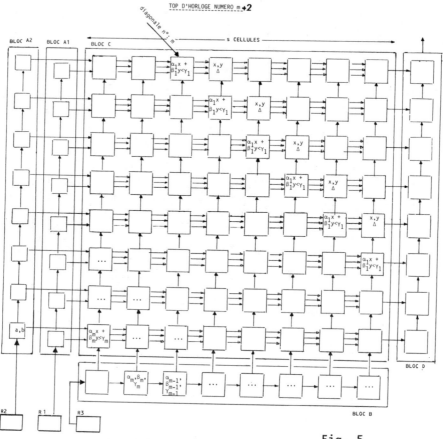

x,y sont les coordonnées de la cellule.

Fig. 5

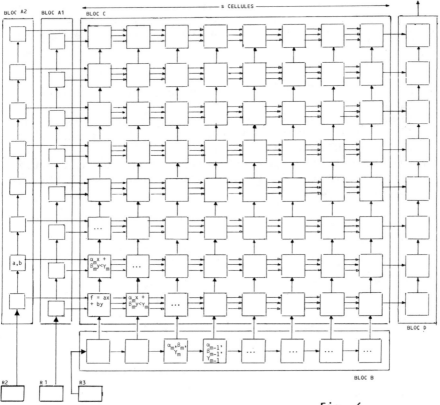

Fig. 6

x,y sont les coordonnées de la cellule.

3. CONCLUSION AND RECOMMENDATIONS

Application of array : The systolic array can be used to
solve multidimensional integer linear programming. An heuristic consists to decompose this problem in collection of
bidimensional linear programs. The resolution of a chain of
subproblems is very efficace by our array.

It can be used also to find an approached solution for
the bidimensional linear programming in real variables. Then
we can use this approach to change two variables at a time
when we pass from a basis to another in the simplex method.

4. REFERENCES

1. Nakechbandi, M. Thèse d'Etat in coure. University of
   Besançon.
2. Nivault, M. and al. (1974). 1er rapport d'évaluation de
   l'optimateur CYBCO. Ecole des Mines de St Etienne, département d'informatique.
3. Vuillemin, (1980). A combinatorial limit to the computing
   power of V.L.S.I. Proceedings of 21-nd Ann. IEEE, Symp.
   on Foun. of Comp. Sc.

THE SYSTEMATIC DESIGN OF
SYSTOLIC ARRAYS

Patrice Quinton
IRISA-CNRS, Campus de Beaulieu
35042 RENNES Cédex
FRANCE

1. INTRODUCTION

The VLSI technology offers us exceptional opportunities to develop parallel computation, for both special-purpose and ge-
neral-purpose devices. Among the several approaches to parallel organization that can take advantage of these new possibilities, the systolic array concept is particularly interesting. As characterized by Kung (1), a systolic array is a
special-purpose parallel device, made out of a few simple
cell types which are regularly and locally connected. Data
circulate through these cells in a very regular fashion and
interact where they meet, giving new results that are sent to
neighbouring cells. A number of systolic arrays have been
designed for implementing efficiently algorithms related to
various areas such as signal processing, numerical analysis,
image processing, etc... However, every systolic array that
has been proposed seems to have been designed in a heuristic
manner.

As the systolic array concept is expected to provide an
efficient framework for the design of special-purpose VLSI
architectures, it seems important to provide designers with
methods that help them to explore various implementations of
the same algorithm. Several alternatives may then be compared
according to various criteria such as the size of the array,
the complexity of the elementary cells, etc...

Attemps to synthesize systolic arrays have been recently
proposed, independently of ours, by Moldovan (2) and Miranker
and Winkler (3). These two approaches start from a program
which defines the problem to be solved. The dependencies

DYNAMICAL SYSTEMS
AND CELLULAR AUTOMATA

347

between the variables of the program, considered as points of Z^n, are extracted. Both methods look for a linear transformation that maps these points onto an a priori given systolic structure.

The method proposed here is based on the possibility of expressing a problem as a set of uniform recurrent equations (4) over a domain of Z^n. When this is the case, it is possible under well-defined conditions to order the computations in such a way that the domain may be mapped onto a systolic array. The main feature of this method is its constructiveness which allows designs to be derived completely automatically from the initial equations. In section 2 we introduce and develop the basic concepts of this approach. In section 3, the method is applied to the convolution product example.

2. FORMAL DESCRIPTION OF THE METHOD

2.1. *Uniform Recurrent Equations*

Consider a subset D of Z^n. To each point z of D is associated a system of equations $E(z)$ given by

$$E(z) \begin{cases} U_1(z) = f(U_1(z+\theta_1), \ldots, U_p(z+\theta p)) \\ U_2(z) = U_2(z) + \theta_2) \\ \ldots \\ U_p(z) = U_2(z+\theta p) \end{cases}$$

where $\theta_1, \theta_2, \ldots, \theta_p$ are vectors of Z^n called *dependence vectors*. Values $U_i(z)$ are assumed to be known outside of D. Such a system is called a *system of uniform recurrent equations* (URE in the following). D is the *domain* of the URE, and the pair (D, Θ), where $\Theta = \{\theta_1, \ldots, \theta_p\}$ is the *dependence graph* of the URE. $x \in D$ is said to *depend strictly* on $y \in D$ if $x = y + \theta_i$ for some dependence vector θ_i. $x \in D$ *depends on* $y \in D$ if

$$x = y + \sum_{i=1}^{p} \alpha_i \theta_i \quad \text{where } \alpha_i\text{'s are non negative integers such that}$$

$$\sum_{i=1}^{p} \alpha_i > 0.$$

We shall suppose that D is the set of integer coordinate points of a convex polyhedral domain of R^n, i.e. a subset of R^n whose elements satisfy a set of linear inequalities. D may then be characterized by its finite set $S = \{S_1, S_2, \ldots, S_u\}$ of *vertices* and its finite set $R = \{r_1, r_2, \ldots, r_v\}$ of *extremal rays* (5) and we have :

$$D=\{x \in Z^n \mid x= \sum_{i=1}^{u} \alpha_i s_i + \sum_{j=1}^{v} \mu_j r_j\}$$

where the α_i's and μ_j's are non negative real numbers and $\sum_{i=1}^{u} \alpha_i = 1$. We suppose moreover, that D has at most one extremal ray r. Notice that vertices s_i and the ray r may always be assumed to be in Z^n.

2.2. Timing functions

A *timing function* for a computation graph (D, Θ), is a mapping t from Z^n to Z such that t is non negative over D, and $\forall x, y \in D$, $t(x) > t(y)$ if x depends on y. If such a function exists, one can solve the URE by computing $E(z)$ at time $t(z)$ since at that time, all input arguments $U_i(z+\theta_i)$ are available. We consider a particular class of timing functions called quasi affine timing functions (QATF) which can be written

$$t(x)= \lfloor \lambda^T x - \alpha \rfloor \qquad \lambda \in Q^n \text{ and } \alpha \in Q$$

where Q denotes the set of rational numbers and $\lfloor u \rfloor$ is the greatest integer lower than or equal to u. We shall denote a QATF as a pair (λ, α). If $\lambda \in Z^n$ and $\alpha \in Z$, t is a strict affine timing function (ATF).

The following theorem gives conditions that must be satisfied by $t=(\lambda, \alpha)$ in order to be a QATF for a dependence graph (D, Θ) :

Theorem 1 : Let (D, Θ) be a dependence graph, and let S and R be respectively the vertices and extremal rays of D. $t=(\lambda, \alpha)$ is a QATF for (D, Θ) if

(i) $-\lambda^T \theta \geq 1$ $\quad \forall \theta \in \Theta$

(ii) $\lambda^T r \geq 0$ $\quad \forall r \in R$

(iii) $\alpha \leq \lambda^T s$ $\quad \forall s \in S$

Moreover, if $\lambda \in Z^n$ and $\alpha \in Z$, the above conditions are also necessary.

The proof of this theorem may be found in (6). Theorem 1 gives a constructive way to find a QATF when there exists one. (i) and (ii) show that the set Λ of possible λ is itself a convex polyhedral domain of R^n. One can thus easily check wether Λ is void or not. For any λ chosen in Λ, (iii) gives the set of values α that are possible.

2.3. *Allocation functions*

Once a QATF $t=(\lambda,\alpha)$ has been found, it remains to map putations onto a finite set of processors in such a way that each processor has at most one calculation to perform at a given instant. Such a mapping will be called an allocation function. More precisely, an allocation function is a mapping a from Z^n to Z^n such that $a(D)$ is finite and $\forall x,y \in D$, $a(x)=a(y) => t(x) \neq t(y)$. In order to obtain designs that have good regularity properties, we shall consider a particular class of allocation functions that we call quasi linear allocation functions (QLAF). Let $a(x)=(a_1(x),\ldots,a_n(x))^T$. Then a is a QLAF if $a_1,a_2,\ldots,a_{n-1}$ are linear and $a_n(x)$ has the form $b(x) \bmod q$ where b is a linear function and q is a positive integer. A simple way to find QLAF's is to "project" the domain D along a conveniently chosen direction, as described in the following theorem :

Theorem 2 :

Let u be a non null vector of Z^n such that $\lambda^T u > 0$ and $u = \dfrac{u}{GCD(u_1,\ldots,u_n)}$. Assume that $u_j \neq 0$. Let a be the mapping defined by $a(x)=(a_1(x),\ldots,a_n(x))$ where

$$a_k = u_j x_k - u_k x_j \qquad \forall k : 1 \leqslant k < j$$
$$a_k = u_j x_{k+1} - u_{k+1} x_j \qquad \forall k : j \leqslant k < n$$
$$a_n(x) = x_j \bmod \lceil 1/\lambda^T u \rceil$$

where $\lceil v \rceil$ denotes the smallest integer greater than or equal to v.

If D has no ray, or if the ray r of D is parallel to u, then a is a QLAF for (D,θ) and $t=(\lambda,\alpha)$.

The proof of Theorem 2 may be found in (6). Notice that if t is an ATF, i.e., $\lambda \in Z^n$ and $\alpha \in Z$, then $\lceil \dfrac{1}{\lambda^T u} \rceil = 1$ and a is a projection of Z^n along the direction defined by u.

The above theorem gives a very simple method for deriving various designs from a unique URE, when D has no ray. When D has a ray r, one can also find an allocation function for a direction u which is not collinear to r by folding D along r first. The interested reader may find in (6) an example of such a transformation for the convolution product.

2.4. Specification of the systolic array

As we shall see in this section, a URE, a QATF $t=(\lambda,\alpha)$ and a QLAF a define completely a systolic array that supports the computation of the URE.

To each point π of $a(D)$ is associated one cell of the systolic array. To each point z of D, let us associate the program $P(z)$ defined by

for j:=1 to p do
  if $a(z+\Theta_j) \in a(D)$
    then read $U_j(z+\Theta_j)$ from cell $a(z+\Theta_j)$
    else read $U_j(z+\Theta_j)$ from memory ;
for j:=1 to p do compute $U_j(z)$ according to the URE ;

for j:=1 to p do
  if $a(z-\Theta_j) \in a(D)$
    then send $U_j(z)$ to cell $a(z-\Theta_j)$
        with a delay $\Delta_j(z)=t(z-\Theta_j)-t(z)$
    else write $U_j(z)$ into memory ;

Cell π of the array has to perform program $P(z)$ at time $t(z)$, for every z lying in $a^{-1}(\pi)$. One can prove that :

Theorem 3 :

(i) $\forall \pi, \forall n \geqslant 0$, there exists at most one z such that $t(z)=n$
(ii) $\forall z \in D, \forall \Theta_j \in \Theta, U_j(z+\Theta_j)$ is available at time $t(z)$.

(iii) $\forall z \in a^{-1}(\pi)$:
$$a(z+\Theta_j)=(\pi_1+a_1(\Theta_j),\ldots,\pi_{n-1}+a_{n-1}(\Theta_j),$$
$$\pi_n+a_n(\Theta_j) \bmod |\frac{1}{\lambda^T u}|)$$

(iv) Suppose D has one ray. Let $(P_n)_{n\in\mathbb{N}}$ be the sequence of programs performed by π. There exists k such that $(P_n)_{n>k}$ is periodic.

Parts(i) and (ii) of the theorem result immediately from Theorem 1 and Theorem 2. Parts (iii) and (iv) may be established by examining the values taken by $a(z+\theta_j)$, $a(z-\theta_j)$ and $\Delta_j(z)$ for every $z \in a^{-1}(\Pi)$.

The consequences of Theorem 3 are the following :

- Each cell has at most one program to perform at a given instant (i), and input values of this program have previously been calculated (ii) ;

- Each cell must only be connected to a finite number of other cells (iii). Moreover, (iii) may be interpreted as a property of locality and extensibility of the design, when one considers a class of URE rather than a unique URE (for example, the convolution product with coefficient vectors of any size).

- Finally, a cell must contain only a finite number of different programs and can determine what program to perform using a finite state control mechanism. (iv)

3. APPLICATION TO THE CONVOLUTION PRODUCT

In the following section, we show how the method described in section 2 may be applied to find various designs for the convolution product. Given a sequence $x(0),x(1),\ldots,x(i)$ and a set of coefficients $w(0),w(1),\ldots,w(K)$, the convolution algorithm consists in computing the sequence $y(0),\ldots,y(i)$ given by

$$y(i) = \sum_{k=1}^{K} w(k)\ x(i-k) \qquad (1)$$

This equation may be rewritten as

$$y(i,-1)=0 \qquad 0 \leqslant i$$
$$y(i,k)=y(i,k-1)+w(k)x(i-k) \qquad (2)$$

From (2), one can find a first URE by noticing that $w(k)$, which is necessary for the calculation of $y(i,k)$, may be provided as a by-product of the calculation of $y(i-1,k)$; in the same way, $x(i-k)$ may be taken from the calculation of $y(i-1,k-1)$. We get :

i: $0 \leqslant i$;   k : $0 \leqslant k \leqslant K$

$$y(i,k)=y(i,k-1)+w(i-1,k)x(i-1,k-1)$$
$$w(i,k)=w(i-1,k) \qquad (3)$$
$$x(i,k)=x(i-1,k-1)$$

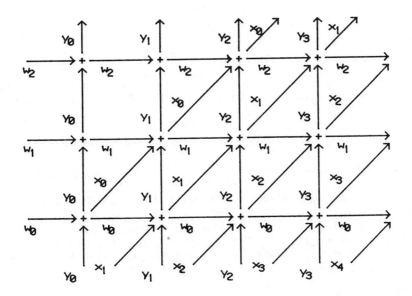

Fig.1: Dependence graph for the convolution product (K=2)

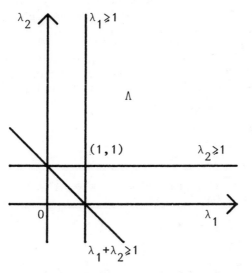

Fig.2: The convex polyhedral domain Λ
defined by Theorem 1

The dependence graph (D,Θ) of (3) is

$$D=\{(i,k) \mid 0{\leqslant}i,\ 0{\leqslant}k{\leqslant}K\}$$

and

$$\Theta=\{(0,-1),(-1,0),\ (-1,-1)\}$$

as shown by Fig.1. D has two vertices $S_1=(0,0)$ and $S_2=(0,K)$, and one extremal ray $r=(1,0)$ (r is defined up to a non-zero scalar factor).

Let $\quad \lambda=(\lambda_1,\lambda_2)^T$. From theorem 1, a QATF (λ,α) must satisfy

$$\lambda_2{\geqslant}1\ ;\ \lambda_1{\geqslant}1\ ;\ \lambda_1+\lambda_2{\geqslant}0\ ;\ \lambda_1{\geqslant}0$$

These conditions define a convex polyhedral domain of Z^n which is depicted by Fig.2. One can for example take $\lambda_1=1$ and $\lambda_2=1$. Condition (iii) of theorem 1 gives then $\alpha{\leqslant}0$. A possible timing function for (D,Θ) is thus $t(i,k)=i+k$. (It turns out that this function is in fact optimal). Fig.3 shows (D,Θ) and $t(i,k)$.

Since D has a ray $r=(1,0)$, the only way to apply theorem 2 is to take the allocation function $a(i,k)=(k,i \bmod 1)$ (since $\lambda^T r=1$). This results in K+1 cells : cell $(k,0)$ executes the computations associated to the points (i,k) of D. The systolic array that results is the well known architecture depicted by Fig.4.

Another solution may be found by returning to equation (2). Notice that value $w(k)$ may be provided by calculation of $y(i-2,k)$ rather than $y(i-1,k)$. We thus find the new URE

$\forall$ i: $0{\leqslant}i$; $\forall$ k : $0{\leqslant}k{\leqslant}K$

$$
\begin{aligned}
&y(i,k)=y(i,k-1)+w(i-2,k)x(i-1,k-1)\\
&w(i,k)=w(i-2,k) \qquad\qquad\qquad\qquad\qquad (4)\\
&x(i,k)=x(i-1,k-1)
\end{aligned}
$$

The dependence vectors are now (Fig.5) :
$$\Theta=\ \{(0,-1),(-2,0),(-1,-1)\}$$

Applying theorem 1 gives
$$\lambda_2{\geqslant}1\ ;\ 2\lambda_1{\geqslant}1\ ;\ \lambda_1+\lambda_2{\geqslant}1\ ;\ \lambda_1{\geqslant}0$$

Domain Λ has a vertex $\lambda=(\frac{1}{2},1)$ from which one can get the timing function $t(i,k)=\lfloor i/2+k \rfloor$ (Fig. 5).

Take $u=r=(1,0)$. Since $\lceil \dfrac{1}{\lambda^T 4} \rceil=2$, we obtain the allocation function

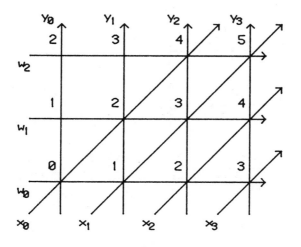

Fig.3 : Timing function t(i,k)=i+k

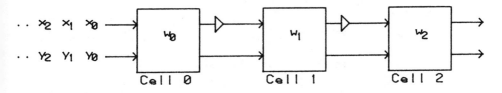

Fig.4 : Systolic array for the convolution product

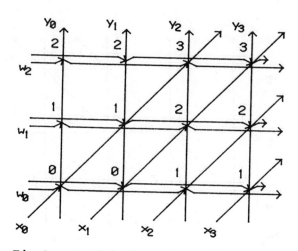

Fig.5 : Another dependence graph for the
convolution product and timing
function $t(i,k) = \left\lfloor \dfrac{i}{2} + k \right\rfloor$

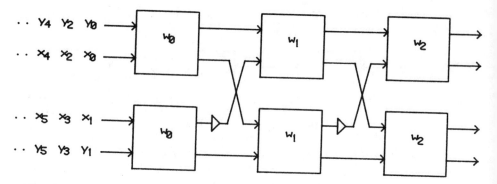

Fig.6 : Systolic array derived from Fig.5.

$a(i,k)=(k,i \bmod 2)$

The systolic array that results(Fig.6) has $2(K+1)$ cells, and computes simultaneously y_{2p} and y_{2p+1} from x_{2p} and x_{2p+1}. This design is twice as fast as the design of Fig.4.

4. CONCLUSION

In this paper, we have described a powerful and fully constructive approach to the design of systolic arrays. This method has been exemplified on the convolution product. Other problems that can be solved by this method include matrix product, recursive filtering, LU decomposition and dynamic programming (these last two problems require a slight extension of Uniform Recurrent Equations). Following this approach a computer aided design system named DIASTOL(7) is currently being implemented. The purpose of this system is to provide a designer with an environment which allows him to find several alternative solutions for a given problem.

REFERENCES

1. Kung, H.T., (1982). Why Systolic Architectures ?, Computer, Vol.15, No.1, pp. 37-46.

2. Moldovan, D.I., (1983). On the Design of Algorithms for VLSI Systolic Arrays, Proceedings of the IEEE, Vol.71, No.1, pp. 113-120.

3. Miranker, W.L., Winkler A., (1982). Spacetime Representations of Systolic Computational Structures, IBM Research Report RC 9775.

4. Karp, R.M., Miller, R.E., Winograd, S., (1967). The Organization of Computations for Uniform Recurrence Equations, JACM, Vol.14, No.3, pp. 563-590.

5. Rockafellar, R.T., (1970). Convex Analysis. Princeton University Press, Princeton.

6. Quinton, P., (1983). The Systematic Design of Systolic Arrays, INRIA Research Report No. 216.

7. André, F., Frison, P., Quinton, P., (1983). DIASTOL : un système de conception assistée pour les architectures systoliques, IRISA Research Report, to appear.

VLSI and Computer Algebra :
The G.C.D. Example

James Davenport* & Yves Robert

Analyse Numérique, Laboratoire I.M.A.G.,

B.P. 68, 38402 SAINT MARTIN D'HERES CEDEX, France

1. INTRODUCTION

In computer algebra systems (such as REDUCE and MACSYMA), the expressions manipulated are large and complicated, for example rational functions in several variables, and hence the algorithms suffer, in general, from a relative slowness (even when they are implemented on the most powerful computers).

Could one hope to increase the capacity or speed (or both) of these systems by implementing certain basic functions in hardware? We are not talking here about completely re-writing a computer algebra system, truly a major task, but adding certain specialised processors to increase the speed of critical algorithms. The progress of VLSI technology means that such implementations are **now** possible.

2. GENERAL REMARKS

Consider, for the moment, the example of multiplication of two polynomials P and Q from $\mathbf{Z}[x]$. The systolic approach introduced by Kung [8], appears attractive, and indeed he proposed a systolic array for this calculation [9]. But the polynomials found in computer algebra are by no means any old polynomials: they are often sparse, and their coefficients are not of the same size. This

---

* Present address: School of Mathematics, University of Bath, Bath, England, BA2 7AY. This work was performed while the first author was a Research Fellow of Emmanuel College, Cambridge.

last is a major drawback, since the size of calculation possible on a circuit is inherently limited; it is necessary to choose between two implementations of integer multiplication:

- small integers, occasionally causing overflows;
- large integers*, almost never completely used, and therefore losing the speed advantage.

In order to achieve a high-performance implementation, it seems that the circuit must use small integers, but that the algorithm must not cause overflow.

The calculation of g.c.d.s is an even more eloquent example: if we choose, after Brown [3], $P(x) = x^8 + x^6 - 3x^4 - 3x^3 + 8x^2 + 2x + 5$ and $Q(x) = 3x^6 + 5x^4 - 4x^2 - 9x + 21$, we find, depending on the exact algorithm chosen:

- 1259333879550074310093114199218750 0 (direct Euclid's algorithm);
- 869224534119352216237500 ("reduced" algorithm);
- 23552630096885020 (sub-resultant algorithm).

We started with two-digit numbers, and finished with 35, 24 or 17 digits, even though the true answer is that the polynomials are relatively prime.

To make matters worse, g.c.d. calculation is a corner-stone of much computer algebra, from the manipulation of fractions to the calculation of integrals. We could cite other examples, from integration [5] or elsewhere, but it is well-known that, in nearly all algorithms of computer algebra, "intermediate expression swell" occurs, and this complicates the search for a hardware implementation as much as it does the existing software. If we examine the methods already developed for use on conventional computers (classical algorithms, modular algorithms and p-adic algorithms), we see that a computer algebra system based on specialised circuits will have to be based on non-classical algorithms, even more than a purely software implementation has to be. For real multi-variate problems, one probably wants to adopt a sparse algorithm, such as [15].

3. A CIRCUIT TO CALCULATE THE G.C.D. OF TWO POLYNOMIALS.

In the rest of this paper, we will describe the implementation of an algorithm for the modular calculation of the g.c.d., solving in the process the problems of size mentioned above. Our choice of this problem is not accidental: this procedure is very often invoked, and can take over 95% of the total time of a calculation [4].

---

* Making the unrealistic assumption that we can bound the size of integers appearing in a calculation before configuring the hardware.

3.1. MODULAR G.C.D.

We are working in the following context: with respect to the original problem (I), we will produce a hardware implementation of step (E), leaving the reduction and reconstruction to be performed classically.

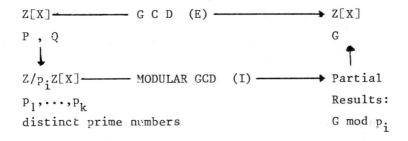

In this approach, it is necessary to do perform several g.c.d. calculations with different choices of p, and to deduce the value of the g.c.d. over the integers from the various modular values by means of the Chinese Remainder Theorem. To explain this, we introduce the operator M_p, which performs reduction modulo p, taking Z to $Z/pZ = Z_p$ and $Z[X]$ to $Z_p[X]$. If p does not divide the leading coefficients $lc(P)$ and $lc(Q)$ of P and Q, then $M_p(\gcd(P,Q))$ divides $\gcd(M_p(P),M_p(Q))$. We say that p is *lucky* if these two polynomials are equal. There are only a finite number of unlucky primes - those that divide the resultant $\text{Res}(P/G,Q/G)$, where G is the g.c.d. of P and Q over the integers [3, 10]. We give below the procedure GCD for this calculation - it is very similar to that of Brown [3] except that we use a bound based on Landau's inequality [12, 13].

```
      Procedure GCD(P,Q)
      Size:=2*Min(Max(|Coeff(P)|),Max(|Coeff(Q)|))
L1:   Base:=a prime number
      SoFar:=ModularGcd(P,Q,Base)
L2:   Degree:=deg(SoFar)
      if degree = 0 then return 1
      While Base < 2**Degree * Size do
L3:     p:=a prime number
      Temp:=ModularGcd(P,Q,p)
      If deg(Temp) > Degree then go to L3
      If deg(Temp) < Degree then
        Base:=p
        SoFar:=Temp
        go to L2
      SoFar:=ChineseRemainder(SoFar,Temp,Base,p)
      Base:=Base * p
      If SoFar does not divide P then go to L1
```

```
If SoFar does not divide Q then go to L1
return SoFar
```

3.2. COMPLEXITY

The aim of the complexity analysis which follows is
a) to specify the total cost of the modular g.c.d. algorithm
b) to discuss the relative cost of step (E), the only step not to be performed on the conventional computer.

Notation

We will suppose that all the prime numbers used are larger than, but close to, 2^β ($\beta = 7$ for example). We write:

$$n_P = \deg(P) + 1, \; n_Q = \deg(Q) + 1,$$

$$l_P = \max \log (P_i), \; l_Q = \max \log (Q_i),$$

$$n = \max(n_P, n_Q), \; l = \max(l_P, l_Q), \; d = \deg(\gcd(P,Q)).$$

We suppose for simplicity that P and Q are monic. All logarithms are to base 2.

Asymptotic Formulae

The following table summarises the various bounds that are relevant

unlucky primes	N_i	$2n(\frac{1}{2} \log 2n + l)/\beta$
lucky primes necessary	N_f	$[(d-1)(\frac{1}{2} \log n + l) + 1]/\beta$
Chinese Remainder cost	C_{CR}	$d\beta N_f O(\log^{3+\varepsilon} N_f + \log^{2+\varepsilon} \beta \log N_f)$

N_i comes from bounding the size of the resultant [10, 13], and improves on that found in [3]. N_f is deduced from theorem 4 of [13] (see also theorem 2 of [12]). The cost of the Chinese Remainder process is deduced assuming a tree-based method (though we wrote a linear method in the program above) - see [1] section 8.11 or [5] chapter 2.

Total cost

We suppose that all calculation modulo p takes a fixed amount of time (we have chosen $\beta = 7$, and most computers use fixed constant time for 16 or 32 bits). This gives, for the three stages:
- reduction modulo p: $(N_f + N_i)(n_P l_P + n_Q l_Q)$
- modular g.c.d.: not performed on the main computer;
- Chinese Remainder Theorem: $d\beta N_f O$(logarithms).

Furthermore, step (E) takes time $O(n)$ (assuming enough hardware is available), and so, before the last reductions are finished, the first inputs will be ready for the Chinese Remainder process. The reduction is bounded by $O(n^2 \log nl, n^2 l^2)/\beta$, and the re-construction by $d^2 lO$(logarithms), and hence we have an overall bound of $O(n^2 l^2, n^2 l$ logarithms).

Some Pragmatic Remarks

The reduction is the most expensive step, and the factor of $1/\beta$ means that progress in VLSI technology, by permitting the treatment of larger numbers on one chip, will diminish N_f and N_i, leading directly to a cheaper implementation.

In practice, unlucky primes are rare, and the bounds given for the coefficients, and hence N_f, are pessimistic. Brown [3], proposes a "heuristic" bound of l, and one could well stop at this point and see if one indeed had a true g.c.d. Of course, l may be larger than the true bound we compute, and a better heuristic bound might be $\min(l_P, l_Q)$.

Finally, we must emphasise that the possibility of unlucky primes, small though it may be, means that we can not pre-determine the primes to use, and hence can not pre-calculate the inverses in the tree-structure for the Chinese Remainder Theorem.

3.3. ALGORITHM AND CIRCUIT SPECIFICATION

For the g.c.d. calculation, we use the method of degree reduction, related to Euclid's algorithm as subtraction is to division:

```
while d(P)*d(Q) > 0 do
    if d(A)≥d(B)
        then A := A - B * x** (d(A)-d(B)) * (lc(A)/lc(B))
        else B := B - A * x** (d(B)-d(A)) * (lc(B)/lc(A))
    if A = 0 then GCD:=B else GCD:=A
```

Each step of the algorithm corresponds to a partial elimination, and it is clear from this form that the algorithm lends itself to a certain parallelism. The execution can be divided into elementary operations performed by identical processors:

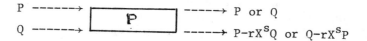

Better still, we can avoid the multiplication by a power x^s (which consists of "shifting" the coefficients of one of the polynomials by s possibilities): if after an elimination step, we have decreased the degree by more than one, we decide to output a zero as the leading coefficient of that polynomial. In this case, the role of the box **P** consists simply in:
- eliminating the zero coefficient;
- decreasing the degree by one;
- transmitting the other polynomial unchanged.

We thus obtain the desired pipe-line implementation in using a vector of identical processors, whose operation is outlined in figure 1.

$$a_0, \ldots, a_n, n \longrightarrow$$
$$b_0, \ldots, b_m, m \longrightarrow$$

CONTROL PART
(n, m, a_n, b_m)

$\downarrow r$

OPERATIVE PART

$\longrightarrow$ new
$\longrightarrow$ values

Figure 1

1) If n or $m = 0$, transmit A and B without modification.
2) Otherwise, if a_n or $b_m = 0$, shift one of the polynomials and transmit the other. A and B are never simultaneously zero. For example, if $a_n = 0$, we suppress a_n, set n to $n-1$, and transmit B unchanged.
3) Otherwise, perform a partial elimination, e.g. if $n \geq m$, $a_i := a_i - r*b_{i+m-n}$, where $r = a_n/b_m$.

We can simplify the presentation somewhat, in remarking that (2) is the same step as (3), but with $r = 0$ similarly, we write $r = 0$ in case 1.

3.4. USE OF MODULAR CALCULATION.

We have seen that calculation modulo a prime number p is indispensable for avoiding the problems of intermediate expression swell. All the calculations are then conducted modulo p, where p is a known *fixed* (the architecture is greatly simplified if p is known when the circuit is designed) prime number. Let us make two preliminary remarks.

i) Modular calculation is about two times as expensive as integer calculation (e.g. the calculation of $a + b$ (modulo p) requires the computation of $a + b$ and $a + b - p$). It is true that the choice $p = 2^n \pm 1$ greatly simplifies the calculations, but we have to be able to treat arbitrary p.

ii) The principal difficulty is in the division. In general one either uses the extended Euclidean algorithm in Z_p, or one uses a table of inverses.

If we are prepared to use a table, though, we can give a quick and elegant solution for the computation of the four elementary operations $+,-,*,/$ modulo p.

An integer modulo p is usually represented as an element of the set $E_p = \{0,1,..,p + 1\}$. Let us fix a primitive root K of p (i.e. a generator of the multiplicative group $Z_p - \{0\}$). Then every non-zero element of E_p can be written uniquely in the form K^i ($0 \le i \le p-2$). If we attach the special representation $*$ to 0, we can represent the element K^i of E_p by i. Multiplication and division are then transformed into addition and division, according to the rules :

$$i \overline{*} j = \begin{cases} * \text{ if } i \text{ or } j = * \\ i + j \bmod p-1 \text{ otherwise} \end{cases}$$

$$i \overline{/} j = \begin{cases} * \text{ if } i = * \\ \text{undefined if } j = * \\ i - j \bmod p-1 \text{ otherwise} \end{cases}$$

For addition (and subtraction) we use *Zech logarithms* [14]: define $Z(u)$ by

$$K^{Z(u)} = 1 + K^u \text{ for } u\varepsilon\{0,1,..,p-2\} - \{\frac{p-1}{2}\},$$

$$Z(\frac{p-1}{2}) = * \text{ and } Z(*) = 0.$$

One deduces the following rules for addition:

$$i \overline{+} j = \begin{cases} j \text{ if } i = * \\ i \text{ if } j = * \\ * \text{ if } j - i = \pm (p-1)/2 \\ i + Z(j-i) \bmod p-1 \text{ otherwise} \end{cases}$$

and similar rules for subtraction.

These representations modulo a prime number p of n bits have $n + 1$ bits, since a special bit is reserved for $*$. The use of this one-dimensional table of Zech logarithms enables us to eliminate all the bivariate operations. In the units λ^2 [11], we have the following costs for the various components:

Component	name	surface
n-bit multiplier	M	$2^{12}n^2$
n-bit adder	A	$2^{12}n$
n-bit table	T	$3n2^{n+6}$

If we assume that division is about the same surface as multiplication, and that mod p units are twice the size of integer ones, we have that the cost of a conventional circuit (multiplier, divisor, adder) is $C(n) = 4M + 2A$, while the cost for a Zech-based solution is $Z(n) = T + 6A$. The ratio is

$$R(n) = \frac{Z(n)}{C(n)} = 3\frac{2^n + 2^7}{n2^8 + 2^7}.$$

This is an increasing function of n, but takes on the value $9/17$ at $n = 8$, meaning that the Zech representation uses just over half the area.

4. CONCLUSION

In [2], Brent and Kung present systolic architectures for the calculation of greatest common divisors of polynomials. The high-level architecture of their circuits is very close to those we present in section 3.3. But our modular approach has a double interest:
-it is the only proposal which solves the problem of "intermediate expression swell", a *sine qua non* of computer algebra algorithms;
-it permits, at the lowest level, an original and parallel implementation of the architecture of each cell.
References [6] and [7] complete this study.

The algorithmic description which we have given shows that it is possible, with today's technology, to build VLSI units specialised for the modular gcd algorithm. Obviously this is a first step - what other VLSI algorithms can be designed for computer algebra? We already know, after [7], that this method can be extended to compute not only the g.c.d. G of P and Q, but also polynomials A and B such that $AP + BQ = G$ (analogous to the extended Euclidean algorithm).

5. REFERENCES

[1] Aho,A.V., Hopcroft,J.E. & UllmanJ.D., "The Design and Analysis of Computer Algorithms". Addison-Wesley, Reading, Mass., 1974.
[2] Brent,R.P. & Kung,H.T., Systolic VLSI arrays for linear-time gcd computation. *In* "VLSI 83" (Eds. F. Anceau & E.J. Aas) pp. 145-154. Elsevier Science Publishers, 1983.
[3] Brown,W.S., On Euclid's algorithm and the computation of polynomial greatest common divisors. *J. ACM 18*(1971) pp. 478-404.

[4] Davenport,J.H., "On the Integration of Algebraic Functions" (Lecture Notes in Computer Science 102). Springer-Verlag, Berlin-Heidelberg-New York, 1981.

[5] Davenport,J.H., "Intégration Formelle", RR IMAG Grenoble 375, June 1983.

[6] Davenport,J.H., "VLSI et Calcul Formel", RR IMAG Grenoble 357, March 1983.

[7] Davenport,J.H.& Robert,Y., "PGCD et VLSI", RR IMAG Grenoble 358, March 1983.

[8] Kung,H.T., "Why systolic architectures", TR Carnegie-Mellon University, 1981.

[9] Kung,H.T., Use of VLSI in algebraic computation: some suggestions. *In* "Proc. SYMSAC 81" (Ed. P.S. Wang) pp. 218-222. ACM, New York, 1981.

[10] Loos,R., Generalized polynomial remainder sequences. *In* "Computer Algebra Symbolic and Algebraic Computation" (Ed. B. Buchberger et al.) pp. 115-137. Springer-Verlag, Vienna, 1982.

[11] Mead,C. & Conway,L., "Introduction to VLSI Systems". Addison-Wesley, Reading, Mass., 1980.

[12] Mignotte, M., Some inequalities about univariate polynomials.

[13] Mignotte, M., Some useful bounds. *In* "Computer Algebra Symbolic and Algebraic Computation" (Ed. B. Buchberger et al.) pp. 259-263. Springer-Verlag, Vienna, 1982.

[14] Zimmer,H.G., "Computational Methods in Number Theory" (Lecture Notes in Mathematics 272) Springer-Verlag, Berlin-Heidelberg-New York, 1972.

[15] Zippel,R.E., Newton's iteration and the sparse Hensel algorithm. *In* "Proc. SYMSAC 81" (Ed. P.S. Wang) pp. 68-72. ACM, New York, 1981.

ITERATION PROPERTIES OF TRANSFORMATIONS OF FINITE SETS WITH APPLICATION TO MULTIVALUED LOGIC

Corina Reischer[*] and Dan A. Simovici[+]

[*]Département de Mathématiques, Université du Québec à Trois-Rivières, Trois-Rivières C.P.500, Qué. G9A 5H7, Canada

[+]Department of Mathematics, University of Massachusetts at Boston, Harbor Campus, Boston, MA 02125, USA

INTRODUCTION

We discuss in this paper several iteration properties of transformations of finite sets. This study, which is technically motivated by the existence of cascade circuits, produces several new structural properties of transformations of finite sets as well as criteria for the existence of cascade realizations of multivalued switching functions.

Our approach is essentially graph-theory oriented ; however, certain results have an algebraic character. We hope that the decomposition result contained in Theorem 4 will enable us to produce more refined versions of the iteration property contained in Theorem 1.

ITERATION PROPERTIES OF TRANSFORMATIONS OF FINITE SETS

We shall designate here the finite set $\{0, \ldots, m-1\}$ by $\langle m \rangle$ (where $m \geq 1$). The monoid of all transformations of the set $\langle m \rangle$ will be denoted by T_m ; the binary operation of this monoid is the composition of mappings, while the unit element is 1_m, the identical transformation of the set $\langle m \rangle$.

We shall attach to a transformation $f \in T_m$ a graph $G_f = (\langle m \rangle, U)$, having $\langle m \rangle$ as set of vertices; the set of edges is $U = \{(x, f(x)) \mid x \in \langle m \rangle\}$. Since the out-degree of each vertex $x \in \langle m \rangle$ is 1, it is clear that G_f consists of oriented cycles to which trees may be attached by their roots. When f is a permutation the graph G_f consists only from cycles.

Let $\kappa(m)$ be the least common multiple of the numbers $1, \ldots, m$.

THEOREM 1. (4) Let k, l be two natural numbers, $k > l > 0$. Then $f^k = f^l$ for *all* transformations $f \in T_m$ if and only if $l \geq m-1$ and $k \equiv l \pmod{\kappa(m)}$.

Proof Let f be a fixed element of T_m. Let a be the

length of the longest attached branch in the graph G_f. If $l \geq a$, for any $x \in \langle m \rangle$, $f^l(x)$ will be a vertex on a directed cycle. Therefore, if b is the least common multiple of the cycles lengths we shall have $f^k(x) = f^l(x)$, for all $x \in \langle m \rangle$, if $k \equiv l \pmod{b}$. Varying the transformations we get the necessity of the statement. The proof of sufficency is similar.   ■ ■ ■

COROLLARY We have $F^k = F^l$, for all p-valued, q-ary switching functions $F : \{0, 1, \ldots, p-1\}^q \rightarrow \{0, 1, \ldots, p-1\}^q$ iff $l \geq p^q-1$ and $k \equiv l \pmod{\kappa(p^q)}$.

Each Boolean function $F : \{0, 1\}^q \rightarrow \{0, 1\}^q$ satisfies the equality

(1) $$F^{2^q-1} = F^{2^q-1+\kappa(2^q)}.$$

We have considered in (3) the *product* of two Boolean functions $f : \{0, 1\}^n \rightarrow \{0, 1\}$ and $g : \{0, 1\}^n \rightarrow \{0, 1\}$ defined by

$$f \circ_i g(x_1, \ldots, x_n) = f(x_1, \ldots, x_{i-1}, g(x_1, \ldots, x_n), x_{i+1}, \ldots, x_n)$$

If we regard $x_1, \ldots, x_{i-1}, x_{i+1}, \ldots, x_n$ as fixed parameters then this operation is essentially a superposition of one-variable functions, hence with respect to this special product we have $f^3 = f$ (by taking $q = 1$ in (1)) as it was established by Lyngholm and Yourgrau (2).

In view of this property only square roots of Boolean functions have sense to be considered with respect to the operation "$\circ_i$". Using the result contained by (1) it makes sense now to study roots of order $2^q-2+\kappa(2^q)$ if the iteration is considered with respect to all the variables (3,7).

We have shown in (4) that $ln \, \kappa(m) = 0(m)$, i.e., $\kappa(m)$ grows exponentially with m. This points that the number of possible different iteration powers of a switching function $F : \{0, 1\}^n \rightarrow \{0, 1\}^n$ grows exponentially with n.

We also gave in (4) a matricial approach of iteration properties of switching functions, providing also a certain limitation on the computational power of the cascades with can be constructed using a switching circuit described by the function F.

ROOTS OF PERMUTATIONS

Let $f \in T_m$. A function $g \in T_m$ is a u-*ary root* of the function f if $g^u = f$. The powers are considered here with respect to the iteration of functions.

In order to establish a criterion for existance of u-ary roots of a function $f \in T_m$, let us consider (6) :

DEFINITION 1. A *simple function* is a function $h \in T_m$,

$h \neq 1_m$, whose graph G_h contains an unique cycle C such that every vertex $x \in \langle m \rangle$ satisfies one of the following two conditions :

  i) either x is a fixed point of h (i.e. $h(x) = x$) or
  ii) there exists $b \in N$ such that $h^b(x) \in C$.

THEOREM 2. (6) For every function $f \in T_m$, $f \neq 1_m$, there exists a family of simple functions $\{h_1, \ldots, h_l\}$ such that $f = h_1 \circ \ldots \circ h_l$ and $h_i \circ h_j = h_j \circ h_i$, for $1 \leq i, j \leq l$.
  We shall refer the functions $h_1, \ldots, h_l$, as the *components* of the function f.

DEFINITION 2. The *spectrum* of a function $f : \langle m \rangle \to \langle m \rangle$, $f \neq 1_m$, is a sequence of natural numbers

$$Spec(f) = (s_1, s_2, \ldots, s_n, \ldots),$$

where s_n is the number of simple functions, components of f, whose unique cycle has length n; $Spec(1_m) = (m, 0, 0, \ldots)$.

  Clearly, since f is a transformation of a finite set, only a finite number of initial components of $Spec(f)$ are non-null.

THEOREM 3. (6) Let $f : \langle m \rangle \to \langle m \rangle$ be a permutation of $\langle m \rangle$ and let u be a prime number. This function has a root of order u iff $l \equiv 0 \pmod{u}$ implies $t_l \equiv 0 \pmod{u}$, where $Spec(f) = (t_1, \ldots, t_l, \ldots)$.
  In the particular case of Boolean functions of n variables, $f : \langle 2 \rangle^n \to \langle 2 \rangle$ and considering the special product "$\circ_i$" (3), the above Theorem permits us to retrieve the necessity of the positivity of the function f in x_i in order to have square roots.
  In (6) we gave also an algorithm to find out the roots of order u of a permutation $f : \langle m \rangle \to \langle m \rangle$, where u is a prime.

STRUCTURAL PROPERTIES OF TRANSFORMATIONS OF FINITE SETS

For $f \in T_m$ we shall consider the *support* of f :

$$Supp(f) = \{x \mid x \in \langle m \rangle, \ x \neq f(x)\}.$$

  The *set of fixpoints* of f is given by

$$Fp(f) = \langle m \rangle \setminus Supp(f).$$

  The *rank* and the *defect* of f are defined by

$$rank(f) = |\{f(x) \mid x \in \langle m \rangle\}|$$

and

$$def(f) = m - rank(f).$$

It is clear that $def(f)$ is given by the number of vertices of G_f whose in-degree is 0.

The *support graph* of $f \in T_m$ is the subgraph of G_f generated by $Supp(f)$.

DEFINITION 3. The transformation $f \in T_m$ is *aperiodical* if for every $x \in Supp(f)$ there is an element $x_0 \in Fp(f)$ and a number k such that for $n \geq k$ we have $f^n(x_0) = x_0$.

The graph G_f of an aperiodical transformation consists only from loops and trees pending on loops.

Let $h \in T_m$ be a simple function, whose unique cycle C is $C = \{x_{i_1}, \ldots, x_{i_k}\}$; the set of vertices of C on which trees are pending is $\{x_{j_1}, \ldots, x_{j_l}\}$, where $\{j_1, \ldots, j_l\} \subset \{i_1, \ldots, i_k\}$. We shall denote by $g_{j_1}, \ldots, g_{j_l}$ the aperiodic functions whose support graphs are the trees pending on $x_{j_1}, \ldots, x_{j_l}$, respectively. Let g be the permutation whose support graph is the cycle C.

LEMMA 1. We have $h = g_{j_1} \circ \ldots \circ g_{j_l} \circ g$.

Proof Let D_j be the set of vertices of the support graph of the aperiodic function g_{j_p}, $p = 1, \ldots, l$. Clearly, $D_{j_p} \cap D_{j_q} = \emptyset$, for $1 \leq p, q \leq l$, $p \neq q$. We remark also that $x \in D_{j_p}$ implies $h(x) \in D_{j_p}$ or $h(x) \in C$. We have

$$g_{j_p}(x) = \begin{cases} h(x), & \text{if } x \in D_{j_p} \setminus \{x_{j_p}\} \\ x, & \text{otherwise} \end{cases}$$

for $1 \leq p \leq l$, and

$$g(x) = \begin{cases} h(x), & \text{if } x \in C \\ x, & \text{otherwise} \end{cases}$$

We can easily verify that $g_{j_p} \circ g_{j_q} = g_{j_q} \circ g_{j_p}$, $1 \leq p, q \leq l$, $p \neq q$.

Moreover, if $x \in C$, then we have

$$(g_{j_1} \circ \ldots \circ g_{j_l} \circ g)(x) = g(x) = h(x)$$

and if $x \in D_{j_p} \setminus \{x_{j_p}\}$, then

$$(g_{j_1} \circ \ldots \circ g_{j_p} \circ \ldots \circ g_{j_l} \circ g)(x) = g_{j_p}(x) = h(x),$$

for $1 \leq p \leq l$. ■ ■ ■

We shall refer to g as the *permutational component* of h;

g_{j_1} , ..., g_{j_l} are the *aperiodic components* of h.
THEOREM 4. (5) Let $f \in T_m$. Then $f = G \circ P$, where G is an
aperiodical transformation from T_m and P is a permutation
from T_m.

DEFINITION 4. A *semitransposition* is a transformation $f \in T_m$ for which $|Supp(f)| = 1$.
   From this Definition it is clear that any two distinct
semitranspositions commute.
   It is well-known the fact that every permutation $f \in T_m$
can be written as a product of transpositions ; by analogy,
any aperiodical function can be written as a product of
semitranspositions. Therefore, any transformation $f \in T_m$
can be written as a product of transpositions and semitranspositions.

CONCLUSIONS AND APPLICATIONS

A combinatorial module M with q input wires and an equal
number of output wires to which we apply signals encoded as
values of the set $\{0, ..., p-1\}$ can be modeled as a function
$F : \langle p \rangle^q \to \langle p \rangle^q$. Thus, a cascade of k such elements (see
Fig. 1) is described by the kth iteration of F. Theorem 1,

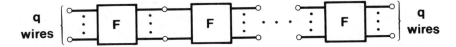

Fig. 1 Cascade of k combinatorial modules

shows that it makes no sense to consider cascades having
length exceeding $p^q - 1 + \kappa(p^q)$ for p-valued switching
devices. For binary two-valued functions, matricial techniques develop by us (4), allow quite good limitations of the
computational power of such cascades.
   In electronic circuits it is often important to separate
the output wire of a circuit from the input wire (see Fig. 2)

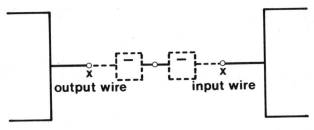

Fig. 2 Separation of an output wire from an input wire

despite the fact that they are logically identical. This
happends when the power demanded by the input wire is significantly larger than the output produced at the input wire.
Instead of joining these wires directly (that is, by a module
described by the identical function 1_p : $\langle p \rangle \to \langle p \rangle$) we inter-
pose two negation modules (in the binary case) which realize
the identity on $\langle p \rangle$. This suggests the technical relevance
of the existence of roots with respect to the iteration power
and our Theorem 3, as well as other several remarks, provide
new insights on this problem.

ACKNOWLEDGEMENTS

The authors wish to acknowledge financial support from the
Natural Sciences and Engineering Research Council (NSERC) of
Canada under Grant A4063.

REFERENCES

1. Davio, M., Deschamps, J.-P. and Thayse, A. (1978).
    "Discrete and Switching Functions". McGraw-Hill
    International Book Company, New York.
2. Lyngholm, C. and Yourgrau, W. (1960). A double-iterative
    property of Boolean functions, *Notre Dame J. of Formal
    Logic* 1, 111-114.
3. Reischer, C. and Simovici, D. (1971). Associative
    algebraic structures in the set of Boolean functions and
    some applications in automata theory, *IEEE Trans.
    Computers*, C-20, 298-303.
4. Reischer, C. and Simovici, D. (1982). Several remarks on
    iteration properties of switching functions, *Proceedings
    of the Twelfth International Symposium on Multipled-
    Valued Logic*, Paris, France, May 25-27, 244-247.
5. Reischer, C. and Simovici, D. (1983). Graph-theoretical
    aspects of properties of transformations of finite sets,
    communication at the Foortheenth Southeastern Conference
    on Combinatorics, Graph Theory and Computing, Atlantic
    Florida University, Boca-Raton, USA, febr. 14-17 (sub-
    mitted for publication).
6. Reischer, C. and Simovici, D. (1983). Roots of n-valued
    switching functions, *Proceedings of the Thertheenth
    International Symposium on Multipled-Valued Logic*, Kyoto,
    Japon, May 24-27, 183-188.
7. Rudeanu, S. (1976). Square roots and functional decom-
    position of Boolean functions, *IEEE Trans. Computers*,
    C-25, 528-532.

THE MULTI-FACETED COMPLETENESS PROBLEM
OF THE STRUCTURAL THEORY OF AUTOMATA

Ivo G. Rosenberg

*Dépt. Math. & Stat. and CRMA, Université de Montréal
C.P. 6128, Succ. "A", Montréal H3C 3J7, Québec, Canada*

1. INTRODUCTION

This mostly expository paper attempts to survey a growing
field of automata theory which, except for a few researchers
(mostly from USSR, GDR and Japan), has received scant attention from the general audience. Many papers in this area
are not easily accessible to western readers (due to language or scarce journals) and so this survey wants to fill
this gap as well as to draw attention to a different approach to automata and the related problems. In the author's view it is perhaps the right time to get involved
since the bridgeheads have been established, the situation
clarified and many problems already located. We wish to
underline the many facets of the completeness problem which
range from the classical two-valued case (solved more than
60 years ago) through many-valued logics, universal algebras,
partial and heterogeneous algebras, delayed gates to the
varied completeness problems for finite and infinite automata.
     The paper starts with a brief introduction to the com-
pleteness for closures which provide a proper framework for
the sequel. Then, in Section 3, automata are introduced as
particular transformations of n-tuples of input sequences
into output sequences over the same alphabet (this out-
sider's stateless view is formally more convenient) and
their natural composition is described in Mal'cev's algebraic formalism (the standard definition of the feedback
loop is, however, deferred to Section 5). It is hoped that
the reader will not get discouraged by the somewhat theoretical framework of Sections 2 and 3 which are most fitting
for the presentation of the sequel. Then, in Section 4 we

375

survey the completeness results for gates, delayed gates and
various kinds of operations. Although this is not strictly
within automata theory, we consider this part important because it serves as the inspirational source, so far has
yielded the best results and often provides useful tools for
the general case. Thus even a reader only interested in
automata may find this the core of the paper. The results
for automata are discussed in Sections 5 and 6. Most of
them are either "negative" in nature (*e.g.*, the non-effect-
iveness) or they only describe the situation (*e.g.*, the
cardinality of the family of maximal sets) without providing
a truly structural insight. Only recently there appeared
some positive results (*e.g.* (2)) but it is hoped that more
are to come soon.

The limited space and the wealth of material forced the
necessity of selecting results and quoting them in a rather
perfunctory way. We tried to give more space to the numerous — and unfortunately not always transparent — definitions than to the results and thus could not even provide
concrete descriptions of the various maximal sets. To our
regret, we had to drop the very important relational aspects and the related Galois theories.

Although a serious attempt has been made to render the
presentation understandable, its condensed nature and the
lack of both proofs and examples made this goal difficult
to achieve. The paper thus will serve its purpose if it
succeeeds in drawing attention to the field. An interested
reader may find more information in Dassow's recent monograph (4) and to some degree in Kudrjavcev's book (23).
The survey lists just references and thus for a more complete bibliography the reader is referred to (4, 23).

2. CLOSURE AND COMPLETENESS

2.1

Closures provide a proper frame for the study of the many-
faceted completeness problem of automata related structures.
Given a set R, a selfmap $X \to [X]_c$ of the set $\underset{\sim}{P}(R)$ (of
subsets of R) is a *closure* on R provided $X \subseteq [X]_c = [[X]_c]_c$
and $[X]_c \subseteq [Y]_c$ whenever $X \subseteq Y \subseteq R$. We say that $X \subseteq R$
is *c-closed* if $[X]_c = X$. The set L_R^c of c-closed subsets
of R forms a complete lattice in which $\inf T = \cap T$ and
$\sup T = [\cup T]_c$ for all $T \subseteq L_R^c$ (moreover, L_R^c determines

$[\ ]_c$ since $[X]_c = \cap\{Y \in L_R^c\colon Y \supseteq X\}$; for more information see *e.g.* (3) II.1).

2.2

A subset X of R is c-*complete* in R if $[X]_c = R$. A c-complete set X in R is a c-*basis* of R is no proper subset of X is c-complete in R, that is, X is irreducible or minimal with respect to c-completeness in R. An element $r \in R$ is c-*Sheffer for* R if $\{r\}$ is c-complete in R. We say that R is *finitely* c-*generated* if it has a finite c-complete set. A c-closed subset X of R is c-*maximal* in R if $X \subset Z \subset R$ for no c-closed set Z, *i.e.* if X is a dual atom of the lattice L_R^c. A family $\underset{\sim}{S}$ of subsets of L_R^c is c-*generic* for R (for finite subsets of R) if each c-incomplete (finite) subset of R extends to a set from $\underset{\sim}{S}$, that is, if $\underset{\sim}{S}$ is cofinal in the poset $L_R^c\diagdown\{R\}$. Let $Y \subseteq R$. We say that $X \subseteq Y$ is c-*complete* in Y if it is c'-complete in Y where $[\ ]_{c'}$ is the induced closure on Y (*i.e.* $[X]_{c'} := [X]_c \cap Y$ for all $X \subseteq Y$). Similarly we modify all the above notions.

2.3

Each c-generic set $\underset{\sim}{S}$ for R provides a c-completeness criterion for R: $X \subseteq R$ *is* c-*complete in* R *if and only if* $X \subseteq Z$ *for* no $Z \in \underset{\sim}{S}$. Here naturally we strive to have $\underset{\sim}{S}$ as irredundant as possible. It is almost immediate that all c-maximal sets in R appear in each c-generic set for R. It may happen that the set $\underset{\sim}{R}^c$ of all c-maximal sets in R forms already a c-generic set for R, that is the lattice L_R^c is dually atomic. Then $\underset{\sim}{R}^c$ is the unique irreducible c-generic set for R providing thus the best universal c-completeness criterion for R. It is indeed universal because it applies in all cases and it only involves the checking of inclusions. However, it requires the knowledge of an explicit list of $\underset{\sim}{R}^c$ with a reasonable description of each $X \in \underset{\sim}{R}^c$ which may be very hard to get. If $\underset{\sim}{R}^c$ is infinite, the c-completeness problem in R may be undecidable even for finite subsets of R (*i.e.* there may exist no algorithm or Turing machine capable of deciding in all cases whether a given

$X \subseteq R$ is c-complete in R. On the other hand even the know-
ledge of a part $\underset{\sim}{S}$ of $\underset{\sim}{R}^c$ may be quite helpful for the c-completeness in R in many cases or at least in several interes-
ting cases. Moreover, a particular $X \subseteq R$ may have properties permitting to rule at once that $X \nsubseteq S$ for S in a
substantial subclass of $\underset{\sim}{R}^c$. Proceeding as in (4) L.1.7,
it is not difficult to see: *Let* $\rho = |R| \geq \aleph_0$ *and let* $\underset{\sim}{R}^c$ *be*
c-generic for R. Then there is a subset T *of* $\underset{\sim}{R}^c$ *of cardi-*
nality at most ρ *which is c-generic for finite subsets of R.*

A particular but important case of $\underset{\sim}{R}^c$ generic is
the following. *If R is finitely c-generated then* (i) *each*
proper c-closed subset of R extends to a c-maximal one,
(ii) *each c-complete set for R contains a c-basis for R and*
(iii) *all c-bases for R are finite.*

2.4

Closures may be composed. If $[\ ]_a$ and $[\ ]_b$ are two closures
on R put $[X]_{ab} := [[X]_a]_b$ for every $X \subseteq R$. In general the
closure properties from 2.2 and 2.3 of $[\ ]_a$, $[\ ]_b$ and $[\ ]_{ab}$
need not be related. Clearly if an a-maximal set X in R
is not b-complete, then X is ab-maximal in R.

The closure b may be defined by an equivalence $\equiv$ on R:
$r \in R$ belongs to $[X]_b$ if $r \equiv x$ for some $x \in X$ (*i.e.*
$[X]_b$ is the union of blocks of $\equiv$ intersecting X).

3. STRUCTURAL THEORY OF AUTOMATA

3.1

We are considering deterministic initial Mealy automata
having one or several input terminals (lines) and a single
output terminal all operating over a fixed alphabet A. For
our purposes we view them as input-output transformations,
i.e. as black boxes formally ignoring their inner states.

We introduce the formal definition. The *alphabet* A is
a set with $|A| > 1$ (in most applications A will be finite).
For a set B and $n \in \mathbb{N} := \{0,1,\dots\}$ let B^n stand for the
set of n-tuples over B (where $B^0 = \{\Lambda\}$ and Λ is the empty
sequence). Further put $B^* := \bigcup_{n \in \mathbb{N}} B^n$ and let $B^{\mathbb{N}}$ denote the

set of infinite sequences $<b_0,b_1,...>$ over B. Thus $(A^n)^{\mathbb{N}}$ is
the set of sequences $<\bar{x}_0,\bar{x}_1,...>$ over A^n with $\bar{x}_t =$
$(x_{t1},...,x_{tn})$ where $t \in \mathbb{N}$ and $x_{t1},...,x_{tn} \in A$. For
$n \in \mathbb{N}$ a map $f:A^n \to A$ is called an n-*ary operation* (other
names connective or function) *on* A. A nullary (or zero)
operation is just a specified element of A.

3.2

Let $n \in \mathbb{N}$. An n-*ary sequential operation* over A is a map
$F:\tilde{x}=(\bar{x}_0,\bar{x}_1,...) \to F(\tilde{x}) = <y_0,y_1,...>$ from $(A^n)^{\mathbb{N}}$ into $A^{\mathbb{N}}$ such
that for all $t \in \mathbb{N}$ we have $y_t = F_t(\bar{x}_0,\bar{x}_1,...,\bar{x}_t)$ where
$F_t:A^{n(t+1)} \to A^n$ is a fixed n(t+1)-ary operation on A $(t \in \mathbb{N})$.
In other words, the output at time t may depend at most on
the "past" and the "present" but there is no "foresight"
i.e. for $\tilde{x}' = <\bar{x}_0',\bar{x}_1',...>$ and $\tilde{x}'' = <\bar{x}_0'',\bar{x}_1'',...>$ with
$\bar{x}_i' = \bar{x}_i'$ for $i=0,...,t$ the corresponding output sequences
$F(\tilde{x}') = <y_0',y_1',...>$ and $F(\tilde{x}'') = <y_0'',y_1'',...>$ have
$y_i' = y_i''$ for all $i=0,...,t$. Note that the output may
react instantaneously to the present input. The set of
n-ary sequential operations over A is denoted $P_A^{(n)}$ and P_A
stands for $\bigcup_{n \in \mathbb{N}} P_A^{(n)}$.

3.3

Each $F \in P_A^{(n)}$ may be identified with a deterministic initial Mealy automaton with n inputs and a single output. For
given $t \in \mathbb{N}$ and $p = (\bar{p}_0,...,\bar{p}_t) \in A^{n(t+1)}$ and an arbi-
trary $\tilde{x} = <\bar{x}_0,\bar{x}_1,...> \in (A^n)^{\mathbb{N}}$ denote $F<\bar{p}_0,...,\bar{p}_t,\bar{x}_0,\bar{x}_1,...>$
by $<y_0,y_1,...>$ and put $\hat{F}_p(\tilde{x}) = <y_{t+1},y_{t+2},...>$. Clearly
$\hat{F}_p$ is obtained from F by fixing the inputs at times $0,...,t$.
Put $\hat{F}_\Lambda = F$ and $Q := \{\hat{F}_p: p \in (A^n)^*\}$. Let Q be the state
set with initial state $\hat{F}_\Lambda$ and let the transition and output
functions $\delta:Q \times A \to Q$ and $\sigma:Q \times A \to A$ be defined by:

$$\delta(\hat{F}_{(\bar{x}_0,\ldots,\bar{x}_{t-1})},\bar{x}_t) = \hat{F}_{(\bar{x}_0,\ldots,\bar{x}_t)}$$

$$\sigma(\hat{F}_{(\bar{x}_0,\ldots,\bar{x}_{t-1})},\bar{x}_t) = F_t(\bar{x}_0,\ldots,\bar{x}_t).$$

A routine check shows that F is the input-output transformation of thus defined automaton. Note, however, that even for A = {0,1} the state set Q may be of cardinality $\aleph_0$.

3.4

Our interest is in the following problem of structural automata theory. If certain basic automata are available, what new automata may be constructed from the basic ones? Here, both "available" and "constructed" need explanation. We assume having at our disposal a reasonably large number of identical copies of each of the basic automata, (*e.g.* they may be commercially available and cheap enough for our budget). We are interested here solely in feasibility ignoring completely optimality (because the latter depends too heavily on the specific "canonical forms", evolves fast with and is quite sensitive to the present technology and is thus worthwhile only as a response to a concrete industrial need).

3.5

We describe in detail the allowable constructions with the exception of the formation of the feedback loop which is defered to Section 5. The description follows Mal'cev's formalism (25) because, short of handwaiving, it takes as much space as any other description. The first three constructions τ, ζ and Δ refer to the manipulation of inputs. First note that, in general, the order of the inputs is relevant while the names or symbols attached to them are not. In most cases it is physically easy to interchange (switch) or identify (fuse) the inputs on a real automaton and therefore we assume that from a given $F \in P_A^{(n)}$ we can create new sequential operations, called *polymers* of F, in this way. Formally for n > 1 and $\bar{a} = (a_1,\ldots,a_n) \in A^n$ put $\tau\bar{a} := (a_2,a_1,a_3,\ldots,a_n)$ and $\zeta\bar{a} := (a_2,a_3,\ldots,a_n,a_1)$. For $\alpha = \tau,\zeta$ put $(\alpha F)(\tilde{x}) :=$ $F\langle\alpha\bar{x}_0,\alpha\bar{x}_1,\ldots\rangle$ for all $\tilde{x} \in (A^n)^{\mathbb{N}}$. Similarly for

$\bar{a} = (a_1,\ldots,a_{n-1}) \in A^{n-1}$ put $\Delta\bar{a} = (a_1,a_1,a_2,\ldots,a_{n-1})$ and $(\Delta F)(\tilde{x}) := F{<}\Delta\bar{x}_0,\Delta\bar{x}_1,\ldots{>}$ for all $\tilde{x} \in (A^n)^{\mathbb{N}}$. For n=0,1 and $\alpha = \tau,\zeta,\Delta$ put $\alpha F = F$. Since it is well known that the transposition (1,2) and the cyclic permutation (1,2,...,n) generate all permutations on {1,...,n}, it is not difficult to see that the repeated applications of τ,ζ and Δ yield all the polymers of F, *i.e.* all $G \in P_A^{(n)}$ obtained from F by manipulating its inputs. For example, $\tau\Delta\zeta F(\tilde{x}) = F(\tilde{y})$ where $\bar{y}_t = (x_{t2},x_{t1},x_{t3},\ldots,x_{t,n-1},x_{t1})$ for all $t \in \mathbb{N}$.

3.6

Two automata can be combined together by attaching the output of the first automaton to one of the inputs of the second. Using ζ and τ we can permute arbitrarily the inputs and thus we assume that the input involved is the first input. For the greatest generality all the inputs of the new automaton are pairwise distinct. Formally let $F \in P_A^{(m)}$ and $G \in P_A^{(n)}$. If $m = 0$ put $F*G = F$. Let $m > 0$ and $\tilde{x} = {<}\bar{x}_0,\bar{x}_1,\ldots{>} \in (A^{m+n-1})^{\mathbb{N}}$. For all $t \in \mathbb{N}$ put

$$\bar{x}_t{}' = (x_{t1},\ldots,x_{tn}),$$

$$\bar{x}_t{}'' = (G_t(\bar{x}_0{}',\ldots,\bar{x}_t{}'),x_{t,n+1},\ldots,x_{t,m+n-1})$$

and define $H := F*G \in P_A^{(m+n-1)}$ by setting $H_t(\bar{x}_0,\ldots,\bar{x}_t)$ $= F_t(\bar{x}_0{}'',\ldots,\bar{x}_t{}'')$ for all $t \in \mathbb{N}$. Clearly * thus defined is a binary operation on P_A; in fact it is associative and $E \in P_A^{(1)}$ defined by $E_t(x_0,\ldots,x_t) = x_t$ for all $t \in \mathbb{N}$ is its neutral element. The algebra $\underline{\underline{P}}_A := {<}P_A;\tau,\zeta,\Delta,*{>}$ is called the *preiterative algebra* on A. A subset $\underline{X}$ of P_A is *s-closed* if $\underline{X}*\underline{X} := \{F*G : F,G \in \underline{X}\} \subseteq \underline{X}$ and $\alpha\underline{X} \subseteq \underline{X}$ for $\alpha = \tau,\zeta,\Delta$. This defines an s-closure $[\ ]_s$ on P_A which is *algebraic* (the set-theoretical union of a chain of s-closed sets is s-closed (3) II.1). Practically nothing is known about the s-completeness in the general case.

3.8

A sequential operation $F \in P_A^{(n)}$ is *finite* if $Q := \{\hat{F}_p :$
$p \in (A^n)^*\}$ (see 3.3) is finite, that is, if the associated
automaton has finitely many states. Let $\hat{P}_A$ denote the set
of finite sequential operations on A. The set $\widetilde{P}_A$ is s-
closed. For A finite we have $|\widetilde{P}_A| = \aleph_0$ (since there are
only finitely many finite automata with n states (4) L.2.6).
For an integer $k > 1$ put $\underline{k} := \{0,\ldots,k-1\}$. It is shown
in (4) Thm. 4.10 that no $\widetilde{P}_{\underline{k}}$ has a finite s-basis, there are
s-complete sets in $\widetilde{P}_{\underline{k}}$ containing no bases while $\widetilde{P}_{\underline{2}}$ has an
s-basis.

   In the next section we review the s-completeness of
certain special s-closed subclasses of $\widetilde{P}_A$.

4. GATES AND DELAYED GATES

4.1

A *gate* (a switching element or device) associated to an
n-ary operation $f:A^n \to A$ is $F \in P_A^{(n)}$ satisfying
$F_t(\bar{x}_0,\ldots,\bar{x}_t) = f(\bar{x}_t)$ for all $t \in \mathbb{N}$ and $\bar{x}_0,\ldots,\bar{x}_t \in A^n$.
Thus a gate corresponds to a one-state memoryless automaton
whose output depends at most on the present inputs. Let O_A
denote the set of gates over A. If A is finite, $|A| = k$,
we may as well assume that $A = \underline{k} := \{0,\ldots,k-1\}$. We review briefly the s-completeless in $O_{\underline{k}}$.

4.2

We start with the simplest case k=2. Operations
$f:\{0,1\}^n \to \{0,1\}$, called *truth* or *Boolean functions* or
logical connectives, were introduced by G. Boole in 1848
(in logics often $\{0,1\}$ is replaced by $\{T,F\}$). Although in
the last century and the early years of the 20th century,
truth functions were studied in depth for logical purposes,
their use for switching circuits was not recognized until
the late 1930's and fully applied only in the fifties.
(Even more recently they found their use in operations
research (7). The s-completeness in $O_{\underline{2}}$ was solved by E.
Post (34) in 1921 (and rediscovered many times since) in
terms of the 5 s-maximal classes in $O_{\underline{2}}$. Twenty years later

Post described all the countably many s-closed subsets of
O_2 (35). In this long fundamental paper he also showed that
each of the countably many s-closed subclasses of O_2 is finitely s-generated.

4.3

For k > 2 the operations on k are called the functions of
the k-*valued logic*. They were introduced by Post and
Lukasiewicz in 1921. Completeless in O_3 was solved by
Jablonskiĭ in 1953 (14,15) and in the general case by the
author in 1965 (38, 39),cf. also (40), by exhibiting an
explicit list of the s-maximal sets in O_k. Their number,
although finite, grows fast with k (from 18 for k=3 (14,15)
to more than 5×10^{12} for k=8 (41)). They come in 6 natural
groups and, in spite of their huge number, many interesting
subsets of O_k have properties permitting to limit the veri-
fication to a few cases (cf. (40,47,42,43)).

Surprisingly for k > 2 the lattice of closed subclas-
ses of O_k has the largest possible cardinality $2^{\aleph_0}$ and in
this respect alone the 2-valued logic stands quite apart
from the many-valued ones and from most of the other s-
closed sets considered here. Our knowledge of the s-closed
subsets of O_k is quite sketchy. These problems are studied
both in universal algebra and multiple-valued logics. There
are some hardware applications in multiple-valued circuits
which are promising because potentially a line may carry
more information and thus increase the operating speed
(cf. (36)).

4.4

Naturally the same questions arise for A infinite. It is
known that there are $2^{2^{|A|}}$ s-closed subsets of O_A and that
many maximal subclasses of O_A (44) but it is an open question whether the lattice is dually atomic. The virtually
untouched problem of the s-completeness for $|A| \geq \aleph_0$ seems
to be hard. A much simpler variant is based on local clo-
sure. For $X \subseteq P_A$ let $F \in [X]_\ell$ if for each finite $B \subseteq A$
there is $G \in X$ such that F and G agree on B (*i.e.* if

$F(x) = G(x)$ for all $x \in (B^n)^{\mathbb{N}}$. The composite $s\ell$-closure (2.4) and the $s\ell$-completeness are intensively studied in universal algebra and to a certain degree parallel to s-closure and s-completeness for A finite. Although the lattice of $s\ell$-closed subsets of O_A is not dually atomic, the work towards a reasonable $s\ell$-generic set in O_A is in progress (45,46).

4.5

Partial operations ($f:D \to A$ with $D \subseteq A^n$) are frequently used and often handy. However, straight-forward concepts for operations usually admit several extensions for partial operations. A natural way to treat partial operations is to adjoin a new element ω to A and extend $f:D \to A$ to an n-ary operation f^0 on $B := A \cup \{\omega\}$ by setting $f^0(x) = f(x)$ for $x \in D$ and $f^0(x) = \omega$ otherwise. This provides an embedding of partial operations on A into the s-closed set R_A of $g \in O_B$ with $g(x_1,\ldots,x_n) = \omega$ whenever $x_i = \omega$ and g depends on its i-th variable. Already in 1966 Freĭvald (5,6) found all 8 s-maximal classes for R_2 and recently Romov (37) obtained partial results for R_k, $k > 2$.

4.6

Heterogeneous operations are special partial operations. We are given the sets $\Sigma_1 = \{A_1,\ldots,A_s\}$ and $\Sigma_2 = \{B_1,\ldots,B_t\}$ of input and output alphabets (where Σ_2 is inclusion free). A heterogeneous operation maps some $A_{i_1} \times \ldots \times A_{i_n}$ into some B_j ($1 \le i_1,\ldots,i_n \le s$, $1 \le j \le t$). Apart from the somewhat degenerate case of $|A_i \cap B_j| \le 1$ for all $i=1,\ldots,s$, $j=1,\ldots,t$ ((23) III.1.1-6) we have that the set $P_{\Sigma_1\Sigma_2}$ of heterogeneous operations is finitely s-generated and has finitely many s-maximal sets. Except for the set O_2 (corresponding to $\Sigma_1 = \Sigma_2 = \{\underline{2}\}$), the set $P_{\Sigma_1\Sigma_2}$ has $2^{\aleph_0}$ s-closed classes. The numbers of s-maximal sets in $P_{\Sigma_1\Sigma_2}$

for the 6 simplest cases are listed in Table 1 (*e.g.* for $\Sigma_1 = \{\{0,1\},\{1,2\}\}$, $\Sigma_2 = \{\{0,1\}\}$ there are 10 s-maximal classes in $P_{\Sigma_1\Sigma_2}$) (23) III. §§2.1-6.

Σ_1	01-12	01-23	01	01	01	012
Σ_2	01	01	01-02	01-23	012	01
#	10	21	6	6	8	6

Tab. 1

4.7

A *delayed gate* associated to an n-ary operation f on A and to delays $\delta_1,\ldots,\delta_n \in \mathbb{N}$ is $F \in P_A^{(n)}$ satisfying
$F_t(\bar{x}_0,\ldots,\bar{x}_t) = f(x_{t-\delta_1,1},\ldots,x_{t-\delta_n,n})$ for all
$t \geq \max(\delta_1,\ldots,\delta_n)$ and $\bar{x}_i = (x_{i1},\ldots,x_{in}) \in A^n$ (i=0,\ldots,t).
We denote this gate by $(f;\delta_1,\ldots,\delta_n)$ and if $\delta_1 = \ldots = \delta_n$
$= \delta$, just by (f,δ). Real switching devices require a certain reaction time which,albeit small,should not be neglected in large or fast switching circuits. Delayed gates model this aspect while still neglecting other transitional phenomena. The s-completeness in the set D_k of delayed gates on k can be easily described (*cf.* (26) Thm.1). For circuits built from delayed gates intuitively completeness should mean that we can represent each $f \in O_A$ with some reasonable delays. Several completeness concepts appear in the literature, *e.g.* (23) IV. §2.1 lists seven variants. Here we discuss one variant based on the following closure on P_A.

For $X \subseteq P_A$ let $F \in [X]_u$ if there is $d \in \mathbb{N}$ and $G \in X$ such that $F_{t+d}(\bar{x}_0,\ldots,\bar{x}_{t+d}) = G_t(\bar{x}_0,\ldots,\bar{x}_t)$ for all $t \in \mathbb{N}$.
Note that an su-complete set in D_A may not be very helpful for the design of circuits since some operations may be only represented with huge delays. In spite of this drawback,

the su-completeness leads to an interesting theory. For
somewhat obscure reasons the literature deals almost exclusively with the su-completeness in the set U_k of uniformly
delayed gates on $\underline{k}$. Already in 1962,   Kudrjavcev
found the su-completeness criterion on U_2 in terms of 9
su-maximal sets and 2 countable families of su-maximal sets
in U_2 (17,18). In subsequent papers together with his collaborators he introduced and studied several other concepts
for delayed gates (see (23) IV. §2). Recently Hikita (8)
determined the su-completeness in U_3. Work is in progress
to solve the su-completeness in U_k for $k > 3$ (12). This
is done by scanning sequences of relations on $\underline{k}$. Using an
ingenious reduction from (11) and the results of (9,10),
the search is now confined to periodic sequences of binary
reflexive polyrelations with strong commuting properties
plus a few odd and probably easy cases.

4.8

Let $U_A^1 := \{(f,1): f \in O_A\}$ denote the set of uniformly
delayed gates with unit delay. Independently of (17,18)
and in a formally slightly different setting Ibuki (13) gave
an su-completeness in U_A criterion for subsets of U_A^1 in
terms of the 6 su-maximal sets in U_A^1.

5.   FEEDBACKS

5.1

Feedback loops play often an important role. The adjunction of a feedback loop means that we connect the output to
one of the inputs. Without loss of generality we may and
will assume that it is the first input. Our model reacts
instantaneously to the present inputs and thus to avoid a
kind of short circuit we may add the feedback loop only if
the dependence on the present first input is fictitious.
Following (4) 2.2, we say that $F \in P_A^{(n)}$ *depends in a de-*
layed way on its first variable if $n > 0$ and
$$F_t(\bar{x}_0,\ldots,\bar{x}_t) = F_t(\bar{x}_0,\ldots,\bar{x}_{t-1},\bar{x}_t') \text{ for all } t \in \mathbb{N},$$
$\bar{x}_0,\ldots,\bar{x}_t,\bar{x}_t' \in A^n$ such that $x_{ti} = x_{ti}'$ for $i=2,\ldots,n$.
Let $a \in A$ and let $F \in P_A^{(n)}$ depend in a delayed way on

its first variable. The operation $\times F$ obtained from F by the adjunction of the feedback loop, is $G \in P_A^{(n-1)}$ defined for every $\tilde{x} = \langle \bar{x}_0, \bar{x}_1, \ldots \rangle \in (A^{n-1})^{\mathbb{N}}$ inductively by setting

$$G_t(\bar{x}_0, \ldots, \bar{x}_t) = F_t(y_0, \bar{x}_0, \ldots, y_{t-1}, \bar{x}_{t-1}, a, \bar{x}_t) = y_t$$

for all $t \in \mathbb{N}$. From the automata point of view, when at time t the signal x_{ti} arrives at the (i+1)-st input terminal (i=1,...,n-1) then instantaneously y_t appears on the output terminal. By the assumption it does not depend at all on the current signal on the first input terminal. At the same moment y_t travels through the feedback loop to the first input terminal which then carries y_t during the time interval (t,t+1). The change of state, which takes place at time t+1, is determined by the present state and the input signals y_t, $x_{t1}, \ldots, x_{t,n-1}$ which appear on the terminals during the whole open time interval (t,t+1). This motivates our definition of $G = \times F$. If F does not depend in a delayed way on its first variable we put $\times F := F$.

5.2

Consider the algebra $\underset{=A}{A} := \langle P_A; \tau, \lessgtr, \Delta, \times, * \rangle$. The corresponding algebraic closure on P_A is denoted $[\ ]_a$. Note that each a-closed set is also s-closed but not vice versa.

The *delay operator* $S \in P_k^{(1)}$ is defined by $S_0(x_0) = 0$ and $S_t(x_0, \ldots, x_t) = x_{t-1}$ for all $t > 0$. The following is well known: *A subset X of the set* $\tilde{P}_k$ (of finite sequential operations) *is a-complete in* $\tilde{P}_k$ *if and only if* $[X]_a$ *contains the set* O_k (of gates) *and* S. (The main idea of the proof proceeds from a coding of the states by elements of k^m, the representation of σ by an (m+n)-ary operation on $\underline{k}$ and of δ by m such operations). It is well known that O_k has a s-Sheffer basis and thus $\tilde{P}_k$ possess a 2-element a-basis. In fact, $\tilde{P}_k$ has an n-element a-basis for all n=1,2,... (cf. (4) Thm. 4.2) and quite interestingly there is an a-Sheffer $F \in \tilde{P}_k^{(2)}$ whose

automaton has only 2 states, clearly the best minimal result (1).

5.3

Since $\tilde{P}_k$ is finitely a-generated, the a-maximal sets in $\tilde{P}_k$
form an a-generic set for $\tilde{P}_k$. However, there are $2^{\aleph_0}$ a-maximal sets in $\tilde{P}_k$ (*i.e.* the set of dual atoms of $L_{\tilde{P}_k}^a$ has the
same cardinality as $L_{\tilde{P}_k}^a$ itself (a non-constructive proof
(18) (**cf.** (4) Thm. 4.14). This number raises questions
about the effectiveness of an a-completeness criterion for
P_k. In fact, the problem has been reduced to the finite
derivation problem of normal Post algorithms ((16) cf. (4)
Thm. 4.16) which is undecidable. In spite of this "pessimistic" result, the knowledge of a large number of a-maximal sets in $\tilde{P}_2$ could prove quite useful (2.3). Thus the
search for such sets is not only far from doomed but presents a rather intriguing challenge.

5.4

Recently A. Nozaki (32) considered the an-completeness in
U_2 for subsets of U_2^1. Here a set X of two-valued gates with
unit delay is an-complete in U_2 if every Boolean function
$f:\{0,1\}^n \to \{0,1\}$ may be constructed from X with some delay
using standard compositions and *feedbacks*. The criterion is
given in terms of the 6 an-maximal sets closely related to
those quoted in 4.8.

6.   SOME OTHER AUTOMATA CLOSURES

6.1

For $t \in \mathbb{N}$ the closure $[X]_{m_t}$ of $X \subseteq P_A$ consists of all
$F \in P_A$ for which there is $G \in X$ such that $F_i = G_i$ for
$i = 0, \ldots, t$, that is, we accept all F undistinguishable from
some $G \in X$ until the time $t+1$. For the composite closure
we have $[\ ]_{am_t} = [\ ]_{sm_t}$ *i.e.* the feedback formation is

dispensable (cf. (4) Thm. 5.2). Recently Buevič (2) gave an sm_t-completeness criterion. He effectively listed the finitely many sm_t-maximal sets in $\widetilde{P}_k$ providing thus one of the few satisfactory positive results in this area.

6.2

Now we can define the closure $[X]_m := \cap_{t \in \mathbb{N}} [X]_{m_t}$. Thus $F \in [X]_m$ iff for every $t \in \mathbb{N}$ the sequential operation F is undistinguishable from some $G \in X$ until the time t+1. For the composite sm-closure we have that the family of sm-maximal sets in $\widetilde{P}_k$ is the union of all sm_t-maximal sets in $\widetilde{P}_k$. Although we know all the sm-maximal sets in $\widetilde{P}_k$, the sm-completeness for finite subsets of $\widetilde{P}_k$ is still undecidable (24).

6.3

The following closure relates to regular sets. Let $F \in P_A^{(n)}$. Given $Y \subseteq A$ put

$$(F)_Y := \{(\bar{x}_0,\ldots,\bar{x}_t) \in (A^n)^t : t \in \mathbb{N}, F_t(\bar{x}_0,\ldots,\bar{x}_t) \in Y\}$$

and say that F *accepts* the subset $(F)_Y$ of $(A^n)^*$. The sets $\{(F)_Y : F \in P_A, Y \subseteq A\}$ are the regular sets ((4) Thm. 5.6). For $X \subseteq P_A$ let $[X]_k$ consist of all $F \in P_A$ such that to each $Y \subseteq A$ we have $(F)_Y = (G)_Z$ for some $G \in X$ and $Z \subseteq A$. Thus $X \subseteq P_A$ is ak-complete in $\widetilde{P}_k$ if every regular set is accepted by some $F \in [X]_a$. Several results concerning ak-completeness and its metric variants are in (4) §5.2. Other completeness definitions-too complex to be mentioned here-are in (4) §§5.4-5.5.

6.4

To conclude we mention a related problem arising for partially recursive operations. Let Z denote the set of partial operations on $\mathbb{N}$. Given an $f:D \to \mathbb{N}$ with $D \subseteq \mathbb{N}^n$ define the partial n-ary operation $\mu f:D' \to \mathbb{N}$ by setting

$$D' := \{(x_1,\ldots,x_n) \in \mathbb{N}^n : (x_1,\ldots,x_{n-1},y) \in D,$$

$$f(x_1,\ldots,x_{n-1},y) = x_n \text{ for some } y \in \mathbb{N}\}.$$

and for every $(x_1,\ldots,x_n) \in D'$ then put $(\mu f)(x_1,\ldots,x_n) = z$
where z is the least solution of $f(x_1,\ldots,x_{n-1},y) = x_n$.

Given $f:D_1 \to \mathbb{N}$ and $g:D_2 \to \mathbb{N}$ where $D_1 \subseteq \mathbb{N}^n$ and $D_2 \to \mathbb{N}^{n+2}$
we define inductively $h = R(f,g)$ by setting

$$h(x_1,\ldots,x_n,0) := g(x_1,\ldots,x_n) \text{ for all } (x_1,\ldots,x_2) \in D_1,$$
$$h(x_1,\ldots,x_n,y+1) := f(x_1,\ldots,x_n,y,h(x_1,\ldots,x_n,y)) \text{ whenever}$$

the right side is defined. In this setting $<2;\tau,\zeta,*,\mu,R>$
is the algebra of *partial recursive operations*. Here we
have a natural example of a modification of the s-closure
from 4.5 obtained by adjoining new operations μ and R.

7. ACKNOWLEDGEMENTS

The author would like to use this opportunity to express his
gratitude to the organizers for inviting him to this exciting colloquium and for the financial support. He also gratefully acknowledges the partial financial support for this
work provided by the NSERC Canada Grant A-5407 and FCAC
Québec Subvention d'équipe EQ 0539.

REFERENCES

1. Buevič, V.A. (1970). The construction of a universal
   bounded deterministic function of two input variables
   having two inner states (Russian). *Problemy Kibernetiki* 22, 75-83.
2. Buevič, V.A. (1980). On the t-completeness in the
   class of automata maps (Russian). *Doklady AN SSSR*
   252, 1037-1041.
3. Cohn, P.M. (1965). "Universal Algebra", Harper and Row,
   2nd edition, D.Reidel Publ.Co. 1981.
4. Dassow, J. (1981). Completeness problems in the
   structural theory of automata, Mathematical research
   B.7, Akademic Verlag, 140pp.
5. Freĭvald, R.V. (1966). A completeness criterion for
   partial functions of logic and many-valued logic
   algebras (Russian). *Doklady AN SSSR* 167, No.6, 1249-

1250. English Translation, Soviet Physics Doklady 11, No.4, 288-289.

6. Freĭvald, R.V. (1966). Functional completeness for not everywhere defined functions of the algebra of logic (Russian). Diskretnyĭ Analiz No.8, 55-68.

7. Hammer, P.L.; Rudeanu, S. (1968). "Boolean Methods in Operations Research and Related Areas". Springer Verlag. French translation: Dunod, Paris, 1970.

8. Hikita, T. (1978). Completeness criterion for functions with delay defined over a domain of three elements. Proc.Japan Acad. 54, 335-339.

9. Hikita, T. (1981). Completeness properties of k-valued functions with delays: Inclusions among closed spectra. Math.Nachr. 103, 5-19.

10. Hikita, T. (1981). On completeness for k-valued functions with delay. Coll. Math. Soc. János Bolyai 28, "Finite Algebra and Multiple-Valued Logic" (Eds. B.Csakany and I.Rosenberg), pp.345-371. North-Holland.

11. Hikita, T. and A. Nozaki (1977). A completeness criterion for spectra. SIAM J. Comput. 6, 285-297. Corrigenda, ibid., 8 (1979) 656.

12. Hikita,T., Rosenberg, I.G. (1983). Completeness for uniformly delayed circuits. Proc. 13th Internat. Symp. Multiple-valued Logics, Tokyo, May 1983, 2-11.

13. Ibuki, K. (1968). Studies on the universal logic circuits (Japanese) Dendenköska Denkitsushin-Kenkyusho Res. rep. 3747.

14. Jablonskiĭ, S.V. (1954). On functional completeness in the three-valued calculus (Russian). Dokl.Akad. Nauk. SSSR 95, 1153-1155.

15. Jablonskiĭ, S.V. (1958). Functional constructions in a k-valued logic (Russian). Trudy Mat.Inst.Steklov 51 5-142.

16. Kratko, M.I. (1964). The algorithmic unsolvability of one problem of the theory of finite automata. (Russian) Diskret.Analiz (Novosibirsk) 2, 37-41.

17. Kudrjavcev, V.B. (1960). Completeness theorem for a class of automata without feedback couplings (Russian). Dokl.Akad.Nauk SSSR 132 (1960), 272-274. English translation: Soviet Math.Dokl. 1 , 537-539.

18. Kudrjavcev, V.B. (1962). A completeness theorem for a class of feedback-free automata (Russian). Problemy Kibernet 8, 91-115. The power of sets of precomplete sets for certain functional systems connected with automata (Russian). Problemy Kibernet. 13 (1965) 45-47

19. Kudrjavcev, V.B.(1971). Concerning S-systems of a k-valued logic (Russian). *Dokl.Akad.Nauk SSSR* 199, 20-22. English translation: *Soviet Math.Dokl.* 121 (1971) 1013-1016. The properties of S-systems of functions of k-valued logic (Russian). *Diskret.Analiz* 19 (1971) 15-47.

20. Kudrjavcev, V.B. (1973). On the functional system P$_\Sigma$ (Russian). *Doklady Akad. Nauk SSSR* 210, No.3, 521-522.

21. Kudrjavcev, V.B. (1973), On properties of S-systems of functions of k-valued logic (Russian, German and English summaries) *Elektron. Informations verarbeit. Kybernetik* 9, 81-105.

22. Kudrjavcev, V.B. (1973). On functional properties of logical nets (Russian). *Math.Nachr.* 55, 187-211.

23. Kudrjavcev, V.B. (1982). Functional systems (Russian). Monograph,Izd.Moskovskogo Univ., 157 pp.

24. Kudrjavcev,V.B., Alešin, S.V., Podkol'zin, A.S. (1978). The elements of the theory of automata (Russian). Moscow.

25. Mal'cev, A.I. (1971). Iterative algebras and Post's varieties (Russian). *Algebra i Logika* 5 (1966) 5-24. English translation: "The Metamathematics of Algebraic Systems" pp. 396-415, North-Holland.

26. Martin, L., Reischer, C., Rosenberg, I.G. (1978). Completeness problems for switching circuits constructed from delayed gates. Proc. 8th Internat. Sympos.Multiple-valued Logic Rosemont 1978, pp.1-10. French version: *Elektron.Inf.Kybernetik* 19, 4/5 (1983) 171-186.

27. Nozaki, A. (1970). Réalisation des fonctions définies dans un ensemble fini a l'aide des organes élémentaires d'entrée-sortie. *Proc.Japan Acad.* 46, 478-482.

28. Nozaki, A. (1972). Complete sets of switching elements and related topics. First USA-Japan Computer Conference, pp. 393-396.

29. Nozaki, A. (1978). Functional completeness of multi-valued logical functions under uniform compositions. *Rep.Fac.Eng.Yamanashi Univ.* 29, 61-67.

30. Nozaki, A. (1981). Completeness criteria for a set of delayed functions with or without non-uniform compositions. Coll.Math.Soc.Janos Bolyai 28, "Finite Algebra and Multiple-Valued Logic (Eds., B.Csákány and I. Rosenberg), pp. 489-519, North-Holland.

31. Nozaki, A. (1982). Maximal *-incomplete sets of functions defined over the set {0,1,2}. Preprint.

32. Nozaki, A. (1982). Completeness of logical gates based on sequential circuits. Preprint No.2, International Christian University, Tokyo.

33. Pöschel, R. and Kalužnin, L.A.,(1979). Funktionen-und Relationenalgebren. VEB Deutscher Verlag der

Wissenschaften, Berlin.

34. Post, E.L. (1921). Introduction to a general theory of elementary propositions. *Amer.J.Math.* 43, 163-185.

35. Post, E.L. (1941). The two-valued iterative systems of mathematical logic. *Annals of Math. Studies,* 5, Princeton.

36. Proceedings of the International Symposia on Multiple-valued Logics 1-13 (1971-1983), IEEE.

37. Romov, B.A., (1981). The algebra of partial functions and their invariants (Russian). *Kybernetik* No.2, March-April 1981, 1-11. English translation: *Cybernetics,* 157-167.

38. Rosenberg, I.G. (1965). La structure des fonctions de plusieurs variables sur un ensemble fini. *C.R.Acad.Sci. Paris, Ser.A-B* 260, 3817-3819. MR 31 # 1185.

39. Rosenberg, I.G. (1970). Uber die funktionale Vollständigkeit in der mehrwertigen Logiken. *Rozpravy Ceskolovenské Akad.Věd., Rada Mat.Přír.Věd.* 80, 4, 1-93. MR45 # 1732.

40. Rosenberg, I.G. (1977). Completeness properties of multiple-valued logic algebras. In "Computer Science and Multiple-Valued Logic, Theory and Applications" (Ed. D.C.Rine) North-Holland, Amsterdam, New York, Oxford, 144-186.

41. Rosenberg, I.G. (1973). The number of maximal closed classes in the set of functions over a finite domain. *J.Combinatorial Theory* 14, 1-7.

42. Rosenberg, I.G. (1970). Complete sets for finite algebras. *Math.Nachr.* 44, 225-258.

43. Rosenberg, I.G. (1980). Functional completeness of single generated or surjective algebras. Proc. Coll. Finite Algebra and Multiple-valued Logic, North-Holland, pp.635-652.

44. Rosenberg, I.G. (1976). The set of maximal closed classes of operations on an infinite set A has card $2^{2^{|A|}}$. *Archiv .d.Math.* 27, 561-568.

45. Rosenberg, I.G. and Schweigert, D. (1982). Local clones. *Elektron.Inf.Kybernet.* 18, 7/8, 389-401.

46. Rosenberg, I.G. and Szabó, L. (1981). Local completeness I. Preprint CRMA-1072. To appear in *Algebra Universalis.*

47. Rousseau, G. (1967) Completeness in finite algebra with a single operation. *Proc.Am.Math.Soc.* 18, 1009-1013.

INDEX